SOIL ECOLOGY RESEARCH DEVELOPMENTS

SOIL ECOLOGY RESEARCH DEVELOPMENTS

TIAN-XIAO LIU
EDITOR

Nova Science Publishers, Inc.
New York

For permission to use material from this book please contact us:
Telephone 631-231-7269; Fax 631-231-8175
Web Site: http://www.novapublishers.com

LIBRARY OF CONGRESS CATALOGING-IN-PUBLICATION DATA

Soil ecology research developments / Tian-Xiao Liu, editor.
p. cm.
ISBN-13: 978-1-60021-971-9 (hardcover)
ISBN-10: 1-60021-971-3 (hardcover)
1. Soil ecology. I. Liu, Tian-Xiao.
QH541.5.S6S643 2007
577.5'7--dc22

2007037984

Published by Nova Science Publishers, Inc. ✦ New York

CONTENTS

PREFACE

Soil ecology is the study of the interactions among soil organisms, and between biotic and abiotic aspects of the soil environment. It is particularly concerned with the cycling of nutrients, formation and stabilization of the pore structure, the spread and vitality of pathogens, and the biodiversity of this rich biological community. This new book presents the latest research in the field rom around the world.

Short Communication A - Bare soil forms an important part of the surface cover in Mediterranean arid and semiarid areas dominated by *Stipa tenacissima* L. tussock grass. Some studies have stressed the relevance of the spatial arrangement of this species in the improvement of sediment dynamic and soil characteristics close to the tussock. Reciprocally, bare soil condition has been considered as a key factor in the physiological performance of *S. tenacissima* stands. On the other side, other studies have neglected the capacity of the root system of this species to explore and consume resources from bare soil surrounding the tussock. In the present work the authors show some evidences that suggest the importance of bare soil near to the tussock in the water status of this species, especially in the high water stress season. The authors have found that water gains in bare soil from water vapour adsorption were highly coincident with the *S. tenacissima* stand transpiration. Furthermore, *S. tenacissima* sub-populations living in rock outcrops with lower connectivity with the surrounded bare soil showed higher signs of physiological water stress than those stands in areas with more soil availability. The authors believe that an interaction of water dynamics between bare soil and *S. tenacissima* tussocks would have to be reconsidered to fully understand ecophysiology of this species.

Short Communication B - The decline in crop production is commonly observed on acid soil, further, a multiple heavy metal pollution except for Al on acid soil is detected in many areas. The present experiment was conducted by using two barley genotypes differing in Al-tolerance, Al tolerant *cv.* Gebeina and Al sensitive *cv.* Shang 70-119, to determine the effect of Al, Cd, Cu individual, binary and ternary combinations on the photosynthetic characteristics and the contents of sucrose and starch in leaves of barley seedlings. The results showed that there were significant decreases in leaf area, leaf fresh and dry weight, chlorophyll content, net photosynthesis (Pn), stomatal conductance (gs), *Fv/Fm* ratio evaluated for chlorophyll fluorescence and sucrose contents, but obvious increases in internal CO_2 concentration (Ci) and starch contents in barley leaves when exposed to the stress medium. These decreases or increases were more marked in Shang 70-119 than in Gebeina. In addition, binary metals combinations of Al + Cd and Al + Cu both resulted in severe changes in these values for two genotypes when compared with the Al alone treatment, indicating the existence of notable synergistic interactions between Al and Cu or Al and Cd, in particular for Shang 70-119. However, ternary metals combination of Al + Cu + Cd produced different interactions in two kinds of genotypes, thus, the slighter toxicity caused by ternary metals mixture than that caused by binary metals mixture was observed in Gebeina, while the reverse results showed in Shang 70-119, suggesting the notable synergistic effect of the three metals existed in sensitive genotype but not in tolerant one. So, it can be concluded that the different responses to acid soil toxicity of two barley genotypes may partly result from the different metals interactions in plant seedlings.

Chapter 1 - The Mediterranean area is characterized by high temperature and low moisture in summer seasons. The aim of this study was to summarize and compare the decomposition dynamics of eight litter types (*Pinus laricio*, *P. pinea, Quercus ilex, Cistus incanus*, *Myrtus communis, Phillyrea angustifolia, Abies alba* and *Fagus sylvatica*) in their natural sites followed for about 2 years. The effect of litter quality was evaluated by comparing lignin, cellulose and nitrogen content. The effect of site characteristics and microclimatic conditions on decomposition was also evaluated by exposing the same litter simultaneously in different stands.

The dominant effect on litter decomposition in Mediterranean area was summer aridity because it highly reduces the activity and growth of microorganisms, affecting also the interactions with litter quality and soil organisms. The decomposition rate was positively related to the Lang aridity index for all the studied litter even if no statistical significance was evidenced. Microbial activity, that was high in autumn and spring, was strongly reduced in summer as evidenced by the high decrease of soil respiration and the activity of cellulases and xylanases. Seasonal fluctuations also occurred for lignin degrading peroxidases, in agreement with the view that the overall microbial community was affected. The overall activity of lignin degrading laccases did not show significant seasonal changes even if its isoenzymes forms were seasonally expressed. N- content on decomposition rate was in line with that reported for temperate areas even if the influence of climatic conditions makes less evident the correlation.

The effect of nutrient, like nitrogen, on decomposition rate was in line with that reported for temperate areas even if the influence of climatic conditions makes less evident the correlations.

The positive relation between lignin decay rate and the initial Mn content support the role of Mn dependent enzymes in such a process. However, the Mn poor *P. pinea* litter was not affected by its low Mn initial content when incubated in the original native stand.

In the litter colonizing microflora pectinolytic and cellulolytic fungi were aboundant from the start of decomposition, followed by the chitinolytic ones that increased after six months of decomposition. Ligninolytic fungi, even if isolated in the first month of decomposition increased when lignin decay started. Considering their taxonomic diversity species of typical genera of the Mediterranean area were mixed to those of ubiquitous genera found also in other regions.

Chapter 2 - In Mediterranean regions climatic seasonality and exerted for millennia human impacts such as grazing and deliberate fire drive the structure and function of ecosystems. Recently, intensive agriculture, invasion of exotic species and atmospheric CO_2 enrichment are also involved. The function of soil subsystem is important for ecosystem maintenance providing nutrients released through microbially mediated decomposition.

Specifically in shallow and nutrient-limited Mediterranean soils, the contribution of soil microbial communities to decomposition is even more important.

The aim of this paper is to examine the effect of climatic and human impacts on characteristics of the soil microbial communities; biomass, composition and activity. Estimates of biomass based on immobilized C and N, total PLFA or counts of bacteria and fungi. For describing structural diversity PLFA and FAME results are used, while for functional diversity BIOLOG data are considered. Results provided by molecular techniques support our discussion of genetic diversity. Finally, data on respiration and enzyme activities are used to assess microbial activity.

The survey shows that vegetation type exerts an influence on soil microbial features. However, in comparison with seasonality the effects of plant species are of lower importance. Grazing, by affecting vegetation patterning and soil compactness, is expected to exert a great influence on soil microbial characteristics. Nevertheless, when seasonality and grazing are compared variations due to season are greater than those induced by grazing. Similar results emerge when seasonality is compared with agricultural practices. The available information concerning the effect of fire and species' invasion on microbial communities is really sparse. Burning reduces microbial diversity and most enzyme activities while invasion by exotic species alters the rhizosphere microbial communities with the magnitude of influence to be increased with time after invasion. CO_2 enrichment, through its effect on plant productivity, affects significantly soil structure. However, no effects on microbial community structure and biomass are detected. Treatment with herbicides, compost or biosolids changes the activity, size and composition of microbes due to physicochemical characteristics of the amendments. Finally, transition from conventional to organic farming does not results in gradual changes in the characteristics of the microbial community.

Chapter 3 - The concept of bacteria living as single isolated cells has been replaced by the vision of these organisms as members of communities that mostly work coordinately, with the aid of communication systems of variable complexity that make possible for every single cell to perceive the rest of bacteria sharing an habitat and talk to them. The languages of bacteria rely on chemical signals, synthesized by the cells and sent out to the surrounding media, where their concentration rises correlated to the increase of cell numbers. When a certain threshold cell density (quorum) is reached, the accumulated chemical signals trigger a coordinated population response, often moved by the need of adaptation to a change in the environment. The term "quorum sensing" was coined to describe this ability of bacteria to monitor the size of their populations before taking the decision of expressing a certain genotype. There are many different molecules released as signals by bacteria, being the best characterized to date acyl-homoserine lactones (AHLs), autoinducing oligopeptides (AIPs), and autoinducer 2 (AI-2). Amongst the known phenotypes regulated by quorum sensing, there are a number of functions related to both beneficial and pathogenic interactions between bacteria and eukaryotic organisms. Bacteria that interact with plants are not an exception, and use sophisticated quorum sensing systems whose involvement on the regulation of important steps for survival and competition in the rhizosphere, root colonization, and establishment of symbiotic or pathogenic associations with plants, has been thoroughly investigated during the last 20 years, yet leaving many questions unanswered. This chapter reviews the basic concepts of the quorum sensing mechanism and its implications, with particular emphasis on the current knowledge of its importance in the interaction of legume plants and nodule-forming bacteria.

Chapter 4 - Soil structure plays a dominant role in the physical protection of soil organic matter by controlling microbial access to substrate, microbial turnover processes, and food web interactions. Good soil structure results in soil productivity, a cornerstone of agricultural sustainability. While there is a wealth of knowledge about soil aggregation, soil microbial biomass and microbial diversity of soil, there is little knowledge of the microbial community ecology of soil aggregates. This review intends to expand upon and examine the microbial nature of soil aggregation: species causal to aggregation and functional groups involved. One focal point will be to examine how a quest for identifying and characterizing key species associated with microaggregates can have implications for management practices to improve soil aggregation and ultimately soil structure. The review will attempt to identify some promising avenues for future research in this area of soil biology that is a central one to soil quality. The authors' goals are to catalyze rigorous, innovative research on current approaches and techniques on the microbial ecology of soil aggregation.

Chapter 5 - The widespread pollution of soils is an increasing urgent problem because of its contribution to environmental deterioration on a global basis. Several toxic compounds, such as heavy metals, often contaminate soils. The main sources of heavy metal pollution are mining, industries and application of metal-containing pesticides, fertilizers and sewage sludge. In recent decades there has been increasing concern with heavy metal, not only because of their toxicity to animals, plants and microorganisms, but also because they are highly toxic, mutagenic and/or carcinogenic to humans.

Due to their small size, which provides a large contact area that can interact with the surrounding environment, microorganisms are the first biota showing the impact of toxic compounds. Microorganisms being in intimate contact with the soil environment are considered to be the best indicators of soil pollution. In general, they are very sensitive to low concentrations of contaminants and provide a rapid response to soil perturbation.

Rhizobium spp. are ubiquitous gram-negative soil bacteria that have a profound scientific and agronomic significance due to their ability to establish nitrogen-fixing symbiosis with legumes, which is of major importance to the maintenance of soil fertility.

There is increasing evidence of the adverse effects of heavy metals on soil microbial processes, including on soil enzymatic activities. Soil enzymes are the driving force behind all the biochemical transformations occurring in the soil. Enzymes catalyse all biochemical reactions and are an integral part of nutrient cycling and soil fertility. Therefore, this chapter evaluated the impact of heavy metals on *Rhizobium* populations isolated from a lead mine which activity ceased 50 years ago. In order to reach this goal some physicochemical parameters that influence metal bioavailability was determined, as well as metal concentrations in soils. Soil enzyme activities are highly affected by soil conditions and their evaluation may provide useful information on soil microbial activity and survival. For this reason soil enzyme activities have been proposed as biological indicators of pollution, specially organic, but information about the influence of heavy metal on soil enzyme activities is scarcer. Thus, it was determined the activity of enzymes such as dehydrogenases, hydrolases, phosphatases, catalase and lipase in heavy metal contaminated soils. Metal tolerance of *Rhizobium* isolates was also screened in artificial media supplemented with different metals (Pb, As, Cd, Cu, Co and Cr) and their tolerance related to soil contamination and enzyme activities.

This chapter can widen the knowledge about the pressure that soil microflora experience under the direct effect of different metals. *Rhizobium* and soil enzyme activities may be useful for the evaluation of agricultural soils pollution, which may be used on the improvement of soil productivity or on the reclaim of contaminated soils.

Chapter 6 - Suppressiveness of pumice, which had been used for 13 years for continuous cropping, was confirmed against bacterial wilt of tomato caused by *Ralstonia solanacearum*, in comparison with unused pumice. Since there were significant differences in some of the chemical and biological properties, suppressive mechanisms of the used pumice were investigated. Contribution of pH, EC and higher amounts of salts, such as Ca, to the suppression mechanisms appeared to be very low. In contrast, microbial biomass and respiration were significantly higher in the used pumice than in the unused pumice and the suppressiveness of the used pumice disappeared after sterilization by autoclaving and gamma irradiation. These results suggested that biological factors rather than chemical factors may be involved in the suppression mechanisms of bacterial wilt in the used pumice. The result of substrate-induced respiration inhibition method indicated that bacteria, rather than fungi, may be the predominant microbial community in the used pumice. Only one isolate, designated as *Burkholderia* sp. W3, showed a suppressive effect on bacterial wilt among 50 dominant bacterial colonies obtained from tomato roots grown in used pumice. When the strain W3 was inoculated into autoclaved used pumice, suppressive effect completely recovered. The bacterial communities of tomato roots evaluated by PCR-DGGE were different between the unused and the used pumice, and roots grown in the used pumice showed more diverse bacterial community. The band corresponding to W3 was not observed in the unused pumice, but there was the band in the used pumice, suggesting that W3 was an initially minor bacterium in tomato roots, but became a major colonizer after repeated cropping. These results may suggest that W3 is involved in one of the major mechanisms in the suppression of bacterial wilt in the used pumice. Unused pumice inoculated with W3 showed higher resistance to bacterial wilt, compared with that without inoculation, and this resistance was further enhanced by the addition of xlyose and glucose. It was further confirmed in small pot and greenhouse experiments that the application of biocontrol agent with substrates, such as lysine, available for the antagonist and not for the pathogen enabled more stable disease suppression in unused pumice. Possible methods to make unused pumice suppressive to bacterial wilt are presented.

Chapter 7 - Phytoextraction, which is the use of plant to extract heavy metals from polluted soils, could be limited by the high toxicity of these elements to plant development. Many data suggest that heavy metals-induced nutritional disturbance is one of the major causes of the plant growth restriction. The present work aims to check the validity of this hypothesis in two halophytes species: *Sesuvium portulacastrum* and *Mesembryanthemum crystallinum* cultivated in the presence of Cd and Ni. Seedlings were grown for 30 days in split-root conditions. Five treatments were applied: in the first two treatments, one half of the root system was immersed in a basal medium (B), while the other half was in the same solution supplemented with 100 µM Cd^{2+} (B/Cd plants) or 50 µM Ni^{2+} (B/Ni plants). In the other treatments, the two halves of the root system were immersed either in a metal-free medium (B/B control plants) or in basal medium containing 100 µM Cd^{2+} (Cd/Cd plants) or 50 µM Ni^{2+} (Ni/Ni plants). At the harvest, dry weight as well as the Cd^{2+}, Ni^{2+}, K^{+}, Ca^{2+} and Fe concentrations in tissues were determined. As compared to Cd/Cd and Ni/Ni treatments, culture on dual medium (B/Cd and B/Ni) alleviated significantly the effects of Cd^{2+} and Ni^{2+} on growth, attenuated the leaf toxicity symptoms and led to appropriate shoot K^{+} and Fe amounts in spite of relatively high Cd^{2+} and Ni^{2+} concentrations. Ca^{2+} status was not modified by Cd^{2+} or Ni^{2+} in *Sesuvium*, but it was decreased by Cd/Cd and Ni/Ni treatments in *Mesembryanthemum*. However, for the latter species, B/Cd and B/Ni plants showed appropriate Ca^{2+} shoot amounts. Hence, the results indicate that nutritional disturbances induced by Cd^{2+} or Ni^{2+}, contributed largely to the growth restriction in both halophytes leading to limitation of heavy metals extracted in the shoots. So, the authors suggest the possibility to enhance the capacity of both species to extract these metals by increasing nutrient availability in soil.

Chapter 8 - Exceptional occurrences of permafrost exist in the forest belt well below the limit of discontinuous alpine permafrost. There are several such sites in the Alps, all steep scree slopes located at the foot of high limestone cliffs. A particularly interesting and important characteristic is that they are vegetated with subalpine vegetation and that some plants display signs of severely limited growth. Research carried out in the last years has demonstrated that the ground is not frozen as a result of a particularly cold microclimate at these locations, but that permafrost is present due to a particular air circulation phenomenon. Under these conditions, soil organic matter accumulates due to the slowing down in degradative processes of plant residues. Consequently, the nutrient availability to plants may be limited and contribute to the reduced plant growth, although scant attention has been devoted to this topic.

To determine vegetation and soil characteristics, tree height, tree ring width measurements and soil analysis were performed in two contiguous areas differently affected by frost conditions in Western Switzerland and characterized by different growth of trees: (I) dwarf trees; (II) reference trees. Near surface soil temperature in the two sites has been collected from a whole winter (November 2001-May 2002). In the same period the N and C dynamics in the organic OH horizon were measured by the buried bag technique and in the undisturbed soils.

Tree growth was significantly lower in the permafrost affected sites, where the winter soil temperature was significantly lower than the reference forest.

During winter an increase of nitrate concentration was recorded in all sites, providing an inorganic N pool ready available for plant growth under the reference forest, but not under the dwarf trees, due to the lower soil temperature which inhibit plant nutrient absorption. Moreover, under the dwarf forest the microbial N immobilization, with a corresponding DON and NH_4^+ decrease, was more evident than in the reference site. The critical conditions under the dwarf trees could have selected a microbial community particularly tolerating cold temperatures, and then more resistant to moderate freeze/thaw events.

Thus, an inorganic N pool, constituted mainly by the leachable NO_3^-, is available in the early growing season in the cold site and in the reference site, but the lower soil temperature under the dwarf trees may inhibit soil nutrient adsorption by plants. Therefore in the cold site there is as asynchronies between the availability of nutrients and their utilization, which may affect the plant growth.

Chapter 9 - The chemical characteristics of soil organic matter (SOM) can be influenced by management and amendment practices which effects can be measured only after long-term experiment. In this long-term study of over 30 years, with a rotation wheat-corn, the authors compared the effects of adding cattle manure (CM) and crop residues (CR) wheat straw or corn-stalks after each crop, on humic substances (HS). Potentiometric titration, thermal analysis (TG-DTA), and spectroscopic methods such as diffuse reflectance infrared Fourier (DRIFT) and liquid nuclear magnetic resonance (^{13}C NMR) spectroscopies were used in order to investigate humic acid (HA) structure. The amendment practices clearly influenced the humic C and the COOH groups content that only increased in CM treatment. The quality of this humic fraction was affected by the different agricultural practices, so that when the soil did not receive any amendment, the aromatic and carboxylic C decreased, whereas the aliphatic C increased as an effect of the crop rotation. With the amendments, in contrast, the aromatic C generally increased, this increase was mainly due to the incorporation of aromatic groups in the structure of HA, arising from the phenolic groups present in the lignin of the crop residues.

Chapter 10 - Soil organic carbon (C) preservation in agroecosystems is crucial point to maintain soil fertility and productivity, and to reduce losses of CO_2 in the atmosphere. Agricultural management practices can differently affect the level of soil organic C (SOC). In this chapter the results of a long-term field experiment (30 years) were investigated to evaluate the effect of mineral fertilization and organic amendments on soil organic C content and on the humic acids (HA) that represent the most important and stable reservoir of soil organic C. The effect of the plant species was also evaluated by comparing wheat and corn monocultures. The amount of corn-derived C in soil and HA at the end of the experiment was calculated by ^{13}C natural abundance measurements. After 30 years of cultivation, the SOC significantly decreased in both unfertilized (Control) cropping systems, especially with continuous corn. Mineral fertilization (Min) and organic amendments (Org) always caused an increase in SOC, especially with Org treatment on continuous corn. The C always increased in HA, except in the unfertilized plots of corn monoculture. Again the highest increase was observed with Org treatment. The amount of corn-derived C in total organic C (TOC) increased in the following order: unfertilized < Min < Org treatments, ranging from 19 to 29%. The turnover time of the older C_3-derived C increased in the same order ranging between 55 and 86 years in the Control and Org treatment, respectively. In the HA the proportion of corn-derived C was similar in the Control and Org treatment (26.4%), lower in the Min treatment (23.7%). Nevertheless if the authors consider the total amount of corn-derived C in soil and the proportion recovered in the HA, the highest was measured in the unfertilized control. In general a proportion ranging from 35% (Org) to 40% (control) of the total corn-derived C in soil was recovered as humic C, confirming the important role of this pool as a C reservoir in soil.

Chapter 11 - Salinity is a major environmental constraint of crop production, and with the climate changes that are being announced for the next decades, like global warming or local reduction of rainfall, this problem will be amplified. Salinity stress negatively impacts agricultural yield throughout the world affecting production whether it is for subsistence or economic gain. *Rhizobium* has considerable scientific, economical and ecological interest because of their ability to establish nitrogen-fixing nodules on leguminous hosts. This feature enables plants to grow in soils with low nitrogen levels, to achieve good crop yields without massive nitrogen fertilization, and, as a consequence, to decrease the contamination of water reservoirs by inorganic nitrogen compounds. Salinity not only affects free-living rhizobia but also considerably restrains the nodulation process and symbiotic nitrogen fixation. To fix nitrogen in saline environments leguminous plants require both free-living rhizobia and hosts tolerant to salt. Therefore, the selection of tolerant phenotypes, which can withstand the negative impact of saline soils, can be of great use to improve nitrogen fixation and productivity in salt-affected soils.

In this work *Rhizobium* was isolated from several locations in Portugal with different environmental conditions. *Rhizobium* isolates were screened for their tolerance to salinity as free-living organisms and for their efficiency to fix N_2 under salt conditions in symbiosis with a legume. Furthermore, *Rhizobium* protein and plasmid profiles were evaluated, in order to identify variability and to relate it with salt tolerance.

To accomplish these goals isolates were grown in YEM supplemented with NaCl (25 to 1800 mM). According to their growth responses isolates were classified in three groups: sensitive (0-50 mM NaCl), tolerant (100-500 mM NaCl) and extremely tolerant (600-1800 mM NaCl). Salt tolerance was a reflex of the conditions experienced in their natural habitats. Forty-one polypeptides were separated by SDS-PAGE. Salt conditions induced differences in protein profiles when compared with controls. Extremely tolerant strains were the less affected, suggesting the putative presence of constitutive mechanisms conferring tolerance to NaCl. Strains displayed different plasmid profiles and classification analysis suggests that the presence of certain plasmids (828, 734, 147 and 82 MDa) can be correlated to salt tolerance. Finally, tolerant strains were demonstrated to efficiently nodulate *Pisum sativum* plants, in the absence of salt, and more importantly in the presence of moderate levels of sodium chloride, which, due to the announced climate changes, could be important in a near future to maintain the present areas of legume cultivation. This is of particular importance because it points out the need of further studies to predict the influence of climate alterations on soil microbial populations. Under this context, because *Rhizobium* is important to several natural and agricultural communities, it may be potential use as an indicator and enabler of agricultural sustainability in affected soils.

In: Soil Ecology Research Developments
Editors: Tian-Xiao Liu, pp. 1-6
ISBN: 978-1-60021-971-9

Expert Commentary

SOIL CHARCOAL AMENDMENTS MAINTAIN SOIL FERTILITY AND ESTABLISH A CARBON SINK – RESEARCH AND PROSPECTS

Christoph Steiner[1]
Institute of Soil Science and Soil Geography, University of Bayreuth,
95440 Bayreuth, Germany

CHARCOAL AS SOIL AMENDMENT

Sustaining soil fertility is a major agricultural constraint in the Amazon Basin (Tiessen, et al., 1994), thus shifting cultivation accompanied with slash and burn agriculture is the prevailing agricultural practice in the humid tropics.

In addition to the predominant and unproductive Ferralsols and Acrilols, an exceptional dark soil is well known by the indigenous people and colonists for its sustained soil fertility. According to its dark color and origin ,the soil was termed *Terra Preta de Indio*. Smith (1879) and Katzer (1903) were among the first who described the *Terra Preta's* properties and presumed its cultural origin. Later their assumptions were strengthened by the soil scientists Sombroek (1966) and Zech, et al. (1990). *Terra Preta* contains significantly more carbon (C), nitrogen (N), calcium (Ca), and phosphorus (P), and the cation exchange capacity (CEC), pH value, and base saturation are significantly higher in *Terra Preta* soils than in the surrounding Ferralsols and Acrisols (Zech, et al., 1990; Glaser, et al., 2000). *Terra Preta* soils contain up to 70 times more black C than the adjacent soils. Due to its polycyclic aromatic structure, black C is chemically and microbially stable and persists in the environment over centuries (Glaser, et al., 2001b). Today, the anthropic origin of *Terra Preta de Indio* is generally accepted and its fertility is most likely linked to an accumulation of P and Ca, associated with bone apatite (Zech, et al., 1990; Lima, et al., 2002) and black C as charcoal (Glaser, et al., 2001a). The evidence that *Terra Preta* was manmade, and thus the proven feasibility to transform one of the most infertile soils into one of the most productive soil inspired charcoal research.

Slash and Char was described as an alternative to slash and burn (Lehmann, et al., 2002) and (Steiner, et al., 2004b) observed that charcoal is currently used by Amazonian settlers to improve soil fertility. If a forest is burned, only around 2-3% of the above-ground C is converted into charcoal (Fearnside, et al., 2001), but charcoal production can capture 50% of the above-ground C. If the charcoal is not used as fuel (e. g., soil amendment) it has a very high recalcitrance against biological or chemical decay and stores the carbon over centuries or millennia. *Slash and char* is an alternative agricultural method producing charcoal out of the aboveground biomass instead of converting it to CO_2 through burning. If re-growing resources are used, *slash and char* could establish as a significant carbon sink and could be an important step towards sustainability and SOM conservation.

On a global scale, the total carbon release flux due to fire is of the order of 4-7 Pg of carbon per year. This flux is almost as large as the rate of fossil fuel consumption (about 6 Pg per year in 1990) (Goudriaan, 1995). Fearnside (2000) calculated a total net emission of carbon from tropical land uses, equivalent to approximately 29% of the total anthropogenic emission from fossil fuels and land-use change. These numbers emphasize the potential for C management if only biomass is utilized being ablaze each year.

As a result of intensive research done in the 1980s, carbonized materials are formally authorized for use as soil amendment material in Japan. Thus Japan used 27% of the total charcoal consumption (50,835 Mg in 1999) for purposes other than fuel. By far the biggest proportion was used for agriculture (30.6%) followed by livestock industries (22.3%), gardening and golf courses (7.6 %) (Okimori, et al., 2003). A Japanese company established charcoal production at an Indonesian tree plantation for pulp production. Their feasibility study with conventional charcoal-making methods showed that 77,000 Mg charcoal could be produced per year, and the carbon emission reductions by the project reaches 62,000 Mg C yr^{-1} (= 230,000 Mg CO_2 yr^{-1}) at an annual wood harvest of 10,000 ha (Okimori, et al., 2003).

SOIL FERTILITY ENHANCEMENT

Antal and Grønli (2003) mentioned that most potting soils, herbicides in carbon-based formulations, and culture media formulations contain charcoal or activated carbons, although the scientific rationale for these applications is absent. Recent studies showed that soil charcoal amendments are indeed capable of increasing soil fertility. Charcoal significantly increased plant growth and nutrition in a pot experiment by Lehmann, et al. (2003) and a field experiment by Steiner, et al., (2007c). The authors proposed that charcoal can improve soil chemical, biological, and physical properties, but could not completely discern the mechanisms of fertility enhancement. Lehmann, et al., (2003) found significantly reduced leaching of applied fertilizer N in charcoal containing pots. This was corroborated by the findings of Steiner, et al., (2007a in submission) and Steiner, et al., (2007b in submission).

Soil respiration and the microbial population growth rate were found significantly altered by charcoal amendments. Steiner, et al., (2004a) found increased microbial activity on charcoal amended plots. *Terra Preta* soils were marked by a very low soil respiration but very high population growth after substrate (glucose) additions. Unmanaged forest soils (Ferralsol) had a higher respiration rate but a very low population growth potential. These results reflect

1 e-mail: Christoph.Steiner@biochar.org.

the relatively high biodegradable soil organic matter (SOM) content of primary forest topsoil but low available nutrients (requirement for microbial population growth), in contrast to refractory *Terra Preta* SOM with high available soil nutrient contents. Thus we conclude that nutrient availability in *Terra Preta* is independent from SOM decomposition. The effects on soil biology seem to be essential as charcoal has the potential to alter the microbial biomass (Steiner, et al., 2004a) and composition (Birk, 2005) and the microbes are able to change the charcoal's properties (Glaser, et al., 2001a). Rondon, et al. (2006) found increased biological N fixation by common beans through charcoal additions and Gehring (2003) increased occurrence of nodulating plants in forests on *Terra Preta* than on adjacent soils.

Biochar Prospects and Essential Research

The global potential of biochar (non fuel use charcoal) reaches far beyond *slash and char*. Inspired by the recreation of *Terra Preta*, most biochar research was restricted to the humid tropics. More information is needed on the agronomic potential of charcoal, the potential to use alternative biomass sources (crop residues) and production of by-products to evaluate the opportunities for adopting a biochar system on a global scale. Biochar as soil amendment needs to be studied in different climate and soil types. Today, crop residue biomass represents a considerable problem as well as new challenges and opportunities.

A system converting biomass into energy (hydrogen-rich gas) and producing charcoal as a by-product might offer an opportunity to address these problems. Charcoal can be produced by incomplete combustion from any biomass, and it is a by-product of the pyrolysis technology used for biofuel and ammonia production (Day, et al., 2005).

The acknowledgement of biochar as carbon sink would facilitate C-trading mechanisms. Although most scientists agree that the half life of charcoal is in the range of centuries or millennia, a better knowledge of the charcoal's durability in different ecosystems is important to achieve this goal. An access to the C trade market holds out the prospect to reduce or eliminate the deforestation of primary forest, because using intact primary forest would reduce the farmer's C credits. Fearnside (1997) estimated the above-ground biomass of unlogged forests to be 434 Mg ha^{-1}, about half of which is C. This C is lost if burned in a slash-and-burn scenario and lost at a high percentage if used for charcoal production. The C trade could provide an incentive to cease further deforestation; instead reforestation and recuperation of degraded land for fuel and food crops would gain magnitude. As tropical forests account for between 20 and 25% of the world terrestrial C reservoir (Bernoux, et al., 2001), this consequently reduces emissions from tropical forest conversion, which is estimated to contribute globally as much as 25 % of the net CO_2 emissions (Palm, et al., 2004).

Today most biomass gasification systems tend to suppress the creation of residuals, like total organic carbon (TOC) and ashes. C-emission trading options and a better knowledge of charcoal as soil additive would add value to these residues. Further, this would facilitate the use of alternative biomass, those which are currently avoided to due their higher TOC residuals. The tarry vapors constitute a significant loss of carbon during carbonization (Antal and Grønli, 2003), although representing another valuable product. Despite a lack of research, these condensed vapors are used for agricultural purposes mainly in Asia and Brazil (GERAIS, 1985; Glass, 2001; Steiner, et al., 2004b).

Japanese researchers attempt to produce charcoal with a specific pore size distribution to favor desired microorganisms (Okimori pers. communication). Pore structure, surface area, and adsorption properties are strongly influenced by the peak temperature during charcoal production (Antal and Grønli, 2003). Increasing porosity is achieved with increasing temperature but the functional groups are gradually lost. In this context, it is also important to discern the mechanisms of nutrient retention (mainly N) due to charcoal applications. The charcoal's low biodegradability (Kuhlbusch and Crutzen, 1995), low nutrient content (Ogawa, 1994; Antal and Grønli, 2003), and high porosity and specific surface area (Braida, et al., 2003) makes charcoal a rather exceptional SOM constituent. *Terra Preta* research has shown that oxidation on the edges of the aromatic backbone and adsorption of other organic matter to charcoal is responsible for the increased CEC, though the relative importance of these two processes remains unclear (Liang, et al., 2006).

Energy from crop residues could lower fossil energy consumption and CO_2-emissions, and become a completely new income source for farmers and rural regions. The biochar by-product of this process could serve to recycle nutrients, improve soils and sequester carbon. A review by Johannes Lehmann (2006) and the article "*Black is the new green*" (Marris, 2006) emphasize the potential of bio-char on a global scale. A global analysis by Lehmann, et al. (2006) revealed that up to 12% of the total anthropogenic C emissions by land use change (0.21 Pg C) can be off-set annually in soil, if slash and burn is replaced by slash and char. Agricultural and forestry wastes add a conservatively estimated 0.16 Pg C yr^{-1}. If the demand for renewable fuels by the year 2100 was met through pyrolysis, bio-char sequestration could exceed current emissions from fossil fuels (5.4 Pg C yr^{-1}). The described mixture of driving forces and technologies has the potential to use residual waste carbon-rich residues to reshape agriculture, balance carbon and address nutrient depletion.

REFERENCES

Antal, M. J. and Grønli, M. *Ind. Eng. Chem. Res.* 2003, 42, 1619-1640.

Bernoux, M.; Graça, P. M. A.; Cerri, C. C.; Fearnside, P. M.; Feigl, B. J. and Piccolo, M. C. In *The biogeochemistry of the Amazon Basin*; McClain, M. E.; Victoria, R. L. and Richey, J. E.; Ed.; Oxford University Press: New York, NY, 2001; pp 165-184.

Birk, J. J. 2005, *Lehrstuhl für Bodenkunde und Bodengeographie*, University of Bayreuth, Diplomarbeit.

Braida, W. J.; Pignatello, J. J.; Lu, Y. F.; Ravikovitch, P. I.; Neimark, A. V. and Xing, B. S. *Environ. Sci. Technol.* 2003, 37, 409-417.

Day, D.;Evans, R. J.;Lee, J. W. and Reicosky, D. *Energy.* 2005, 30, 2558-2579.

Fearnside, P. M. *Climatic Change.* 1997, 35, 321-360.

Fearnside, P. M. *Climatic Change.* 2000, 46, 115-158.

Fearnside, P. M.; Lima, P. M.; Graça, A. and Rodrigues, F. J. A. *Forest Ecol. Management.* 2001, 146, 115-128.

Gehring, C. 2003, University of Bonn, *PhD theses.*

GERAIS, F. C. T. D. M. STI/CIT, Brasilia, *Documentos* 23, 1985, pp 18.

Glaser, B.; Balashov, E.; Haumaier, L.; Guggenberger, G. and Zech, *W. Org Geochem.* 2000, 31, 669-678.

Glaser, B.; Guggenberger, G.; Haumaier, L. and Zech, W. In *Sustainable management of soil organic matter*; Rees, R. M.; Ball, B. C.; Campbell, C. D. and Watson, C. A.; Ed.; CABI Publishing: Wallingford, UK, 2001a; pp 190-194.

Glaser, B.; Haumaier, L.; Guggenberger, G. and Zech, *W. Naturwissenschaften.* 2001b, 88, 37-41.

Glass, V. *Globo Rural.* 2001, vol. 188.

Goudriaan, J. In Carbon sequestration in the biosphere. Processes and prospects.; Beran, M. A.; Ed.; *Global environmental change*; Springer Verlag: Heidelberg, Berlin, 1995; 33, pp 3-18.

Katzer, F. *Grundzüge der Geologie des unteren Amazonasgebietes* (des Staates Pará in Brasilien). M. Weg: Leipzig, Germany, 1903.

Kuhlbusch, T. A. J. and Crutzen, P. *J. Global Biochem. Cy.* 1995, 9, 491-501.

Lehmann, J.; da Silva Jr, J. P.; Rondon, M.; Cravo, M. d. S.; Greenwood, J.; Nehls, T.; Steiner, C. and Glaser, B. *17th World Congress of Soil Science*, 2002, Bangkok, Thailand; The International Union of Soil Sciences.

Lehmann, J.; da Silva Jr., J. P.; Steiner, C.; Nehls, T.; Zech, W. and Glaser, B. *Plant Soil.* 2003, 249, 343-357.

Lehmann, J.; Gaunt, J. and Rondon, M. *Mitigation and Adaptation Strategies for Global Change*. 2006, 11, 403-427.

Liang, B.; Lehmann, J.; Solomon, D.; Grossman, J.; O´Neill, B.; Skjemstad, J. O.; Thies, J.; Luizão, F. J.; Petersen, J. and Neves, E. G. *Soil Sci. Soc. Am.. J.* 2006, 70, 1719-1730.

Lima, H. N.; Schaefer, C. E. R.; Mello, J. W. V.; Gilkes, R. J. and Ker, J. C. *Geoderma.* 2002, 110, 1-17.

Marris, E. *Nature*. 2006, 442, 624-626.

Ogawa, M. *Farming Japan*. 1994, 28-5, 10-30.

Okimori, Y.; Ogawa, M. and Takahashi, F. *Mitigation and Adaptation Strategies for Global Change*. 2003, 8, 261-280.

Palm, C.; Tomich, T.; Noordwijk, M. v.; Vosti, S.; Gockowski, J.; Alegre, J. and Verchot, L. *Environment, Development and Sustainability*. 2004, 6, 145-162.

Rondon, M. A.; Lehmann, J.; Ramírez, J. and Hurtado, M. *Biol Fertil Soils*. 2006, DOI 10.1007/s00374-006-0152-z.

Smith, H. H. Brazil, *the Amazons and the coast*; C. Scribner's Sons: New York, NY, 1879.

Sombroek, W. G. 1966, Wageningen, The Netherlands, PhD theses.

Steiner, C.; Garcia, M. and Zech, W. In *Terra preta nova*: *A tribute to wim sombroek;* Woods, W. I.; Teixeira, W.; Lehmann, J.; Steiner, C. and WinklerPrins, A.; Ed.; Springer: 2007a; in preparation.

Steiner, C.; Glaser, B.; Teixeira, W. G.; Lehmann, J.; Blum, W. E. H. and Zech, W. J *Plant Nurtr. Soil. Sc.* 2007b, in submission.

Steiner, C.; Teixeira, W. G.; Lehmann, J.; Nehls, T.; Macêdo, J. L. V. d.; Blum, W. E. H. and Zech, *W. Plant Soil.* 2007c, 291, 275-290.

Steiner, C.; Teixeira, W. G.; Lehmann, J. and Zech, W. In *Amazonian Dark Earths: Explorations in space and time*; Glaser,. B. and Woods, W. I.; Ed.; Springer Verlag: Heidelberg, 2004a; pp 195-212.

Steiner, C.; Teixeira, W. G. and Zech, W. In *Amazonian Dark Earths: Explorations in space and time*; Glaser, B. and Woods, W. I.; Ed.; Springer Verlag: Heidelberg, 2004b; pp 183-193.

Tiessen, H.; Cuevas, E. and Chacon, P. *Nature.* 1994, 371, 783-785.

Zech, W.; Haumaier, L. and Hempfling, R. In *Humic substances in soil and crop sciences; selected readings*; Ed.; ASA-SSSA: Madison, WI, 1990; pp 187-202.

In: Soil Ecology Research Developments
Editors: Tian-Xiao Liu, pp. 7-17
ISBN: 978-1-60021-971-9

Short Communication A

WATER INTERACTIONS BETWEEN BARE SOIL AND VEGETATION IN SEMIARID MEDITERRANEAN STEPPES: SOME NEW EVIDENCES

David A. Ramírez[1,2], Francisco Domingo[3,4] and Juan Bellot[1]
[1] Departamento de Ecología, Universidad de Alicante Ap. 99-E-03080, Alicante, Spain.
[2] Laboratorio de Ecología de Procesos, Dpto. Biología, Universidad Nacional Agraria La Molina, Ap.456, Lima, Perú
[3]Estación Experimental de Zonas Áridas, Consejo Superior de Investigaciones Científicas, 04001 Almería, Spain
[4] Departamento de Biología Vegetal y Ecología, Escuela Politécnica Superior, Universidad de Almería, 04120 Almería, Spain

ABSTRACT

Bare soil forms an important part of the surface cover in Mediterranean arid and semiarid areas dominated by *Stipa tenacissima* L. tussock grass. Some studies have stressed the relevance of the spatial arrangement of this species in the improvement of sediment dynamic and soil characteristics close to the tussock. Reciprocally, bare soil condition has been considered as a key factor in the physiological performance of *S. tenacissima* stands. On the other side, other studies have neglected the capacity of the root system of this species to explore and consume resources from bare soil surrounding the tussock. In the present work we show some evidences that suggest the importance of bare soil near to the tussock in the water status of this species, especially in the high water stress season. We have found that water gains in bare soil from water vapour adsorption were highly coincident with the *S. tenacissima* stand transpiration. Furthermore, *S. tenacissima* sub-populations living in rock outcrops with lower connectivity with the surrounded bare soil showed higher signs of physiological water stress than those stands in areas with more soil availability. We believe that an interaction of water dynamics between bare soil and *S. tenacissima* tussocks would have to be reconsidered to fully understand ecophysiology of this species.

INTRODUCTION

The observation and enjoyment of the structural complexity of semiarid landscapes has stimulated a considerable amount of research aimed at describing and cataloguing the processes that support such variability. As it is well known to ecologists, the description of the structure of the landscape is fundamentally a matter of scale. For instance, landscapes dominated by semiarid Mediterranean steppes can be described using at least two biologically meaningful spatial scales, namely a coarsed-grained scale defined by the slopes as a fine-grained scale defined at the individual plant level within slopes. One source that justifies and connects both scales is the behavior of hydrological fluxes and its relation to the abundance and distribution of soil. Indeed hydrological fluxes can be regarded as landscape-modelling forces that condition the disposition of the vegetation in the terrain and the ecophysiological performance of individual plants. A fine example of such interactions can be found in semiarid Mediterranean steppes dominated by *Stipa tenacissima* L. - a perennial, rhizomatous and tussock grass, which represent a prominent vegetation type in the Mediterranean basin (Le Houérou 2001). These ecosystems are characterized by being subjected to drastic environmental conditions, notably hot summers with severe rain shortages that cause an important light and water stress to the vegetation (Joffre et al. 1999). Characteristically, rangeland Mediterranean soils are often shallow, discontinuous, poorly structured and with rock outcrops, which represent critical constraints for the settlement and even the restoration of the vegetation in this region (Vallejo et al. 2000). Such environmental conditions result in low plant productivity, low plant cover and high susceptibility (sometimes an actual trend) to degradation and desertification (Cortina et al. 2004). The amount of bare-ground areas on *S. tenacissima* grasslands represent an important total figure in the Iberian Peninsula (between 40 - 82%, Maestre et al. 2007). The dynamic interaction between these areas and vegetated clumped patches is one important trait of these ecosystems. For instance, Puigdefábregas and Sánchez (1996) found that interaction occurs at different scales. At the micro-site scale, the overland flow of water and sediments is retained in the uphill part (or terrace) of the *S. tenacissima* tussock. This soil under the tussock has more infiltration (Cerdà 1997), abundance and stability of organo-mineral of aggregates (Puigdefábregas et al. 1999; Bochet el al. 1999), soil moisture (Puigdefábregas and Sánchez 1996), and organic matter content (Sánchez 1995; Bochet et al. 1999; Maestre et al. 2001) than bare soil areas, which helps create "resource islands" in this ecosystems (*sensu* Reynolds et al. 1999) that facilitate the establishment of other vascular plants (Maestre et al. 2001; Maestre 2006). In turn, the intensity of the accumulation of soil sediments influences the mortality of stems and the branching rate, which ultimately has a negative effect on the space occupation of the tussock (Sánchez and Puigdefábregas 1994). At the stand scale, the spatial pattern of *S. tenacissima* tussocks depends on topographic conditions because slope angle and length, catchment area influence the circulation of sediments (Puigdefábregas and Sánchez 1996). Thus, in hillsides with moderate sediment circulation the spatial pattern of the tussocks results in contour vegetated lines parallel to the hillside that allow the accumulation of sediment. However, under conditions of circulation of sediment such as those promoted by high slope angles or slope-length the tussock structure is broken, forming downhill strips. On the other hand, recent studies (Maestre and Cortina 2006) have proved that bare soil condition is a key factor to consider in order to understand the physiological performance of the *S. tenacissima* stands.

Thus the coarsed-grained scale can be briefly characterized and we can now focus in the "within stand" processes, that is, the interactions between water and soil at the individual plant level.

Some authors have measured (Puigdefábregas and Sánchez 1996) and pointed out (Maestre and Cortina 2006) the importance of water transported via runoff to the tussock terrace because this peculiar type of water notably improved morphological (leaf length and number per stem) and physiological performance (assessed using foliar $\delta^{13}C$, $\delta^{15}N$, nitrogen concentration and carbon to nitrogen ratio) in *S. tenacissima*. The lower root fines (Ø < 5mm) weight of this species found in bare soil (compared to those found under tussock) have been the main argument to play down the importance of water use via the tussock root system from bare soil (Puigdefábregas and Sánchez 1996; Puigdefábregas et al. 1999). However, the occurrence of water inputs from non-rainfall water phenomena and some ecophysiological traits that we shall show in this communication question the classical paradigm about the use of soil water in semiarid steppes.

THE PROCESS

In order to better address the role of soil water in semiarid steppes we firstly bring about the physical phenomenon product of the energy exchange between the soil and the atmosphere known as water vapour adsorption by soil (WVA). This phenomenon is likely to occur when the humidity of the soil pores is less than that in the adjacent atmosphere (see Agam and Berliner 2006, for a detailed explanation of this process). Kosmas et al. (1998) have highlighted the importance of water from WVA in semiarid environments. Indeed, these authors calculated that the total WVA was +26.25% of the rainfall water during February – August, 1996 period in Athens-Greece. Moreover, Kosmas et al. (2001) estimated that water gains from WVA can supply the 55.7 to 70.7% of the soil water evaporation demands. Agam and Berliner (2004, 2006) proved the paramount relevance of WVA, superior even to that of dew formation in the soil water gains in the Negev Desert - Israel. The main argument of these last authors was that the soil surface temperature did not reach the dew point temperature. Due to the importance of this type of energy exchange between the soil and the atmosphere, a brief account of the environmental factors that facilitate and limit the occurrence of WVA in arid and semiarid areas is shown in Table 1. It is noteworthy that some recent works have presented evidences of the occurrence of WVA in southern and southeastern areas of the Mediterranean Spain (Verhoef et al. 2006; Ramírez et al. 2007a). Moreover, Kosmas et al. (1998, 2001) suggested that WVA is an important water resource for vegetation in semiarid Mediterranean areas mainly during periods of high water stress. Below we provide with further evidences that support this hypothesis in a *S. tenacissima* steppe in SE Spain.

EVIDENCES

The evidences we aim to present in this review were obtained from ongoing studies carried out in *S. tenacissima* stands belonging to a South-facing micro-basin (19 ha, altitude and slope range from 479 to 800 m a.s.l. and 37 to 73% respectively) of the "El Ventós" Alicante, SE-Spain (38° 28' N, 0° 37' W). The yearly average rainfall in this area is 291 mm. The soil is a calcareous regosol (FAO-UNESCO) silt loam developed over marls and calcareous bedrocks (Chirino 2003).

Table 1. Main environmental characteristics positively (+) and negatively (-) related with water vapour adsorption by soil (WVA) occurrence in arid and semiarid zones (information from: Kosmas et al., 1998, 2001; Agam and Berliner 2004, 2006; Verhoef et al., 2006). The physical attributes affected by the environmental characteristics in the WVA process *were taken from* Verhoef et al. (2006) study

Environmental Factors	Effect	Physical Attributes Affected
Soil wetness	-	Surface soil air humidity
Among of potential evaporation during the period preceding the WVA period	+	
Clay content	+ *	
Minimum value of atmospheric humidity	-	Diurnal course of atmospheric vapour density
Diurnal amplitude of atmospheric humidity	+	
Proximity of sea	+	
Surface cover: mulch, stones, vegetation	-	Roughness parameters and localized wind speed

* Kosmas et al. (2001) found an inverse relationship between accumulated WVA and clay content. However these author suggested a combined effect with the macro-porosity which affects water vapor diffusion between atmosphere and soil mass.

1. Stipa Tenacissima's Water Responses under Water Stress Conditions

In August 22, 2003 our team recorded a surprisingly fast response of dawn leaf water potential (ψ) after a light rainfall event (1.59 mm day^{-1}) in our study area during the only precipitation event recorded in the summer season of 2003 (Ramírez et al., 2007a). Benefiting from such small amount of water, *S. tenacissima* was able to escape of its turgour loss point (ψ = -3.8 MPa; according to Pugnaire et al. 1996) to move on to a condition of reduced water stress (ψ = -2.7 MPa) after such a light rainfall. The amount of water that prompted the response was the lowest available in the literature. For instance, studies like Sala and Lauenroth (1982) and Ivans et al. (2003) have characterized plant water responses to "short" (*sensu* the referred authors) rainfalls in summer. However, the range of rainfall values assessed in these works was between 5 – 15 mm day^{-1}, figures notably higher than our assessed light rainfall value (1.59 mm day^{-1}). Pugnaire et al. (1996) studied the water response of *S. tenacissima* after watering in summer season and still the simulated rainfall assessed was 31 mm day^{-1}, a figure much higher than our assessed summer rainfall figure.

The architectural structure of *S. tenacissima* is an interesting factor to conjure up at this point of our reasoning, trying to explain the rapid response of *S. tenacissima* to a very small amount of rainfall. The plant is characterized by a thick tussock caused by a vertical leaf green arrangement (Valladares and Pugnaire 1999) and accumulation of litter from dead leaves (Domingo et al. 1996). This architectural configuration facilitates high interception rates mainly under light rainfall events, as shown by Domingo et al. (1998). Shortly, they demonstrated, using a comparative approach, that *S. tenacissima* has lower canopy water drainage (when the storage water is low) than *Retama sphaerocarpa* and *Anthyllis cytisoides* (other prominent species in semiarid areas in Spain). They also highlighted that the spatial structure of *S. tenacissima* promotes low stemflow and throughfall. Due to its high sensitivity to atmospheric conditions, namely a high evaporation demand such as that characteristic of summer in our study area, the intercepted water is quickly evaporated: the daily average total potential evaporative energy in August 2004 in our study area was 5.18 mm day^{-1} (Ramírez 2006). These results would help explain the rapid response of *S. tenacissima* to very low soil water inputs via summer light rainfall events. Taking into consideration that the tussock structure of *S. tenacissima* promotes great interception of light rainfalls, a root water harvest from bare soil close to the tussock can be expected.

One of the mechanisms developed by *S. tenacissima* that help avoid high water stress in the summer season is widespread stomatal closure (Pugnaire and Haase 1996; Balaguer et al. 2002). Indeed, the obtained figures for stomatal conductance during this water-stressed season were very low (between 0 and 0.08 mmol H_2O m^{-2} s^{-1}). However, Ramírez et al. (2007a) found that our assessed *S. tenacissima* grassland showed atypical high stomatal conductance values (between 21.8 and 43.1 mmol H_2O m^{-2} s^{-1}) in the 2004 summer. Therefore the question arises as to why the water status of this *S. tenacissima* stand does not correspond with other stands in the summer season.

2. Stand Evapotranspiration and WVA

Ramírez et al. (2007b) estimated stand evapotranspiration using two methods (scaling procedures and a multi-source evaporation model) in the spring and summer seasons in our study area. In the latter season, the evapotranspiration value assessed during a 10-day period was very similar to the water gains from the WVA process (Figure 1, Ramírez et al. 2007a). This result suggests the importance of the WVA process in our study area as a resource for the evaporative demands from evapotranspiration during the season with the highest water stress. Ramírez et al. (2007a) highlighted the importance of the WVA of the bare soil above soil under vegetation in the water demands from the *S. tenacissima* stand evapotranspiration, this arguments is supported due to the fact that, like to other semiarid Mediterranean environments (Kosmas 1998, 2001; Verhoef et al. 2006), the WVA values in bare soil were higher than in soil under vegetation.

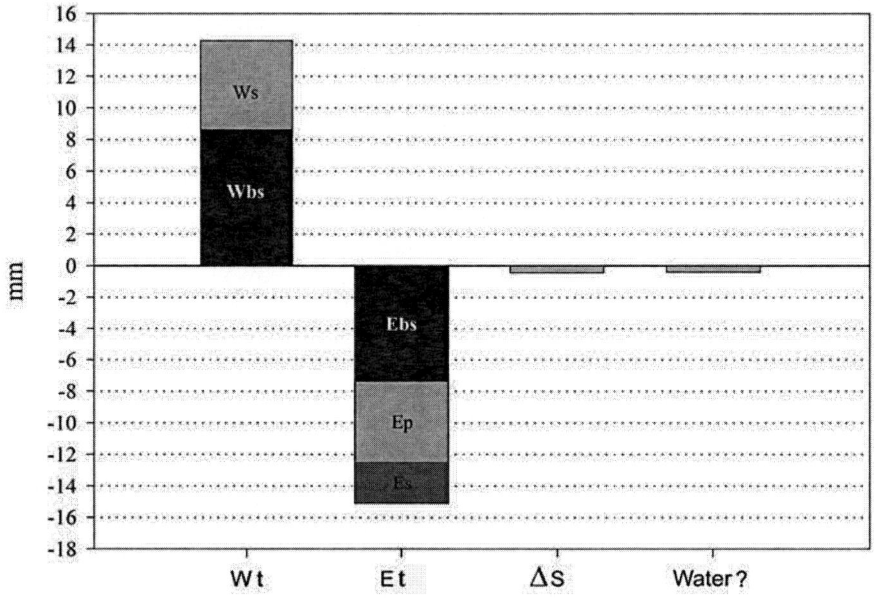

Figure 1. Total summer water balance assessed in the period from 08th to 18th August, 2004 in our study area. Ebs, Es and Ep = total evaporation from bare soil, soil under vegetation and *S. tenacissima* stand transpiration respectively, calculated from a multi-source evapotraspiration model (Clumped model). Wbs and Ws = total water gains from water vapour atmospheric adsorption process in bare soil and under vegetation soil respectively. ΔS = difference of the soil water storage at a depth of 15 cm, Water ? = water not accounted for by the water balance. Adapted from Ramírez et al. (2007a).

3. Soil Availability

We carried out an ecophysiological assessment of three *S. tenacissima* stands characterized by three gradients, namely a gradient in the cover of *S. tenacissima*, in average soil depth and altitude (see Table 2). The study was carried out in summer 2004 (season with high water stress). The stand with lower soil depth and higher variability of this variable (hereafter stand III, Table 2) showed the greater water stress symptoms. Accordingly, this stand obtained lower average gas exchange and F_v/F_m (maximum photosystem II photochemical efficiency) at dawn and at midday (0.33±0.03 and 0.18±0.02, respectively) than stands I and II (Table 2). Stand III was located in the upper part of the micro-basin, a place characterized by the presence of rock outcrops where the tussocks are into "soil pockets", and where the *S. tenacissima* stand cover is spatially clumped because it depends on the clumped distribution of the soil available (Figure 2). The root system of *S. tenacissima* tussocks into "soil pockets" is limited only to the soil under vegetation (Figure 3). These tussocks have no connection with the bare soil, and therefore the root water harvest from the important bare soil water gains detected in summer by Ramírez et al. (2007a) cannot happen. On the other hand, the stand with higher soil depth (stand I) showed higher gas exchange values (Table 2). In this micro-basin's sector the *S. tenacissima* cover shows a random distribution pattern again following the pattern of soil availability (Figure 2). The higher soil availability and bare soil connection with the root system in this stand allowed the *S. tenacissima* tussocks to make use of the bare soil water gains from WVA as explained above.

Table 2. Structural and ecophysiological traits of three *Stipa tenacissima* stands in a South-facing micro-basin located in Alicante, SE - Spain. The ecophysiological assessment was carried out in August - 2004 (summer season) under high water stress conditions. Gas exchange measures: E = average transpiration rate, A = average net photosynthesis, g = average stomatal conductance; chlorophyll fluorescence measure: F_v/F_m = maximum photosystem II photochemical efficiency. CV = variability coefficient. Different letters mean significant differences detected by Tukey HDS test ($p < 0.05$) from one-way ANOVA in each assessed ecophysiologial variable

Traits	Variables	Stand I	Stand II	Stand III
Structure	Altitude (m a.s.l.)	415	640	675
	S. tenacissima cover (%)	41.9	8.9	18.1
	Average soil depth (cm) n = 36	15.5	11.2	9.1
	CV soil depth (%)	65.5	62.0	122.8
Ecophysiology (± standard error)	E (mmol H_2O m^{-2} s^{-1})	1.1 ± 0.03 a	0.6 ± 0.01 b	0.3 ± 0.01 c
	A (μmol CO_2 m^{-2} s^{-1})	2.0 ± 0.1 a	1.2 ± 0.06 b	0.7 ± 0.1 c
	g (mmol H_2O m^{-2} s^{-1})	25.4 ± 1.3 a	15.0 ± 0.6 b	6.1 ± 0.4 c
	Green LAI (m^2 m^{-2})	2.0 ± 0.1 a	3.1 ± 0.3 b	1.3 ± 0.1 c
	Dawn F_v/F_m (relative unities)	0.51 ± 0.03 a	0.54 ± 0.01 a	0.33 ± 0.03 b
	Midday F_v/F_m (relative unities)	0.23 ± 0.02 a	0.35 ± 0.01 c	0.18 ± 0.02 b

CONCLUSION

The evidences presented here suggest an important relationship between the water status of *Stipa tenacissima* and bare soil water dynamic promoted by non-rainfall water (WVA) and very light precipitation. We acknowledge that specific studies (using isotopic assessment) are required to thoroughly determinate and quantify a possible physiological integration as the main mechanism responsible for the transport of the soil-adsorbed water (mainly from bare soil). Hopefully our study will encourage further works focusing in the dynamics of water, bare soil and vegetation, a paramount topic in semiarid Mediterranean regions.

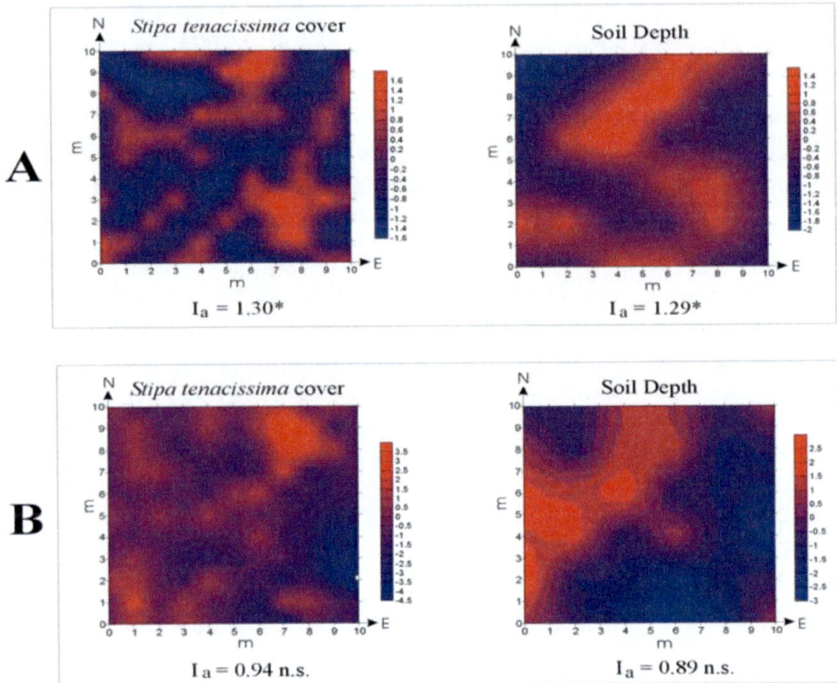

Figure 2. Map of the SADIE index of clustering of the *Stipa tenacissima* tussock cover and soil depth assessed in a 100 m^2 plot (positive and negative values mean patch and gaps zones respectively, see Perry et al. 1999). The spatial pattern of both variables is assessed with the SADIE index of aggregation. (I_a) in the plot (the reader is referred to Maestre and Cortina 2004, for a detail description of the SADIE method in semiarid steppes) in zones close to the *S. tenacissima* stands I (A) and III (B) (see Table 2). I_a = SADIE index of aggregation, where the clumped, random and regular spatial pattern is defined when: $I_a > 1$(*), $I_a \approx 1$ (n.s.) and $I_a < 1$ (*) respectively. * and n.s. mean significant values (at $p < 0.05$) and no significant values ($p > 0.05$) respectively assessed by a permutation test.

Figure 3. *Stipa tenacissima* tussocks belonging to stand III in the upper part of the assessed micro-basin. The arrow underlines the rock outcrop in this area.

ACKNOWLEDGEMENTS

Financial support was provided by the "Efecto de la cubierta vegetal en el balance hídrico y en la disponibilidad y calidad del agua propuesta para aumentar la recarga de acuíferos" Project, funded by the CICYT (ref.: REN2000-0529HID, CGL2004-03627), by the European Community "AQUADAPT: Energy, Environment and Sustainable Development" (EVK1-2001-00149) and the PROBASE (ref.: CGL2006-11619/HID) and CANOA (ref.: CGL2004-04919-C02- 01/HID) projects funded by the Spanish Ministry of Education and Science.

REFERENCES

Agam (Ninari), N and Berliner, PR (2004). Diurnal water content changes in the bare soil of a coastal desert. *J. Hydrometeorol., 5,* 922 – 933.

Agam N and Berliner, PR (2006). Dew formation and water vapor adsorption in semi-arid environments-A review. *J. Arid Environ., 65,* 572-590.

Balaguer, L; Pugnaire FI; Martinez-Ferri, E; Armas, C; Valladares, F and Manrique, E (2002). Ecophysiological significance of chlorophyll loss and reduced photochemical efficiency under extreme aridity in *Stipa tenacissima* L. *Plant Soil, 240,* 343-352.

Bochet, E; Rubio, JL and Poesen, J (1999). Modified topsoil islands within patchy Mediterranean vegetation in SE Spain. *Catena, 38,* 23-44.

Cerdà, A (1997). The effect of patchy distribution of *Stipa tenacissima* L. on runoff and erosion. *J. Arid Environ., 36,* 37-51.

Chirino, E (2003). Influencia de las precipitaciones y de la cubierta vegetal en el balance hídrico superficial y en la recarga de acuíferos en clima semiárido. PhD thesis, University of Alicante, Spain.

Cortina, J; Bellot, J; Milagrosa, A; Catarla, RN; Maestre, FT; Rubio, E.; Ortiz de Urbina, JM and Bonet, A (2004). Restauración en semiárido. In: Vallejo VR and Alloza JA, editors. *Avances en el estudio de la gestión del monte mediterráneo.* Valencia – Spain; Fundación Centro de Estudios Ambientales del Mediterráneo, pp. 345 - 406.

Domingo, F; Van Gardingen, PR and Brenner AJ (1996). Leaf boundary layer conductance of two native species in southeast Spain. *Agr. Forest Meteorol., 81,* 179-199.

Domingo F; Sánchez G; Moro, MJ; Brenner, AJ and Puigdefábregas, J (1998). Measurement and modelling of rainfall interception by three semi-arid canopies. *Agr. Forest Meteorol., 91,* 275-292.

FAO-UNESCO (1988) *Soil map of the world.* Revised legend. Rome. World Soil Resources, Report 60.

Ivans, CY; Leffler, AJ; Spaulding, U; Stark, JM; Ryel, RJ and Caldwell, MM (2003). Root responses and nitrogen acquisition by *Artemisia tridentate* and *Agropyron desetorum* following small summer rainfall events. *Oecologia, 134,* 317-324.

Joffre, R; Rambal, S and Damesin, C (1999). Functional attributes in Mediterranean-type ecosystems. In: Pugnaire FI and Valladares F, editors. *Handbook of functional plant ecology.* New York; Marcel Dekker, Inc.; pp. 347-379.

Kosmas, C; Danalatos, NG; Poesen, J and van Wesemael, B (1998). The effect of water vapor adsorption on soil moisture content under Mediterranean climatic conditions. *Agric. Water Manage., 36,* 157-168.

Kosmas, C; Marathianou, M; Gerontidis, St; Detsis, V; Tsara, M and Poesen, J (2001). Parameters affecting vapor adsorption by the soil under semi-arid climatic conditions. *Agric. Water Manage., 48,* 61-78.

Le Houérou, HN (2001). Biogeography of the arid steppeland north of the Sahara. *J. Arid Environ., 48,* 103-128.

Maestre, FT; Bautista, S; Cortina, J and Bellot, J (2001). Potential for using facilitation shrubs on a semiarid degraded steppe. *Ecological Applications, 11(6),* 1641-1655.

Maestre, FT and Cortina, J (2004). Insights into ecosystem composition and function in a sequence of degraded semiarid steppes. *Restoration Ecology,12(4),493-501.*

Maestre, FT (2006). Linking the spatial patterns of organisms and abiotic factor to ecosystems function and management: Insights from semi-arid environments. *Web Ecology, 6,*75-87.

Maestre, FT and Cortina, J (2006) Ecosystem structure and soil-surface condition drive the variability in the foliar $\delta^{13}C$ and $\delta^{15}N$ of *Stipa tenacissima* in semiarid Mediterranean steppes. *Ecol. Res., 21,* 44-53.

Maestre, FT; Ramírez, DA and Cortina J (2007). Ecología del esparto (*Stipa tenacissima* L.) y los espartales en la Península Ibérica. *Ecosistemas 2007/2* (URL: http://www.revistaecosistemas.net/articulo.asp?Id=480andId_Categoria=19andtipo=portada)

Perry, JN; Zinder, L; Holland, JM and Alston, RD (1999). Red-blue plots for detecting cluster in count data. *Ecol Lett, 2,* 106-113.

Puigdefábregas; J and Sánchez, G (1996). Geomorphological implications of vegetation patchiness on semi-arid slopes. In: Anderson MG and Brooks SM, editors. *Advances in hillslopes processes.* Vol 2; New York; John Wiley and Sons Ltd; pp. 1027-1060.

Puigdefábregas, J; Sole, A; Guitierrez, L; del Barrio, G and Boer, M (1999). Scales and processes of water and sediment redistribution in drylands: results from the Rambla Honda field site in Southeast Spain. *Earth-Sci. Rev., 48,* 39-70.

Pugnaire, FI and Haase, P (1996). Comparative physiology and growth of two perennial tussock grass species in a semi-arid environment. *Ann. Bot., 77,* 81-86.

Pugnaire, FI; Haase, P; Incoll, L and Clark, SC (1996). Response of tussock grass *Stipa tenacissima* to watering in a semi-arid environment. *Funct. Ecol., 10,* 265-274.

Ramírez, DA (2006). Estudio de la transpiración del esparto (*Stipa tenacissima* L.) en una cuenca del semiárido alicantino: un análisis pluriescalar, PhD thesis, University of Alicante, Spain.

Ramírez, DA; Bellot, J; Domingo, F and Blasco, A (2007a). Can water responses in *Stipa tenacissima* L. during the summer season be promoted by non–rainfall water gains in soil?. *Plant Soil, 291,* 67-79.

Ramírez, DA; Bellot, J; Domingo, F and Blasco, A (2007b). Stand transpiration of *Stipa tenacissima* grassland by sequential scaling and multi-source evapotranspiration modelling. *J Hydrol.* (*in press*, doi: 10.1016/j.jhydrol.2007.05.018).

Reynolds, JF; Virginia, RA; Kemp PR; de Soyza, AG and Tremmel, DC (1999). Impact of drought on desert shrubs: effects of seasonality and degree of resource island development. *Ecological Monographs, 69,* 69-106.

Sala, OE and Lauenroth, WK (1982). Small rainfall events: an ecological role in semiarid regions. *Oecologia, 53 (3),* 301-304.

Sánchez, G (1995). Arquitectura y dinámica de las matas de esparto (*Stipa tenacissima* L.), efectos en el medio e interacciones con la erosión. Ph.D. thesis, Autónoma de Madrid University, Spain.

Sánchez, G and Puigdefábregas, J (1994). Interactions of plant growth and sediment movement on slopes in a semi-arid environment. *Geomorphology, 9,* 243-260.

Valladares, F and Pugnaire, FI (1999). Tradeoffs between irradiance capture and avoidance in semi-arid environments assessed with a crown architecture model. *Ann. Bot., 83,* 459-469.

Vallejo, VR; Serrasoles, I; Cortina, J; Seva, JP; Valdecantos, A and Vilagrosa, A (2000). Restoration strategies and actions in Mediterranean degraded lands. In: Enne G, Zanolla Ch. and Peter D, editors. *Desertification in Europe: mitigation strategies, land-use planning.* Brussels; EC DGXII, Environment and Climate Program, European Commission; pp. 211-233.

Verhoef, A; Díaz-Espejo, A; Knight, JR; Villagarcía, L and Fernández, JE (2006). Adsorption of water vapour by bare soil in an olive grove in Southern Spain. *J. Hydrometeorol., 7(5),* 1011-1027.

In: Soil Ecology Research Developments
Editors: Tian-Xiao Liu, pp. 19-29

ISBN: 978-1-60021-971-9

Short Communication B

EFFECT OF AL, CD AND CU INTERACTIONS ON PHOTOSYNTHETIC CHARACTERISTICS AND SUGAR CONTENTS IN BARLEY LEAVES DIFFERING IN AL TOLERANCE

Guo Tian Rong*[1], *Mi Zhong Xiang,
Zhang Yan Hua and Tian Run Gang
Life Science College, Shaoxing University, Shaoxing, 312000,
People's Republic of China

ABSTRACT

The decline in crop production is commonly observed on acid soil, further, a multiple heavy metal pollution except for Al on acid soil is detected in many areas. The present experiment was conducted by using two barley genotypes differing in Al-tolerance, Al tolerant *cv.* Gebeina and Al sensitive *cv.* Shang 70-119, to determine the effect of Al, Cd, Cu individual, binary and ternary combinations on the photosynthetic characteristics and the contents of sucrose and starch in leaves of barley seedlings. The results showed that there were significant decreases in leaf area, leaf fresh and dry weight, chlorophyll content, net photosynthesis (Pn), stomatal conductance (gs), *Fv/Fm* ratio evaluated for chlorophyll fluorescence and sucrose contents, but obvious increases in internal CO_2 concentration (Ci) and starch contents in barley leaves when exposed to the stress medium. These decreases or increases were more marked in Shang 70-119 than in Gebeina. In addition, binary metals combinations of Al + Cd and Al + Cu both resulted in severe changes in these values for two genotypes when compared with the Al alone treatment, indicating the existence of notable synergistic interactions between Al and Cu or Al and Cd, in particular for Shang 70-119. However, ternary metals combination of Al + Cu + Cd produced different interactions in two kinds of genotypes, thus, the slighter toxicity caused by ternary metals mixture than that caused by binary metals mixture was observed in Gebeina, while the reverse results showed in Shang 70-119, suggesting the notable synergistic effect of the three metals existed in sensitive genotype but not in

[1] Corresponding author: Tel: +86 75 88345007. *E-mail address*: guotr@163.com .

tolerant one. So, it can be concluded that the different responses to acid soil toxicity of two barley genotypes may partly result from the different metals interactions in plant seedlings.

Keywords: Aluminum, Copper, Cadmium, Toxicity, Barley (Hordeum vulgare L.), Leaf traits, Chlorophyll contents, Photosynthesis, Chlorophyll fluorescence, Starch, Sucrose.

ABBREVIATIONS

Al, -aluminum;
Cd,- cadmium;
Cu,- copper;
Pb,- lead;
Zn, -zinc.

1. INTRODUCTION

Acid soils have extended over 40% to 70% of the world's arable soils (Rengel, 1992). A multi-element pollution except for aluminum (Al) on acid soil is detected in many areas though Al is considered as the major factor constraining crop performance when soil pH is lower than 5.5 (Chen et al., 2000). Many reports showed that Al can disturb cell metabolism by inhibiting nutrients absorption, stomatal function, photosynthesis, respiration and altering the activity and quantity of the key enzyme of various metabolic pathways of the treated-plants. Nevertheless, there have been few reports about the combined effects of Al and other heavy metals on plants on acid soil up to now except our previous study suggesting the existed synergistic effect between Al and low amount of Cd in barley seedlings (Guo et al., 2004)

Numerous authors suggest that the inhibition of plant growth detected in many plant species subjected to Al or heavy metals stress is often associated with a reduction in net photosynthesis, which in turn is closely associated with the decreases in stomatal conductance (gs), green leaf area, and leaf chlorophyll contents. The stomatal conductance is one of the main factors regulating CO_2 flow and leaf transpiration (Kozlowski and Pallardy, 1997; Jones, 1998), while the ability to maintain the rate of photosynthetic CO_2 uptake under stress treatment is fundamental for plant growth and production. On the other hand, a simple, non-intrusive way to monitor the performance of the photosynthetic apparatus is to measure chlorophyll fluorescence and the changes in the characteristics related to fluorescence, such as initial fluorescence (Fo), maximum fluorescence (Fm), variable fluorescence (Fv), and the ratio between them (Lichtenthaler and Rinderle, 1988). The inverse relationship between the photosynthetic activity and the *in vivo* chlorophyll fluorescence was detected in many situations (Krause and Weis, 1991). Furthermore, starch and sucrose are two main fixed photosynthetic carbohydrates. During active periods of photosynthesis, starch is formed as temporary storage form of fixed carbon and is deposited as starch granules in the chloroplast, while the sucrose is transported to different organs and is the most commonly used photoassimilate by plants (Verma and Dubey, 2001). Thus, the contents of starch and sucrose are the important indicators of plants photosynthetic capacity. In addition, the great differences in Al, Cu or Cd tolerance have been found among plant species and genotypes within a species (Wu and Zhang, 2003; Guo et al., 2004; Nikookar et al., 2005).

Therefore, this study was carried out to analyze the effects of Al, Cd, and Cu interactions on several photosynthesis related characteristics, chlorophyll fluorescence and the accumulations of photosynthetic carbohydrate in two barley genotypes differing in Al tolerance. Further, the present investigation was designed to determine whether the growth reductions in the Al-stressed barley plants observed in previous study (Guo et al., 2007) resulted from a decline in total leaf area or a reduction in Pn, or both. Also, chlorophyll a, b contents, PS □activity, contents of sucrose and starch in leaves of stressed plants were examined to determine if these factors contributed to the growth response and the different Al sensitivity of two barley genotypes.

2. Materials and Methods

2.1. Plant Material, Growth Conditions and Treatments

The experiment was carried out in 2006 at Shaoxing University, Shaoxing, China. Two 2-rowed winter barley (Hordeum vulgare L.) genotypes differing in Al tolerance were used, Al tolerant cv. Gebeina and Al sensitive cv. Shang 70-119. The seeds were surface sterilized in 0.2% NaClO for 20 min, rinsed with distilled water 4 times and germinated in moist quartz sand in a culture room. When seedlings grew the second leaf (10-day old), they were selected for uniformity and transplanted to a modified Hoagland nutrient solution with the photoperiod of 8/16h (day/night) and the temperature of 23±2°C throughout the experiment. One week after transplanting to the basic solution culture, Al as $AlCl_3.6H_2O$, Cd as $CdCl_2$, and Cu as $CuCl_2.2H_2O$ were added to the nutrient solution, and the solution pH was adjusted with HCl to following 6 treatments: (1) pH 6.5 (control); (2) pH 4.5; (3) Al (100 μmol L^{-1} Al, pH 4.5); (4) Al + Cu (100 μmol L^{-1} Al + 10 μmol L^{-1} Cu, pH 4.5); (5) Al + Cd (100 μmol L^{-1} Al + 10 μmol L^{-1} Cd, pH 4.5); (6) Al + Cu + Cd (100 μmol L^{-1} Al + 10 μmol L^{-1} Cu + 10 μmol L^{-1} Cd, pH 4.5). The solution pH in each container was adjusted every other day with HCl or NaOH as required. The experiment was laid out in a completely randomized design with five replicates. The nutrient solution in the growth container was continuously aerated with pumps and renewed every 5 days.

2.2. Measurements and Statistical Analysis

The upper second fully expanded leaves were sampled for analysis of net photosynthesis (Pn), stomatal conductance (gs), internal CO_2 concentration (Ci) and Fv/Fm, at 30 days after treatments. The Pn, gs, Ci were determined by using a portable carbon dioxide infrared analyzer, model CIRAS-1 (Analytical Development Co. Ltd., England). The measurements were taken between 09:00 and 11:00 hours for the intact plants. The measurements of the chlorophyll fluorescence emission kinetics were taken using a portable fluorometer, model FMS2CIRAS-1 (Plant Efficiency Analyser, PEA, Hansatech Instruments Ltd., England). The leaves used for the measurements were pre-conditioned in the dark for 30 minutes, and illuminated for 5 seconds to induce fluorescence. The initial fluorescence (F0) and the maximum fluorescence (Fm) were measured, further, the variable fluorescence (Fv = Fm – F0) and the Fv/Fm ratio were calculated. Leaf area (LA) per plant were measured with copy method described by Tao (2006), and leaf chlorophyll contents for the upper second fully expanded leaves were estimated according to Zhang (1992). At the same time, starch and sucrose contents in leaves were determined according to the description by Verma and Dubey (2001) and Hashimoto el al. (2004) respectively. Then, plants of each treatment were harvested and washed thoroughly with distilled water for 2 hours, leaves were separated and weighted as leaf fresh weight (LFW), then dried at 80°C four 48h and weighted as leaf dry weight (LDW). Then, leaf water content (LWC = (LFW - LDW) / LFW) were calculated.

All data presented are the mean values of three replicates. Statistical assays were carried out by one-way ANOVA using Student's t-test to evaluate whether the means were significantly different.

3. Results and Discussion

3.1. Leaf Traits

Leaf traits in terms of leaf area (LA), leaf fresh and dry weight (LFW and LDW), and the estimated data leaf water content (LWC) all significantly decreased when plants were exposed to stress medium including low pH alone, especially for the ternary metals combination of Al + Cd + Cu in Shang 70-119 (table 1). Inhibition of leaf expansion in stressed plants was in agreement with the previous report by Florence et al. (2002) who detected a reduction in cucumber leaf area of young expanding leaves under 10 $\mu g\ g^{-1}$ Cu supplementary in sand, but not for mature leaves. Similarly, Gichner et al. (2006) showed that the average tobacco leaf area significantly reduced in plants growing on the combined heavy metals (Cd, Cu, Pb and Zn) polluted soil. Reduction in leaf expansion in this investigation is likely related to direct metals inhibition of cell growth as observed in Cu-stressed bean seedlings (Maksymiec et al., 1995). In addition, the noticeable decreases in fresh and dry weight detected in stressed barley plants were in good coincidence with the results obtained on tobacco plants (Gorinova et al., 2007). Further, compared with the Al alone treatment, binary metals combinations resulted in a severe inhibition than that of Al alone treatment for both barley genotypes, indicating the existence of notable synergistic interactions between Al and Cu or Al and Cd. However, for ternary metals combination of Al + Cu + Cd, it produced different toxic results in two genotypes. In Gebeina, the toxicity caused by ternary metals mixture was slighter than that caused by binary metals combinations, but severer than that caused by Al alone treatment, while for Shang 70-119, noticeable synergistic effect was detected in the three metals. Our previous study (Guo et al., 2007) also demonstrated that whole plant growth of barley seedlings was dramatically inhibited by the treatments of Al, Cd, Cu individual, binary and ternary combinations, and the toxicity was highly correlated with Al, Cd and Cu contents in tissues. Likewise, the inhibition was more severe under Al + Cd and Al + Cu treatments than Al alone treatment, but the ternary metals combination induced the different toxic results in two genotypes.

Table 1. Effect of the different Al, Cd and Cu treatments on leaf traits of barley seedlings

Genotypes	Treatments	LA ($cm^2 \cdot plant^{-1}$)	LDW ($g \cdot plant^{-1}$)	LFW ($g \cdot plant^{-1}$)	LWC (g/g)
	pH 6.5	65.2	0.171	0.553	0.691
	pH 4.5	63.1	0.163	0.516	0.684
	Al	59.4	0.131	0.351	0.627
Gebeina	Al + Cu	56.2	0.121	0.306	0.601
	Al + Cd	56.1	0.120	0.312	0.603
	Al + Cu + Cd	58.7	0.130	0.343	0.623
	pH 6.5	66.1	0.169	0.538	0.688
	pH 4.5	63.4	0.160	0.497	0.678
Shang 70-119	Al	57.9	0.124	0.314	0.605
	Al + Cu	54.6	0.112	0.255	0.561
	Al + Cd	54.1	0.110	0.248	0.557
	Al + Cu + Cd	51.4	0.101	0.207	0.532
LSD0.05 Between genotypes		1.2	0.003	0.033	0.021
LSD0.05 Between treatments		2.3	0.008	0.039	0.019

LA, leaf area; LDW, leaf dry weight; LFW, leaf fresh weight; LWC, leaf water content =(LFW –LDW)/LFW.

3.2. Chlorophyll Contents

Chlorophyll absorbs radiant energy, which is used for photosynthesis, so chlorophyll content in the cell has been used as a parameter of following plant growth. In the present investigation, the total chlorophyll contents, chlorophyll a and b contents all decreased under stress condition, while the decreases in chlorophyll a contents is more obvious than that in chlorophyll b content expressed as the ratios of chlorophyll a/b (table 2), which was in agreement with the reports by Woolhouse (1974) who showed that chlorophyll content decreased as leaf was senescent, and chlorophyll a content decreased noticeably as compared with chlorophyll b. Khudsar et al (2001) found that the contents of chlorophyll a and b were reduced under the effect of Cd toxicity. In addition, Stobart et al. (1985) suggested that heavy metals could inhibit chlorophyll biosynthesis at the level of photochlide reductase. Metals combinations induced the further reduction in chlorophyll content when compared with the Al alone treatment, especially for Shang 70-119. However, ternary metals combination produced different effects in two barley genotypes, thus, the more reductions were observed in Shang 70-119, while the adverse results were detected in Gebeina when compared with that of the binary metals combinations.

Table 2. Effect of the different Al, Cd and Cu treatments on chlorophyll contents of barley leaves

Genotypes	Treatments	Chl content ($mg . g^{-1}$)	Chl a content ($mg . g^{-1}$)	Chl b content ($mg . g^{-1}$)	Chl a/b
	pH 6.5	1.18	0.79	0.39	2.03
	pH 4.5	1.15	0.76	0.39	1.94
	Al	1.09	0.71	0.38	1.87
Gebeina	Al + Cu	1.04	0.67	0.37	1.81
	Al + Cd	1.03	0.66	0.37	1.78
	Al + Cu + Cd	1.08	0.70	0.38	1.84
	pH 6.5	1.19	0.79	0.40	1.98
	pH 4.5	1.15	0.75	0.40	1.88
Shang 70-119	Al	1.03	0.67	0.36	1.86
	Al + Cu	0.99	0.64	0.35	1.83
	Al + Cd	0.98	0.62	0.36	1.72
	Al + Cu + Cd	0.95	0.59	0.36	1.64
LSD0.05 Between genotypes		0.04	0.02	0.01	0.06
LSD0.05 Between treatments		0.03	0.03	0.01	0.05

Chl, chloropgyll.

3.3. Photosynthesis

The presence of Al, Cd, and Cu individual, binary and ternary in the culture medium all decreased the net photosynthesis (Pn) and stomatal conductance (gs), but increased the internal CO_2 concentration (Ci) in the stressed leave. The changes in Pn, gs and Ci were maximum when plants were exposed to the ternary metals combination for Shang 70-119, but the binary metals combinations for Gebeina (table 3). For instance, Pn under Al, Al + Cu, Al + Cd, and Al + Cd + Cu treatments was 87.3%, 84.1%, 84.1%, 86.2% and 83.4%, 79.1%, 79.7% 74.9% of the control for Gebeina and Shang 70-119, respectively. Thus, the decline in Pn was associated with a parallel decrease in gs and increase in Ci, which was in agreement with previous observations on cucumber under copper toxicity (Florence et al., 2002). Similarly, Khudsar et al. (2001) showed that Pn and gs reduced significantly at 5-50 $\mu g.g^{-1}$ Cd treatments, but Ci also decreased obviously at 10 $\mu g.g^{-1}$ Cd onwards. Up to a 53% decrease in Pn was also found in the presence of 550 $\mu mol.L^{-1}$ Al in two *Triticosecal* cultivars (Nunes et al., 1995). However, Lidon et al. (1997) found an increase of 52% in gs in maize plants under Al concentrations of up to 1 mmol. L^{-1}. Simon et al. (1994) also observed a similar response in tomato plants, where Al at a 25 $\mu mol\ L^{-1}$ concentration caused a 38% increase in the gs value compared with the control. But at the highest concentration evaluated (50$\mu mol\ L^{-1}$), a 62% decrease in the gs value compared with the control. Further, for Ci, Moustakas et al. (1996) found a 10 % decrease of Ci in *Thynopyrum bessarabicum* in the presence of 1 mol L^{-1} Al. Nevertheless, Dong et al. (2005) found that 1 $\mu mol\ L^{-1}$ Cd and 10 $\mu mol\ L^{-1}$ Cd toxicity both increased Ci in tomato plants. In the present study, the increase in the metal toxicity in growth medium decreased Pn and gs, but increased Ci, indicating that inhibition in Pn by metal toxicity did not result from a stomatal closure.

Table 3. Effect of the different Al, Cd and Cu treatments on photosynthesis of barley leaves

Genotypes Treatments	Gebeina	Shang 70-119	Gebeina	Shang 70-119	Gebeina	Shang 70-119
	Pn (CO_2 μmol m^{-2} s^{-1})	g_s (mol m^{-2} s^{-1})	Ci (μl L^{-1})			
pH 6.5	18.9 a	18.7 a	0.43 a	0.44 a	259 c	256 e
pH 4.5	17.8 b	17.3 b	0.41 a	0.40 b	268 c	284 d*
Al	16.5 c	15.6 c*	0.38 b	0.37 c	292 b	306 c*
Al+ Cu	15.9 d	14.8 d*	0.35 b	0.33 d *	316 a	339 b **
Al + Cd	15.6 d	14.9 d*	0.35 b	0.33 d *	323 a	341 b *
Al + Cu + Cd	16.3 c	14.0 e**	0.37 b	0.30 d **	301 b	362 a **

abc, different letters within columns indicate significant differences ($P < 0.05$) between treatments.

* and **, represent significant differences at the 0.05 and 0.01 probability levels, respectively, between cultivars.

3.4. Chlorophyll Fluorescence

Pn in stress treated plants was inhibited mainly as a result of low chlorophyll content and disturbing in chloroplast structure, on the other hand, inhibition in PS □activity showed in *Fv/Fm* ratio, an indicator of structural damage and affecting the photosynthetic transport of electrons, could also play an important role. In this study, a significant decline in *Fv/Fm* ratio evaluated for chlorophyll fluorescence was detected in barely seedlings in the presence of Al, Cd and Cu individual, binary and ternary combinations (Figure1).

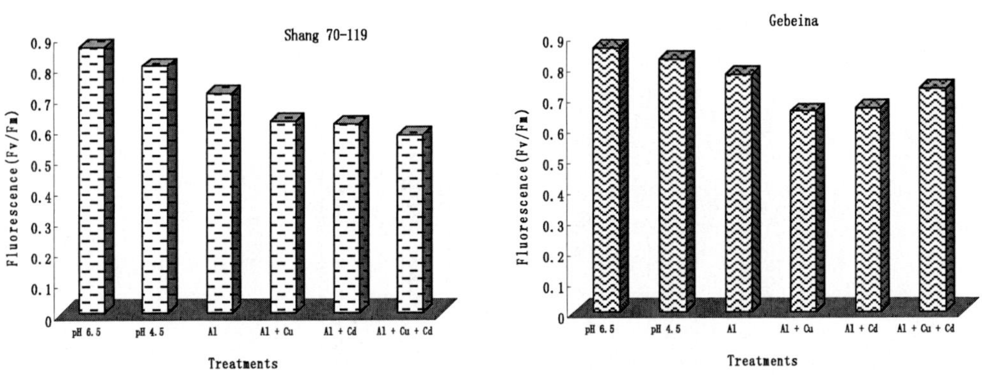

Figure 1. Effect of the different Al, Cd and Cu treatments on chlorophyll fluorescence (Fv/Fm) of barley leaves.

This decrease was more marked in Shang 70-119 than in Gebeina, especially for the metal combinations treatments. The parameter significantly decreased to 0.576 (66.9% of the control) at the ternary metals combination in Shang 70-119, where for Gebeina, a marked decrease was observed at binary metals combinations, 84.7% of the control. Zhou et al. (2006) found that a significant decrease of F_v/F_m value in cyanobacterium *Microcystis aeruginosa* cells treated with 4 μmol. L^{-1} Cd concentration after 48 h, and who suggested that this could result from the decrease of the rate constant for energy trappings of PSII reaction centers. However, Panković et al. (2000) showed that F_v/F_m was not significantly influenced by toxic concentrations of Cd treatment, while 5 mmol. L^{-1} Cd decreased quantum efficiency of PSII electron transport (Φ_{II}) by 30%.

3.5. Starch and Sucrose Contents

There was an increase in starch concentration but a decrease in sucrose concentration when plants were exposed to the stress medium (Figure 2)

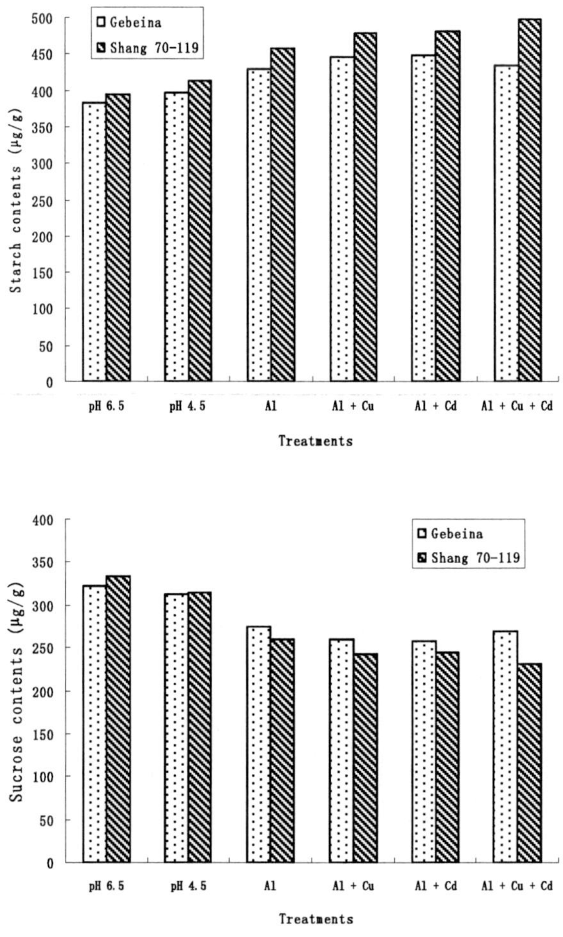

Figure 2. Effect of the different stress treatments on starch and sucrose contents in barley leaves.

These changes were more obvious in Shang 70-119 than that in Gebeina. Binary metals combinations induced further changes in these values for both barley genotypes, but ternary metals combination resulted in different effects in two kinds of genotypes, more obvious change was seen in Shang 70-119, but the adverse results occurred in Gebeina when compared to the binary combinations. The present results were not completely with one accord with the report by Florence et al. (2002) who showed that significant increases both in sucrose and starch contents in cucumber mature leaves under 10 $\mu g\ g^{-1}$ Cu supplementary in sand. Similarly, Moya et al. (1993) found that 0.1 mmol. L^{-1} Cd and 0.5 mmol. L^{-1} Ni induced the decrease in net photosynthesis of treated plants, while, the total carbohydrate content in the shoots of these plants was higher than in controls, thus, the starch and sucrose content in the shoots of 0.5 mmol. L^{-1} Ni treated plants were respectively up to 2.6 and 4.0 times greater compared to controls. Authors suggested that the influence of Al and heavy metals on sucrose concentration is very variable, depending mainly on the plant species or variety, and on the Al and heavy metal activity and the combinations of metals in the nutritive solution.

4. Conclusion

The toxicities of Al, Cd, and Cu separately and in combinations and the metals interactions in this study were assessed by estimating the leaf trait, chlorophyll content, photosynthesis, chlorophyll fluorescence and sugar content in barley leaves after 5-week supply of the Al, Cd, and Cu excess in the nutrient solution. LA, LFW, LDW, LWC, chlorophyll content, especially chlorophyll a content, Pn, gs, PS □activity showed in *Fv/Fm* ratio, and sucrose content all decreased, while Ci and starch content increased significantly in barley leaves when exposed to stress medium including low pH alone, especially for Shang 70-119. Binary metals combinations induced the further changes in these values when compared with the Al alone treatment for both genotypes, indicating the existence of notable synergistic interactions between Al and Cu or Al and Cd. However, ternary metals combination produced different effects in two barley genotypes, thus, when compared with the binary metals combinations, the more obvious changes were observed in Shang 70-119, while the adverse results were detected in Gebeina. Further, compared to other leaf traits, leaf fresh weight was the more sensitive endpoint, and consequently it would be a good parameter to assess acid soil toxicity for barley plants. The analysis of chlorophyll content and fluorescence, gs and Ci showed that the inhibition of Pn in barley seedlings by Al, Cd, and Cu toxicity *in vivo* partly resulted from the reductions of chlorophyll content, the active PSII reaction centers and gs, while the uncoupled changes of Pn and Ci indicated the reduction of Pn by metal toxicity did not result from a stomatal closure.

Acknowledgements

First author acknowledge the support provided by the Educative Department of Zhejiang Province, Chinese and Australian Bureau of Cereal Research and Development (UT-8), respectively.

References

Chen, HM, Zheng, CR, Wang, SQ, Tu, C. (2000). Combined pollution and pollution index of heavy metals in red soil. *Pedosphere,* 10: 117-124.

Dong, J, Wu, FB, Zhang, GP. (2005). Effect of cadmium on growth and photosynthesis of tomato seedlings. *J. Zhejiang U. Sci. B.* (6): 974-980.

Florence, VD, Daniel, E, Badr, AS, Badot, PM. (2002). Effects of copper on growth and on photosynthesis of mature and expanding leaves in cucumber plants. *Plant. Sci.,* 163: 53-58.

Gichner,T, Patková, Z, Száková J, Demnerová, K. (2006). Toxicity and DNA damage in tobacco and potato plants growing on soil polluted with heavy metals. *Ecotox. Environ. Safe,* 65: 420-426.

Gorinova, N, Nedkovska, M, Todorovska, E, Simova-Stoilova, L, Stoyanova, Z, Georgieva, K, Demirevska-Kepova, K, Atanassov, A, Herzig, R. (2007). Improved phytoaccumulation of cadmium by genetically modified tobacco plants (Nicotiana tabacum L.): physiological and biochemical response of the transformants to cadmium toxicity. Environ Pollut, 145: 161-170.

Guo, TR, Zhang, GP, Zhang, YH. (2007). Physiological changes in barley plants under combined toxicity of aluminum, copper and cadmium. *Colloid Surface B,* 57: 182-188.

Guo, TR. Zhang, GP. Zhou, MX, Wu, FB, Chen, JX. (2004). Effect of aluminum and cadmium toxicity on growth and antioxidant enzyme activities of two barley genotypes with different aluminum tolerance. *Plant Soil,* 258: 241-248.

Hashimoto, A, Yamanaka, A, Kanou, M, Nakanishi, K, Kameoka, T. (2004). Simple and rapid determination of metabolite content in plant cell culture medium using an FT-IR/ATR method. *Bioprocess Biosystems Eng.*, 27: 115-123.

Jones, HG. (1998). Stomatal control of photosynthesis and transpiration. *J Exp Bot,* 49: 387-398,

Khudsar, T, Mahmooduzzafar, Iqbal, M, (2001). Cadmium-induced changes in leaf epidermes, photosynthetic rate and pigment concentrations in *Cajanus cajan. Biol. Plant,* 44: 59-64.

Kozlowski, TT, Pallardy, SG. Physiology of woody plants. San Diego: Academic Press, 1997.

Krause, GH, Weiss, E. (1991). Chlorophyll fluorescence and photosynthesis:the basics. *Annu Rev. Plant Physiol. Plant Mol. Biol.,* 42: 313–349.

Lichtenthaler, HK, Burkart, S, Schindler, C, Stober, F. (1992). Changes in photosynthetic pigments and *in vivo* chlorophyll fluorescence parameters under photoinhibitory growth conditions. *Photosynthetica,* 27: 343-353.

Lidon, FC, Ramalho, JC, Barreiro, MG. (1997). Aluminum modulation of the photosynthetic carbon reduction cycle in *Zea mays. Photosynthetica,* 34: 393–400.

Maksymiec, W, Bednara, J, Baszynski, T. (1995). Response of runner bean plants to excess copper as a function of plant growth stages: effects on morphology and structure of primary leaves and their chloroplast ultrastructure. *Photosynthetica,* 31: 427-435.

Moustakas, M, Ouzounidou, G, Eleftheriou, EP, Lannoye R. (1996). Indirects effects of aluminum stress on the photosynthetic apparatus. *Plant Physiol. Biochem.,* 34: 553–560.

Moya, JL, Ros R, Picazo, I. (1993). Influence of cadmium and nickel on growth, net photosynthesis and carbohydrate distribution in rice plants. *Photosynth. Res.,* 36: 75-80.

Nikookar, K, Moradshahi, A, Hosseini, L. (2005). Physiollogical responses of *Dunaliella salina* and *Dunaliella tertiolecta* to copper toxicity. *Biomol. Eng.,* 22: 141-146.

Nunes, MA, Matos, CM, Azinheira, HG, Quartin, VL. From light to biosphere. In: Mathis P Photosynthesis. Netherland: Kluwer Academic Publisher, 1995.

Panković, D, Plesničar, M, Arsenijević-Maksimović, I, Petrović, N, Sakač, Z, Kastori, R. (2000). Effects of Nitrogen Nutrition on Photosynthesis in Cd-treated Sunflower Plants. *Annal. Bot.*, 86: 841-847.

Rengel, Z. (1992). Role of calcium in aluminum toxicity. *New Phytol.,* 121: 499–513.

Simon, L, Kieger, M, Sung, SS, Smalley, TJ. (1994). Aluminum toxicity in tomato. Part 2. Leaf gas exchange, chlorophyll content, and invertase activity. *J. Plant Nutri.,* 17: 303-317.

Stobart, AK, Griffiths, WT, Ameen-Bukhari, I. (1985). The effect of Cd^{2+} on the biosynthesis of chlorophyll in leaves of barley. *Physiol. Plant*, 63: 293-298.

Tao, HB, Lin, S. (2006). Comparison on disc method with copy method and length-width method for measuring leaf area of rice. *Plant Physiol. Commun.* 42: 496-498. (in chinese)

Verma, S, Dubey, RS. (2001). Effect of cadmium on soluble sugars and enzymes of their metabolism in rice. *Biol. Plant*, 44:117-123.

Woolhouse, HW. Longevity and senescence in plant. Oxford: Sci Prog, 1974.

Wu, FB, Zhang, GP, Dominy, P. (2003). Four barley genotypes respond differently to cadmium: peroxidation and activities of antioxidant capacity. *Environ. Exp. Bot*, 50: 67-78.

Zhang, XZ. Research methodology of crop physiology. Beijing: Agriculture Press, 1992. (in Chinese).

Zhou, WB, Juneau, P, Qiu, BS, (2006). Growth and photosynthetic responses of the bloom-forming cyanobacterium *Microcystis aeruginosa* to elevated levels of cadmium. *Chemosphere*, 65: 1738-1746.

In: Soil Ecology Research Developments
Editors: Tian-Xiao Liu, pp. 31-61

ISBN: 978-1-60021-971-9

Chapter 1

LEAF LITTER DECOMPOSITION DYNAMICS IN MEDITERRANEAN AREA

A. Fioretto, S. Papa, A. Pellegrino and A. Fuggi
Dipartimento di Scienze della Vita, Seconda Università di Napoli. Via Vivaldi n 43, 81100 Caserta ,Italy

ABSTRACT

The Mediterranean area is characterized by high temperature and low moisture in summer seasons. The aim of this study was to summarize and compare the decomposition dynamics of eight litter types (*Pinus laricio, P. pinea, Quercus ilex, Cistus incanus, Myrtus communis, Phillyrea angustifolia, Abies alba* and *Fagus sylvatica*) in their natural sites followed for about 2 years. The effect of litter quality was evaluated by comparing lignin, cellulose and nitrogen content. The effect of site characteristics and microclimatic conditions on decomposition was also evaluated by exposing the same litter simultaneously in different stands.

The dominant effect on litter decomposition in Mediterranean area was summer aridity because it highly reduces the activity and growth of microorganisms, affecting also the interactions with litter quality and soil organisms. The decomposition rate was positively related to the Lang aridity index for all the studied litter even if no statistical significance was evidenced. Microbial activity, that was high in autumn and spring, was strongly reduced in summer as evidenced by the high decrease of soil respiration and the activity of cellulases and xylanases. Seasonal fluctuations also occurred for lignin degrading peroxidases, in agreement with the view that the overall microbial community was affected. The overall activity of lignin degrading laccases did not show significant seasonal changes even if its isoenzymes forms were seasonally expressed. N- content on decomposition rate was in line with that reported for temperate areas even if the influence of climatic conditions makes less evident the correlation.

The effect of nutrient, like nitrogen, on decomposition rate was in line with that reported for temperate areas even if the influence of climatic conditions makes less evident the correlations.

The positive relation between lignin decay rate and the initial Mn content support the role of Mn dependent enzymes in such a process. However, the Mn poor *P. pinea* litter was not affected by its low Mn initial content when incubated in the original native stand.

In the litter colonizing microflora pectinolytic and cellulolytic fungi were aboundant from the start of decomposition, followed by the chitinolytic ones that increased after six months of decomposition. Ligninolytic fungi, even if isolated in the first month of decomposition increased when lignin decay started. Considering their taxonomic diversity species of typical genera of the Mediterranean area were mixed to those of ubiquitous genera found also in other regions.

INTRODUCTION

The cycle of organic matter and related mineral elements play a key role in the relationships between the soil, the vegetation and the surrounding environment.

The primary net productivity of forest vegetation is dependent on exogenous environmental factors such as soil and climate, and to endogenous factors such type of tree cover and age (Santa Regina et al., 1997). Trees retain a part of their production in perennial structures, whose nutritive elements form the mineralomass of the phytocenosis. In the forest ecosystems, the largest fraction of net primary production is delivered every year to the upper layers of soil and, consequently, provides the main above-ground contribution of carbon and nutrients to the forest floor. Litter deposit depends on the productivity of plant communities, which, in turn, is affected by plant species composition, soil fertility, water retention and climate (Meentemeyer et al. 1982; Kouki and Hokkanen, 1992; Pausas et al., 1997). An inverse linear relationship between total litter production and latitude has been found (Vogt et al., 1986). Leaf tissues account for about 70% of aboveground litterfall (Meentemeyer et al. 1982), even if the deposition of woody litter tends to increase with forest age (Harmon et al., 1987).

Most of detritus is delivered to the upper layers soil where it is subject to degradation. Litter disappearance is dependent on physical and chemical degradation, heterotrofic consumption and decomposition. The last process, carried out by bacteria and fungi with the help of pedofauna, involves the mineralization and humification of organic matter with the progressive release and/or immobilization of nutrients (Aber and Melillo, 1991; Coûteaux et al., 1995). In this view it is an important factor controlling nutrient cycling and soil humus formation, in particular, in forest ecosystems.

The decomposition process is affected by litter quality, climatic factors and soil organisms (Virzo De Santo et al., 1993; Berg et al. 1993, 1995 a; Rutigliano et al. 1996; Moore et al., 1999; Osono e Takeda, 2001; Fioretto et al., 2001 a and b). The quality depends on chemical and biochemical features of the litter, in terms of relative abundance of water soluble substances, polymer carbohydrates including pectin, hemicellulose and holocellulose, lignin and other aromatic compounds, lipids and waxes as well as nutrient availability. C/N ratio, lignin/N, holocellulose/lignin, cellulose/lignin/N are indices that reflects the fraction of labile compounds (carbohydrates, proteins) and recalcitrant compounds (mainly lignin, but also suberins, resins, fats and waxes) (Mac Claugherty and Berg, 1987; Taylor et al., 1989; Vallejo 1993; Rovira and Vallejo, 2002). As a consequence, the litter quality affects the amount and chemical features of humus as well as the microclimate of the soil surface.

Water soluble substances and other labile compounds are degraded in the early stage of decomposition by fast growing microorganisms that require high concentration of nitrogen (Swift et al., 1979). Cellulose and lignin, the most abundant components of litter, constituting 70-80% of fallen litter, are slowly decomposed.

Nevertheless, because lignin physically protects most of cellulose and hemicellulose from enzymatic hydrolysis, neither group of compounds decompose independently (Cooke and Whipps, 1993). Each fraction has characteristic exponential kinetics of decomposition, so that the total mass loss of litter is the sum of a number of exponential functions (Minderman, 1968).

Decomposition process consists of 2 main phases: I) the former phase, that lasts 1-2 years, in which the decomposition rate is regulated by the C and N availability, as well as litter morphological characteristics (Melillo et al., 1989); II) the latter phase, lasting more than 3 years, in which the rate is low and regulated by lignin and Mn concentration, being such element essential for some lignin-degrading enzymes, such as Mn peroxidases (Berg et al., 1995 b). On the other hand high lignin content has a rate reducing influence (Berg et al., 1996; Rutigliano et al., 1996; Coûteaux et al., 1998; Tian et al., 2000; Fioretto et al., 2005 a) in particular, when associated with high N content, because new and stable complexes are formed (Berg and Ekbohm, 1991; Coûteaux et al., 1995).

STUDIES ON LITTER DECOMPOSITION: A PERSPECTIVE

The studies on litter decomposition, considered by Odum (1983) an emergent property of ecosystems being sensitive to changes in ecosystem function, started more than 30 years ago.

Numerous researchers studied the decomposition dynamics in relation to climatic condition and substrate quality (Virzo De Santo et al., 1993; Coûteaux et al., 1995; Cortez et al. 1996; Coûteaux et al., 1998; Berg et al., 1998), to nutrient release (Laskowski et al., 1995; Berg and Laskowski 1997; Fioretto et al., 2001a), to nature and abundance of decomposer organisms and their interactions with the fauna (Cortez 1998; Cortez and Bouché 1998; Andrén et al. 2001).

Many data have driven to conclude that the climate is more important than substrate quality as a predictor of decomposition rates over large scales (Meentemeyer and Berg, 1986). In particular climate is a dominant factor in areas subjected to unfavourable weather conditions. Litter quality, instead, largely prevails under favourable climatic conditions (Couteaux et al., 1995).

Soil microrganisms play a key role in the decomposition of organic matter. In boreal forests, microbial communities mainly fungi and bacteria transform more than 90% of the plant litter carbon, leaving only 5% of it to soil animals (Berg and Laskowski 2006). In temperate forest soil the microorganisms are also the dominating primary decomposers.

Fungi play the main role among the decomposers for the degradation of plant organic material. Saprophytic microfungi are ubiquitous decomposers and respond rapidly to the addition of new substrate, because of their hyphal growth pattern, production of vegetative spores, specific survival strategies and capacity to produce a great variety of enzymes important in the decomposition process (Kjøller and Struwe, 2002). They contribute up to 90% of the total respiration of soil organisms (Kjøller and Struwe, 1982) and a lot of them (Basidiomycetes) can attack the lignocellulose matrix that other organisms are unable to metabolize. Due to their different versatility the composition of the fungal community changes during decomposition establishing a microbial succession (Frankland, 1998).

Bacteria also contribute to litter decomposition by releasing a large variety of enzymes in the environment.

The knowledge of the interaction between fungi and bacteria is important to understand the decomposition process. Nevertheless, generally, the fungal biomass exceeds the bacterial component mainly in the initial stage of cellulose, lignin and chitin decomposition (Swift et al., 1979).

In recent years, a greater attention has been turned to microbial degrading capacity by considering the extracellular enzymatic activities (Fioretto et al., 2000; Sinsabaugh et al., 2002; Kjøller and Struwe 2002; Di Nardo et al., 2004). The study of the decomposition dynamics in relation to the enzymes that degrade the major structural constituents of plant material, may provide information on specific metabolic and functional aspects of microbial communities (Sinsabaugh et al., 1991).

Microbial communities release hundred of different enzymes into the environment. Those directly involved in the degradation of lignocellulose (cellulases and phenoloxidases) and in N, P and S cycles are of primary interest. The degradation of the major structural constituents of plant organic matter requires multi-component enzyme systems, involving different microbial taxa. Cellulose and lignin degradation, for example, requires glucose oxidase and cellobiose dehydrognase that oxidize glucose and cellobiose, respectively. Glucose oxidase generates H_2O_2, which is required by ligninolytic peroxidases, while cellobiose dehydrogenase reduces quinones generated by the oxidative depolymerization of lignin. Glucose and cellobiose are produced by cellulose hydrolysis and, both, are feedback inhibiters of cellulolysis (Sinsabaugh et al., 1991). The substrates also interact as well as the enzymes. Cellulose, lignin and hemicellulose are covalently linked and physically intercalated in plant cell wall (Marsden and Gray, 1986). So, extensive degradation of litter needs of the concerted action of a variety of enzymes.

Laccases appear to be the best lignin degrading agents (Reid, 1995) and are released by many fungi (Tuor et al., 1995), especially white-rot fungi. These enzymes act in synergy with other polyphenoloxidases such as Fe-peroxidases or Mn-peroxidases. Cellulases are enzymes that degrade cellulose. They are divided at least in 2 enzyme classes: endoglucanases (endocellulase), which exhibit different affinities to cristalline cellulose and cellobiohydrolases (exocellulase). In principle, the level of activity of cellulases should be related to the rate of cellulose degradation in litter. However, the relationship is complex and only an index of cellulase interactions (the product of exocellulase and endocellulase activities) correlated with the rate of disappearance of cellulose from decomposing leaf litter, both within and between species (Linkins et al., 1990; Sinsabaugh et al., 1994).

Chitinases catalyze hydrolysis of chitin that is the main structural component of cell walls of most fungi and arthropods. They are produced by many species of bacteria (streptomycetes and other actinomycetes), fungi as well as plants and play an important physiological and ecological role in ecosystems as recyclers of chitin, by releasing C and N nutrients (Chernin and Chet, 2002).

By considering the large number of factors involved and interacting, litter decomposition is a complex process that requires an interdisciplinary approach.

Decomposition Studies in Mediterranean Area

The Mediterranean area constitutes a geographic zone critical for litter decomposition being characterised by summer seasons with high temperature and low moisture. These factors limit the grow and activity of microorganisms. Some organisms of pedofauna are also limited by such conditions being reported that the animal mortality is high during the dry season. However, high diversity and activity of microflora lead to high decomposition rates during the milder, wetter periods (Couteaux et al., 1995). In addition, the sclerophylly of the leaves in some ecosystems reflect a high content of structural compounds that slow down decomposition.

Numerous studies on the litter decomposition have been reported in temperate and boreal coniferous and deciduous forests (Berg and Lundmark, 1987; Blair, 1988; Wise and Schaefer, 1994; Berg and Laskowski, 1997; Berg, 1998; Chadwick et al., 1998; Berg et al, 1998; Berg, 2000; Dilly et al., 2001). Only a few data are available for Mediterranean ecosystems (Escudero et al., 1987; Ferran and Vallejo, 1992; Arianoutsou, 1993; Stamou et al., 1994; Gallardo et al., 1995; Kurz et al., 2000). Nevertheless, few studies have been performed on decomposition by comparing for the same litter mass loss rate, nutrient dynamics, qualitatively changes and microbial succession.

Here, the attention is focused on decomposition dynamics studies in Mediterranean area, and in particular in the South-Italy. The mass loss pattern and the qualitative changes of different litter types incubated in their natural environments is compared. Moreover, the influence of litter quality on decomposition rate is also considered by discussing the results obtained on different leaf litter incubated in the same plots. The influence of environmental conditions on decomposition rate are also evaluated by exposing the same litter type simultaneously in different stands. The relationships between lignin and cellulose decay and relative enzyme activities is also discussed.

Decomposition Dynamics and Changes in Substrate Composition

The decomposition dynamics of eight litter types (*Pinus laricio*, *P. pinea*, *Quercus ilex*, *Cistus incanus*, *Myrtus communis*, *Phillyrea angustifolia*, *Abies alba* and *Fagus sylvatica*) in their natural sites of South Italy followed for around 2 years were compared.

The stands of litter collection and incubation (Figure 1) were located within: a) the Fossiata forest (Golia site) (39°24'N; 16°34'E) in the Sila Mountains (Calabria) for *P. laricio*, b) the Natural Reserve of Castel Volturno (40°57' N, 13°55' E) in particular, in an area with low maquis, for *P. angustifolia*, *M. communis* and *C. incanus*, c) the Natural Reserve Tirone Alto Vesuvio in locality Terzigno (40°49' N, 14°28' E) for *P. pinea*; d) the WWF Oasis "Bosco di S.Silvestro" (41°30'N, 13°50'E) for *Q. ilex* litter; e) Taburno-Camposauro Mountain complex (41°06' N, 14°36' E) for *F. sylvatica* and *A. alba* litters. The sites differed in climatic and edaphic characteristics and some of these are summarized in Table 1.

The studies were performed by litter bag methods (mesh size 1 mm^2) with the only exception of pine litters, that were enclosed in large shallow containers of wood (100x100x5 cm) with top and bottom of terylene net (mesh size 1x1.5 mm).

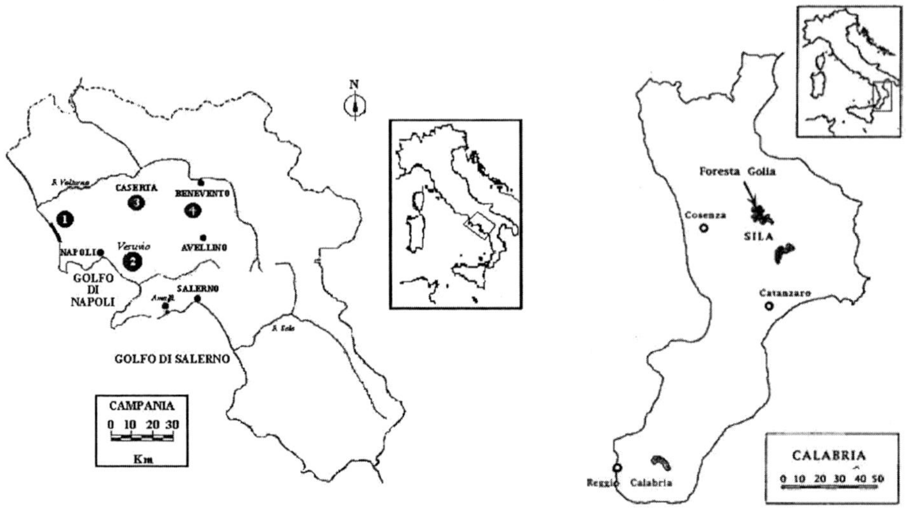

Figure 1. Localization of the litter incubation sites in the South Italy. A - sites in Campania region: 1, Natural Reserve of Castel Volturno; 2, Natural Reserve Tirone-Alto Vesuvio; 3, WWF Oasis "Bosco di San Silvestro"; 4, Taburno-Camposauro Mountains Complex. B – site in Calabria region: Natural Reserve "La Fossiata", Sila Mountains.

Table 1. Some climatic and edaphic characteristics of the studied sites

	Golia Forest	Castel Volturno Reserve	Taburno-Camposauro	WWF Oasis S.Silvestro	Tirone Alto Vesuvio Reserve (Terzigno)
Dominant plant species	*Pinus laricio*	*Cistus incanus* and *Myrtus communis*	*Fagus sylvatica*	*Quercus ilex*	*Pinus pinea*
Altitude (m a.s.l.)	1300	6-9	1100	200	250
Mean annual temperature (°C)	9.0	18.6	7.9	16.3	13.2
Mean temperature of the coldest month (°C)	1.6	10.6	-0.3	9	5.9
Mean temperature of the hoddest month (°C)	18	28	18	26	23
Annual precipitation (mm)	1225	680	2166	1190	960
Substrate type	siliceous	Siliceous-calcareous	pyroclastic material	calcareous	lapillus
	sandy	sandy	sandy	sandy	sandy

Table 1. Some climatic and edaphic characteristics of the studied sites (Continued)

	Golia Forest	Castel Volturno Reserve	Taburno-Camposauro	WWF Oasis S.Silvestro	Tirone Alto Vesuvio Reserve (Terzigno)
Soil pH	5.5	8* 8.5** 7.8***	6.2	7.0	6.0
Soil organic matter content (%)	8.7	2.2* 4.1** 13.6***	19	13.5	6.1
Nitrogen content (%)	0.22	0.19* 0.21** n.d.***	0.9° 0.65°°	0.5	0.3
C/N ratio	21.5	6.9* 11.4** n.d.***	16° 14°°	26	24

* under *C. incanus* shrubs; ** under *M. communis* shrubs; *** under *P. angustifolia* shrubs; ° Beech stand, °° Beech-fir stand

n.d. no determined

Data from: Virzo De Santo et al., 1993; Fioretto et al., 2001; De Marco et al., 2004; Papa et al., 2002

The litters showed different litter mass loss rates (Figure 2). The leaf litter of *M. communis* and *C. incanus* showed the highest accumulated mass loss at about 2 years of decomposition. For these two litters the degradation went on and after 3 years (data not shown) the masses lost were about 90% and 70%, respectively (Fioretto et al., 2005 a). Lower mass loss rates were observed in the litters of *F. sylvatica*, *A. alba*, *Q. ilex* and finally by *P. laricio* and *P. pinea*. The two latter litters lost only about 20% of the initial mass after two years. *P. angustifolia* litter showed a different pattern of mass loss. In fact, during the first 100 days of exposure it lost about 25 % of the initial mass, but subsequently further losses were not recorded, suggesting that it quickly reached limit values of decomposition.

The differences in decomposition rates among the species really emerge by comparing the decay constants (Table 2). For all species, apart from *P. angustifolia*, the decay constant values suggested that the decomposition could be divided in two intervals: the first year of decay and the remaining period of about 400-480 days. *M. communis* had the highest decay constant in both periods. *C. incanus*, that among the species showed a middle constant value in the first year around 59% of that of *M. communis*, increased its decomposition rate in the second year retaining about the same difference with *M. communis*. *F. sylvatica* and *A. alba*, that had the highest decay constants in the first year, even if lower than that of *M. communis*, reduced such rate to 1/3 and 1/4, respectively, in the remaining period. *Q. ilex* and *P. pinea* litter showed the same constant values in the first year and in the second year, even if they were reduced to 50%. Finally, *P. laricio* had the lowest decay constant among the tested species in the first year but not after.

The mass loss of all the litters at the end of the first year of decomposition showed a positive correlation, even if not statistically significant, with the Lang aridity index (annual precipitation/annual mean temperature). One of the main limiting factors in Mediterranean environments is the summer aridity, therefore an increase of precipitations or a decrease of temperature promotes the decomposition rate. By comparing the measured mass loss with those of the other litter types in the Mediterranean area (Table 3), it emerges that, because the high variability showed by different species in different environments, all our litters showed comparable values.

Climatic or microclimatic differences among the experimental sites did not justified the decomposition patterns of the eight studied litter species. Therefore a combination of climatic influence and chemical composition effects was suggested.

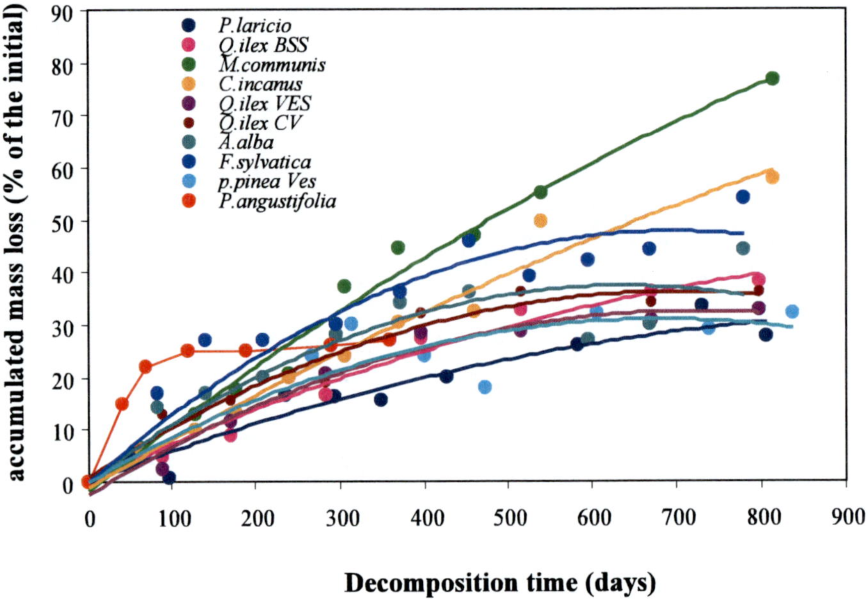

Figure 2. Accumulated mass loss versus decomposition time of *Myrtus communis*, *Cistus incanus*, *Phillyrea angustifolia*, *Quercus ilex*, *Fagus sylvatica*, *Abies alba*, *Pinus laricio* and *Pinus pinea* leaf litters in their native sites in South Italy. Standard deviation was around 10%. Data from Fioretto et al., 2000; Papa, 2000; Virzo De Santo et al., 1985, 1993; Fioretto et al. 1998; De Marco et al., 2004.

Table 2. Decay constants of the eight studied litters during the first year of decomposition (K_1) and during the remaining period concerning the second year and the beginning of the third one (K_2). The decay constants were calculated by the formula $X/X_0 = e^{-kt}$ (X=mass remaining at time t, X_0=initial mass) (Olson, 1963)

	K_1 y^{-1}	K_2 y^{-1}
Myrtus communis	0.58	0.71
Cistus incanus	0.34	0.40
Phillyrea angustifolia	0.31*	----
Quercus ilex	0.29	0.14
Fagus sylvatica	0.43	0.16
Abies alba	0.40	0.10
Pinus laricio	0.17	0.13
Pinus pinea	0.29	0.10

* 1.58 by considering only the first 100 days of decomposition, when the mass losses were clearly evident.

Nitrogen, as well as P and S, whose their concentrations are often highly correlated (Taylor et al., 1991; Berg and Laskowski, 2006) because contained in defined ratios in compounds such as protein and nucleic acid, are rate-regulating in the early stage of decomposition. The initial concentration of these nutrients affects the ability of the litter to their net accumulation, immobilization or release. Nitrogen is often limiting for the growth of decomposer populations in forest litters (Berg et al., 1987) and, as a consequence, is immobilized by microflora mainly fungi (Arianoutsou, 1993; Laskowski et al., 1995). Accumulation with respect to its initial amount have been also recorded, probably due external sources such as throughfall, microbial nitrogen fixation and, in particular, translocation by fungi from deeper soil layers. On the contrary, nitrogen is released from litters during decomposition when its initial concentration exceeds the need of microorganisms. Nevertheless, to predict its dynamics by its initial concentration, it must be considered that only a part of the total amount can be really bioavailable (Berg and Laskowski, 2006).

In our litters *F. sylvatica* had the highest N content, followed by *A. alba*, *Q. ilex*, *C. incanus* and finally by *M. incanus*, *P. angustifolia*, *P. pinea* and *P. laricio* (Table 4). As concern its dynamics during decomposition (Figure 3), *F. sylvatica*, *C. incanus* and *M. communis* litters released nitrogen from the early stage of decomposition (Virzo De Santo et al., 1998; Fioretto et al., 2005 a). After about 3 years of decomposition their residual amount was 70%, 60% and 30%, respectively. Among these three, *M. communis* litter released the greatest amount of nitrogen, in spite of the lowest initial N content and a C-to-N ratio higher than 25 (Table 3), a limit above that it has been supposed a net accumulation. *A. alba* and *P. pinea* litter accumulated nitrogen during the first year of exposure but subsequently a slow release occurred (Virzo De Santo et al., 1985; 1998) and their residual amounts were 80% and 100%, respectively, at the end of study period. Similarly, *Q. ilex* litter immobilized nitrogen for about 2 years of decomposition and only during the third year slowly released it. In *P. laricio* litter nitrogen increased as decomposition progressed so that its residual amount after 2 years of decomposition was about 150% of the initial (Figure 3) (Fioretto et al., 2001 a). In *P. angustifolia*, where unfortunately the N dynamics was followed for only 250 days of decomposition, its amount did not change during the first 4 months, but subsequently it started to increase (data not shown in Figure 3) (De Marco et al., pers. com.).

The content of lignin, a complex aromatic polymer resistant to degradation, also differed among the litters studied. *F. sylvatica* and *A. alba* had the highest content while *M. communis* and *Q. ilex* the lowest (Table 4).

Initial N content affects the early stage of decomposition while lignin concentration the later phase. In this view relationships between decay constants of the litters for the first year (K_1) and for the remaining period (K_2) with their N and lignin contents, respectively, were tried. A positive trend between N initial contents and K_1 while a significant ($P<0.05$) negative correlation between initial lignin contents and K_2 were found.

Generally, the initial concentrations of lignin affects its degradation. In rich lignin litter, lignin disappearance begins at early stage of decomposition; on the contrary, in poor lignin litter there is accumulation of lignin and/or like substances as decomposition proceeds (Berg and Laskowski, 2006). A negative influence of initial N concentration on lignin degradation has been reported (Hermann et al., 1977; Berg and Ekbohm, 1991; Berg and Laskowski, 2006), due to the repression of fungal lignolytic enzymes by low molecular recalcitrant N compounds (Keyser et al., 1978), derived by chemical reactions between ammonium or amino acids and phenolic groups in lignin or partly degraded lignin (Nömmik and Vahtras, 1982; Stevenson, 1982).

Table 3. Mass loss and nitrogen content of different litters in Mediterranean area

	Mass loss (% of the initial)	N %	Study area	Altitude (m a.s.l.)	Authors
Quercus rotundifolia	60	1.12	Near Salamanca (Spain)	820-880	*Escudero et al. 1987*
Quercus pyrenaica	75	1.01	Near Salamanca (Spain)	850	"
Quercus lanuginosa	38	n.d.	Causse Mejean (France)	1000	*Martin et al. 1994*
Quercus pyrenaica	36	n.d.	Fuenteguinaldo, Sierra de Gata (Spain)	870	"
Quercus pyrenaica	38	n.d.	Navasfrias, Sierra de Gata (Spain)	1000	"
Quercus rotundifolia	46	0.92	Tierra de Campos (Spain)	688	*Hernandez et al. 1995*
Pinus pinaster	35	0.76	Tierra de Campos (Spain)	688	"
Pinus pinea	41	0.98	Tierra de Campos (Spain)	688	"
Arbutus unedo	37	0.72	Stavros (Northern Greece)	20	*Arianoutson 1993*
Quercus coccifera	33	0.73	Stavros (Northern Greece)	20	"
Castanea sativa	25	0.88	Sierra de Bejar (Spain)	1150	*Gallardo et al. 1995*
Quercus pyrenaica	28	1.00	Sierra de Bejar (Spain)	1350	"
Pinus sylvestris	21	0.60	Sierra de Bejar (Spain)	1550	"
Pinus sylvestris	17	n.d.	Pirenees (Spain)	1540	*Pausas 1997*
Pinus sylvestris	43	1.25	Sierra de la Demanda (Spain)	1250	*Santa Regina and Tarazova 2001*
Castanea sativa	28	n.d.	Anduze, Cevennes (France)	380	*Cortez 1998*
Castanea sativa	42	n.d.	Le Vernet, Cevennes (France)	520	"
Castanea sativa	42	n.d.	Salidès, Cevennes (France)	860	"
Quercus petraea	18	n.d.	Anduze, Cevennes (France)	380	"
Quercus petraea	33	n.d.	Le Vernet, Cevennes (France)	520	"
Quercus petraea	30	n.d.	Salidès, Cevennes (France)	860	"
Quercus ilex	17	n.d.	Anduze, Cevennes (France)	380	"
Quercus ilex	25	n.d.	Le Vernet, Cevennes (France)	520	"
Quercus ilex	20	n.d.	Salidès, Cevennes (France)	860	"
Fagus sylvatica	11	n.d.	Anduze, Cevennes (France)	380	"
Fagus sylvatica	32	n.d.	Le Vernet, Cevennes (France)	520	"
Fagus sylvatica	24	n.d.	Salidès, Cevennes (France)	860	"

Lignin degradation is also affected by climate. The comparison of the lignin decay of a standard *Pinus sylvestris* litter within a large range of climatic conditions evidenced that it was more quickly degraded under a cold and dry climate than under a wet and warm one (McTiernan et al., 2003; Saiya Cork et al., 2002).

In our studies most of litters showed a lignin decay in the early stage of decomposition. In particular, lignin degradation, even if at different rates, began earlier in *F. sylvatica*, *A. alba* (Rutigliano et al., 1996) and *C. incanus* (Fioretto et al., 2005 a) litters with a lignin content higher than 30% (Table 4). Such value has been considered a threshold over which the decay of lignin in the early stage of decomposition occurs in boreal and temperate forests (Mc Claugherty and Berg, 1987). After three years of litter exposure, about 70%, 47% and

60% of the initial lignin amount were lost in *F. sylvatica*, *A. alba* and *C. incanus*, respectively (Table 5). In *P. agustifolia*, with an initial lignin concentration lower than such threshold, the lignin decay started early but at a slow rate (De Marco et al., 2004). On the contrary, in *M. communis*, this process was delayed in the time (its degradation began after about eight months) but subsequently rapidly decomposed (Fioretto et al., 2005 a).

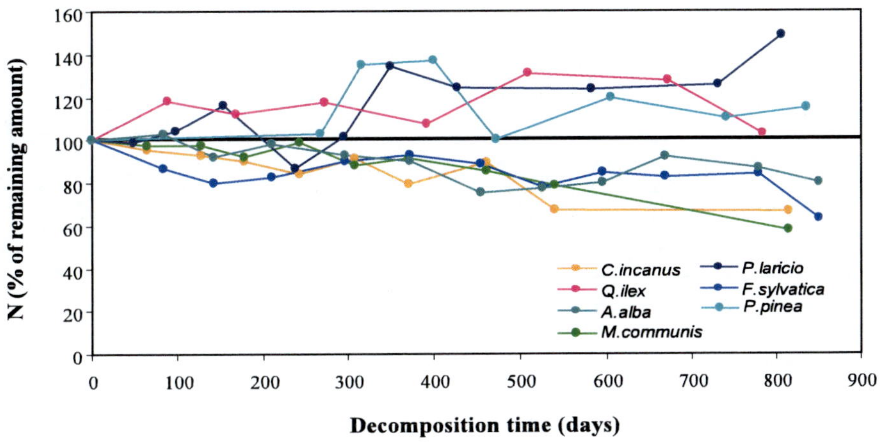

Figure 3. Nitrogen dynamics during leaf litter decomposition of *Myrtus communis*, *Cistus incanus*, *Quercus ilex*, *Fagus sylvatica*, *Abies alba*, *Pinus laricio* and *Pinus pinea* in their native sites in South Italy. Standard deviation was around 10%. Data from Fioretto et al., 2000; Papa, 2000; Virzo De Santo et al., 1998; Fioretto et al. 1998.

Table 4. Litter quality of Quercus ilex, Myrtus communis, Cistus incanus, Pinus laricio, Pinus pinea, Phillyrea angustifolia, Fagus sylvatica and Abies alba in their native stands

A.alba		*Q. ilex*	*M. communis*	*C. incanus*	*P. laricio*	*P. pinea*	*P. angustifolia*	*F. sylvatica*
13.2	N content (mg g^{-1} d.wt.)	11.1	7.8	10.1	5.96	6.24	6.9	16.8
37	C/N	46	38,26	38.33	86	78	-	25
48	Lignin (%IOM)	15	18	34	27	37	27	41
-	Cellulose (%IOM)	39	37	36	-	-	-	-
36	Lignin/N	13	23	31	45	59	40	24
60	Mn (μg g^{-1} d.w.)	40	50	70	490	53	16	50
-	Zn (μg g^{-1} d.w.)	30	20	80	10	16	-	-
4.7	K (mg g^{-1} d.w.)	3.12	3.18	4.98	3.5	3.3	0.16	1.2
1.03	Mg (mg g^{-1} d.w.)	0.89	3.05	1.56	1.28	2.09	0.17	1.86
-	Na (mg g^{-1} d.w.)	0.33	1.72	3.05	-	-	0.07	-

Data from: Papa et al., 2007; Fioretto et al. 2003; Fioretto el al. 1997; De Marco et al., 2004; Rutigliano et al. 1996; Berg et al. 2003

After about three years the 78% of the initial lignin litter was degraded (Table 5). In *Q. ilex* litter, lignin had the lowest concentration and was not significantly degraded during the whole study period similarly to *P. laricio*. Unfortunately, no data are available for *P. pinea* in Terzigno site.

In addition, in *C. incanus* and *M. communis* litter the decay constants of lignin in the first year was higher than that of organic matter suggesting that lignin was more quick degraded than other litter components (Table 2 and Table 5).

Table 5. Residual lignin amount, as per cent of the initial (T_0) after about 1 year (T_1), 2 years (T_2) and 3 years (T_3) of decomposition in the eight litters. The lignin decay constant (*K* per year) for the first year (K_1), for the second (K_2) and third year (K_3) were also reported

	Residual amount (% of the initial)			Lignin decay rate Ky^{-1}		
	T_1	T_2	T_3	K_1	K_2	K_3
Fagus sylvatica	66	41	31	0.41	0.47	0.39
Abies alba	68	63	53	0.38	0.08	0.21
Myrtus communis	44	37	22	0.80	0.14	0.59
Cistus incanus	68	39	31	0.38	0.82	0.31
Quercus ilex	100	~100	~100	0	0	0
Phillirea angustifolia	90	-	-	0.10	-	-
Pinus laricio	~100	~100	87	0	0	0.07
Pinus pinea	-	-	-	-	-	-

In this view, *F. sylvatica* and also *C. incanus* showed a lignin decay constant higher than that of organic matter in the second year. The other litters showed lower values.

To degrade lignin, Mn, among other nutrients, seems to be essential. High level of Mn can enhance the production of Mn peroxidase, a lignin degrading enzyme having Mn as cofactor (Perez and Jeffrey, 1992; Archibald and Roy, 1992).

Among the studied species *P. laricio* had the highest initial Mn concentration and *P. angustifolia* the lowest (Table 4). For the other litters the values ranged from 40 to 70 $\mu g\ g^{-1}$.

The lignin decay constants of our litters appeared positively related to initial Mn concentration (K_1-Mn for $P < 0.05$; K_2-Mn for $P < 0.01$). However, in the advanced decay phase, when lignin degraded very slowly, Mn appeared no a stimulating factor but rather an inhibitor because a negative relationship was found between lignin decay constant during the third year and Mn initial content ($P < 0.05$). The lignin decay constants were also negatively related to initial N content (K_1-N for $P < 0.001$; K_2-N n.s.; K_3-N for $P < 0.01$). On the contrary, lignin decomposition rate was not correlated to climatic conditions.

Quality Effect on Decomposition Rate

Quality effect on decomposition rate was tested incubating different litter types in a same stand.

In this view, the dead needles of *P. laricio* and *P. pinea*, collected in their woods in the Fossiata Forest and Tirone Alto-Vesuvio Reserve, respectively, as well as of *P. sylvestris* collected in Jädraås (Sweeden) were simultaneously incubated in the *P. laricio* and in the *P. pinea* woods.

Pine litters differed in their chemical composition. The highest content of N, S, Mg and Ca, among the macronutrients, were found in the *P. pinea* litter (Table 4) while the lowest ones in the *P. sylvestris* litter (4.9 mg g^{-1}; 0.66 mg g^{-1}; 0.30 mg g^{-1}; 3.44 mg g^{-1}, respectively).

P and K had the highest content in the litter of *P. laricio* (Table 4) while the lowest ones in the *P. sylvestris* (0.23 mg g^{-1}; 1.32 mg g^{-1}, respectively). Among the micronutrients, Mn, Zn and Cu were high in the *P. sylvestris* (1026 μg g^{-1}; 49μg g^{-1}; 5.5 μg g^{-1}) while, as said before, *P. pinea* litter contained the lowest Mn amount (Table 4). As concern the lignin content, *P. pinea* showed the highest value compared to that of *P. laricio* (Table 4) and *P. sylvestris* (37%) ones.

The different litter quality caused different rate and pattern of mass loss (Fioretto et al., 1998; Virzo De Santo et al., 1993).

In the *P. laricio* wood, *P. pinea* showed a delay in the start of decomposition and a low decomposition rate throughout the whole study period (2.5 y). It resulted in a mass loss of 18% of the initial mass as compared to 33%-34% of *P. sylvestris* and *P. laricio*, respectively, (Table 6) after 2.5 years of incubation, in spite of the highest initial N concentration that normally stimulate the decomposition rate (Witkamp, 1966). Probably, as suggested by Berg (1986) the high N content associated to high lignin content determined the synthesis of new recalcitrant N compounds in the litter that inhibited the decomposition. Moreover, *P. pinea* exhibited the lowest Mn initial content and it is probable that this concentration was limiting for microbial activity. The Mn dynamics in this litter evidenced an increase of absolute amount during decomposition (Fioretto et al., 2001a). The decay started at the end of the first year of exposure when the Mn immobilized reached 86 μg g^{-1}, value probably sufficient to sustain decomposition, that in the second year had a rate higher than that of the other two litters (Table 6). By comparing the lignin decay dynamics, it emerges that in *P. laricio* and *P. pinea* only 10% of the initial lignin was degraded during about 2 years of decomposition. However, in *P. sylvestris* no lignin was degraded in the same period (data not shown).

Table 6. Decay constant (*K*) of needle litter of *P. laricio*, *P. pinea*, *P. sylvestris* calculated by the formula $X / X_0 = e^{-kt}$ (X = mass remaining at time t, X_0 = initial mass) (Olson, 1963)

	First year ($K\,y^{-1}$)	Second year ($K\,y^{-1}$)	Total period (2 y) ($K\,y^{-1}$)
P. laricio	0.18	0.22	0.17
P. pinea	0.004	0.26	0.08
P. sylvestirs	0.21	0.22	0.16

In the *P. pinea* stand, the native needle litter, had in the early stage of decomposition a mass-loss dynamics similar to those of *P. laricio* and *P. sylvestris* litters (Virzo De Santo et al., 1993), suggesting that the initial low Mn content did not influence negatively the decomposition rate in the native stand probably because of microflora specialized to work at low Mn level or of a translocation of the element from soil.

ENVIRONMENTAL EFFECT ON DECOMPOSITION RATE

The influence of environment on decomposition rate and pattern was tested incubating a same litter type in different woods.

The leaf litter decomposition of *Q. ilex* was studied in the WWF Oasis "*Bosco di S. Silvestro*" (BSS), where senescent leaves were collected at litterfall time, and in the Nature Reserves of Castel Volturno (CV), Caserta (sea level) and Tirone-Alto Vesuvio, Mt Vesuvius (VES), Naples (500 m a.s.l.).

The studied sites differed in soil characteristics, in such as the distribution of particle sizes that, in turn, affects water holding capacity, pH, N content and C/N ratio (Table 7) as well as by pluviometric regime that appeared different. Referring to the nearest meteorological stations, the VES wood appeared to receive the greatest amount of precipitations during the studied period (1999-2001), and BSS the smallest (with the only exception of the 1999 year). It changed the soil moisture content (Papa et al., 2002).

In spite of these differences among the three oak woods chosen for the comparative study, the leaf litter of *Quercus ilex* did not show significant differences in decomposition rate and at the end of the incubation period (about 2 years) the remaining litter weight ranged from about 50 to 60 % of the initial mass (Figure 4 a).

The degradation of the main recalcitrant material, holocellulose and lignin, however some little differences among the sites. Cellulose was degraded from the first sampling and continued at a constant rate in Castel Volturno holm-oak wood. Its initial mass was reduced to 50% after 2 years of incubation (Figure 4 b). In this stand, a significant increase of fungal biomass occurred after about 300 days of exposure during the wet autumn season (Figure 5). In the other two sites, cellulose degradation began after 6 months of exposure and continued to the end of the study period. Lignin, on the contrary, did not show any significant degradation during the decomposition period (Figure 4 c).

Table 7. Soil characteristics of the three oak woods located within the Natural Reserve Tirone Alto-Vesuvio, WWF Oasis "Bosco di San Silvestro" and Natural Reserve of Castel Volturno (Campania Region, South Italy)

		Natural Reserve of Tirone-Alto Vesuvio	Natural Reserve of Castel Volturno	Oasis WWF Bosco di S.Silvestro
Substrate		Volcanic	siliceous-calcareous	Calcareous
Fraction*	Designation	Distribution of particle sizes (mass %)		
1 cm- 2 mm	Gravel	4.5	0	3.0
2 – 0.2 mm	Coarse sand	41.2	76.8	17.8
0.2 – 0.02 mm	Fine sand	40.0	22.8	51.4
< 0.02 mm	Clay and silt	14.3	0.4	27.8
Water holding capacity (H_2O g/100 g d.wt)		71.7	62	156.7
pH		6.5	8.0	7.0
Potential pH		5.0	6.8	6.4
C org (%)		21.7	13.8	13.5
N (%)		1.47	0.51	0.50
C/N		14.8	27.1	26.0

* Gravel >1 cm was removed before fraction separation.

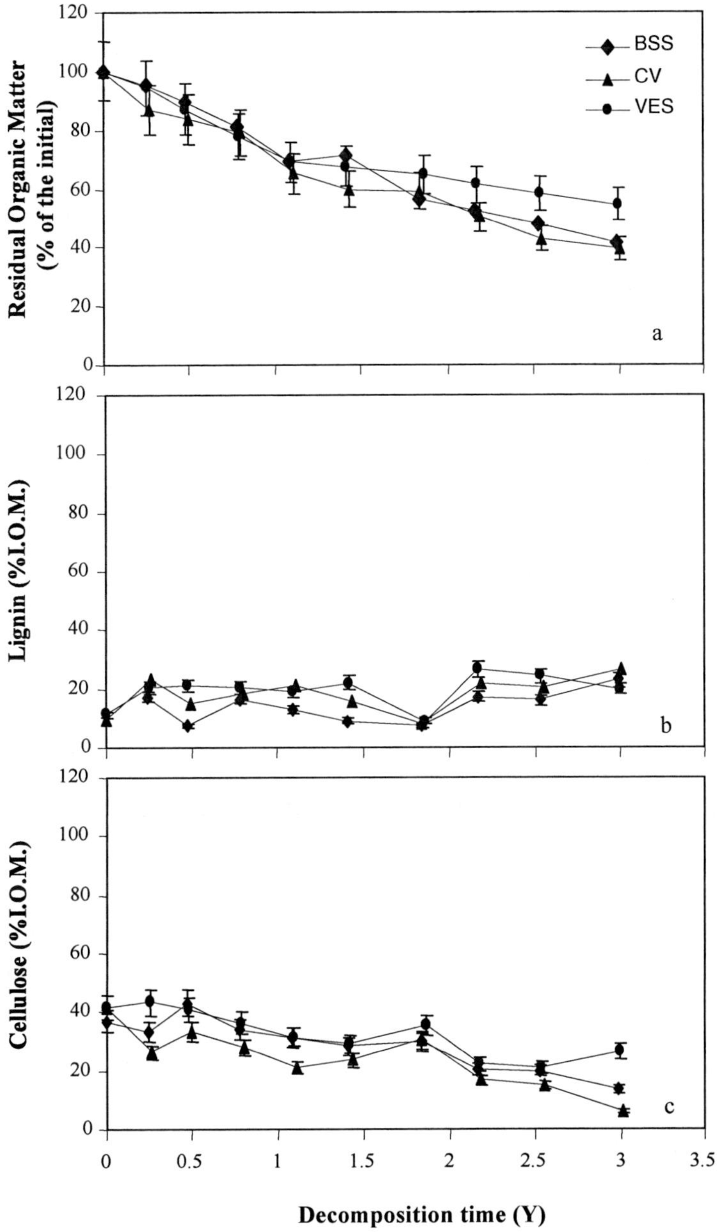

Figure 4. Decay dynamics of organic matter (A), lignin or lignin like substances (B) and holocellulose (C) during decomposition of *Quercus ilex* leaf litter in three woods of Campania Region (South Italy). Residual organic matter (R.O.M.), holocellulose and lignin are reported as % of the initial mass (I.O.M.). Data from Papa, 2000.

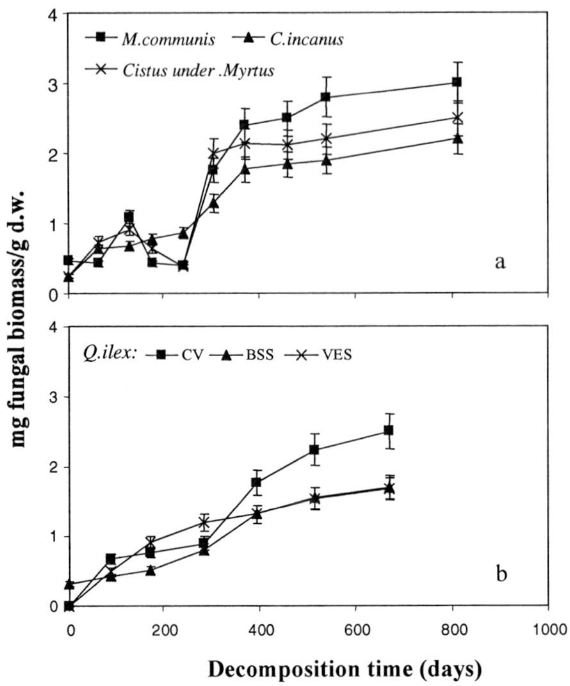

Figure 5. Fungal biomass during leaf litter decomposition of *Cistus incanus* and *Myrtus communis* on top soil under their shrubs as well as of *Cistus* under *Myrtus* shrubs in a Mediterranean maquis (A) and of *Quercus ilex* in three holm-oak woods (B) of South Italy. Data from Fioretto et al., 2000; Papa, 2000.

By comparing the mass-loss dynamics of *P. laricio* and *P. pinea* litters in the both incubation stands, generally the decomposition rate occurred at a lower rate in the *P. laricio* than in *P. pinea* wood. This suggests that in *P. laricio* stand the dryness (P/T = 73) superimposed the relatively low temperature, in particular in the winter (Virzo De Santo et al., 1993).

On the other hand, the same litter of *P. sylvestris* incubated in the *P. laricio* and *P. pinea* stands as in other coniferous forest along a climatic transect North-South Europe (Berg et al., 1993; 1995 a and b; Virzo De Santo et al., 1993) evidenced that the scarcity of precipitation along the year and in particular in the hot summers negatively influences the decomposition rate in Mediterranean area.

Microclimatic Effects on Decomposition Rate

Microbial community composition in the rhizosphere is affected by a complex interaction between soil type, plant species and root zone location (Marschner et al., 2001). It is affected by plant species (Pinzari et al., 1999; Miethling et al., 2000; Rutigliano et al., 2004; Fioretto et al., 2005 b) depending on the composition of root cells and root exudates which, in turn, is affected by root zone, plant age, N supply and mycorrhizal fungi infection (Merbach et al., 1999). In addition, soil aeration and physical-chemical characteristics result in different microbial communities and spatial variability (Gelsomino et al., 1999; Carelli et al., 2000).

Cistus incanus is a summer deciduous species while *Myrtus communis*, an evergreen species. This two type of shrubs create different microclimatic conditions on the top soil because of their different habits. For example, soil temperature at 5 cm dept before midday, similar in winter, were higher over 2-4°C under *Cistus* than under *Myrtus* shrubs in summer. Moreover, even if there was no significant change in water availability at the sampling time (about 3 h after dawn) also in summer, however, some dew, was formed by the high nocturnal humidity, and it moistened the litter under the bare *Cistus* but not under *Myrtus* at sunrise. These microclimatic conditions affected decomposition: both litters incubated under *Myrtus* showed reduced decomposition rate during the dry summer of the first year of decomposition when the litter bags were more exposed to air because less covered by new litter. The mass loss rate of *Cistus* litter incubated under *Cistus*, instead, did not change (data not shown) in the same period (Fioretto et al. 2000). Probably under *Cistus* the daily drying and rewetting cycles in summer can favour mechanical fragmentation and decomposition, as reported for sweet chestnut and oak leaf litters (Anderson, 1973; Witkamp and Olson, 1963).

In addition, the distribution of soil enzyme activities suggested a different microbial community in the soil under the two shrub types. The activities, in particular of xylanase, cellulase and invertase, were almost twice in the soil under *Cistus* during the wet periods, clearly indicating a different functionality of the community (Fioretto et al., 2005 b; Papa et al., 2007).

The litters of *Cistus*, incubated under *Cistus* and *Myrtus* shrubs, as well as of *Myrtus*, incubated under *Myrtus* plants, showed similar decomposition rate during the first 8 months of incubation (Figure 6) (the average decay constant (k) ranged between 0.29±0.03 y^{-1} and 0.33±0.03 y^{-1}) (Fioretto et al. 2000). Subsequently, it increased only for litters incubated under *Myrtus* shrubs in line with the increase of the fungal biomass (Figure 5).

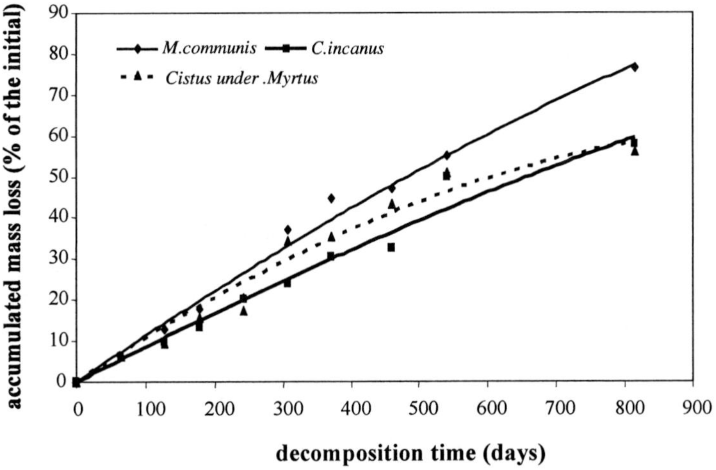

Figure 6. Accumulated mass loss versus decomposition time of Cistus incanus litter incubated on top soil under Cistus shrubs and Myrtus shrubs in a Mediterranean maquis. As comparison, also accumulated mass loss of Myrtus litter under Myrtus shrubs is reported. Data from Fioretto et al., 2000.

The enzyme activities (discussed below), on the contrary, differenced only between the two litter types (always higher in *Myrtus* than in *Cistus*) independently from incubation microsites (under *Cistus* or *Myrtus* shrubs). Nevertheless, the effect of different rhizosphere microbial populations could not be excluded and the contrasting data of mass loss dynamic and microbial activity could depend on complex interaction among litter quality, microbial communities colonizing litters as well as rizhosphere communities.

The decomposition rate of *Fagus sylvatica* litter showed similar results. It was higher in the beech stand than in a beech-fir stand of Monte Taburno during the first 15 months of exposition (Figure 7). Its higher decay rate could be related to different microclimatic condition as well as to interaction among beech, fir and decomposers (Virzo De Santo et al., 1985). Similar differences were also found by Herlitzius (1983) for beech leaf litter exposed in beech and *Picea abies* woods.

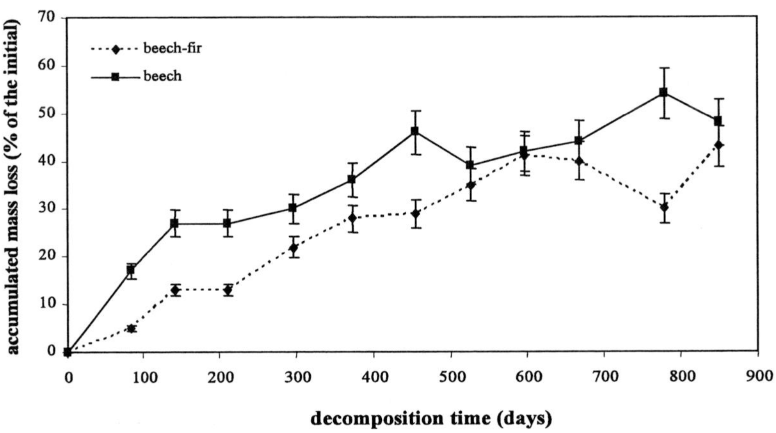

Figure 7. Accumulated mass loss versus decomposition time of *Fagus sylvatica* litter incubated in a beech stand and a fir-beech stand on Monte Taburno (Campania Apennines, South Italy). Data from Virzo De Santo et al., 1985.

MICROBIAL ACTIVITY DURING DECOMPOSITION

Water availability is the main factor affecting microorganism activities in the Mediterranean area as suggested by the strong correlation between soil and litter water content and microbial respiration (Papa et al., 2002; Fioretto et al., 1998; 2000; 2001). So, even if microbial respiration generally increases with temperature, it was at the lowest level in hot and dry summer and at highest level in spring and autumn for the needle litter of the three pine types in *P. laricio* wood either as well as for *Q. ilex* litter in the three relative stands and for *M. communis* and *C.incanus* litters in the low maquis (Fioretto et al., 1998; 2000; 2001 b).

Similar seasonal trend was also evidenced by litter enzyme activities (Figure 8). The extractable activities of cellulase (EC 3.2.1.4) and xylanase (E.C 3.2.1.8), degrading cellulose and hemicellulose, respectively, as well as of peroxidase (EC 1.11.1.7), lignolytic enzymes, showed seasonal variations in *Cistus*, *Myrtus* and *Quercus* litters during the decay process (Figure 8) (Fioretto et al 2000, 2001 b; Papa, 2000). This pattern was independent from site or microsite of incubations; therefore, as said before, no significant differences of activity were found in the *Cistus* litter incubated under *Cistus* and under *Myrtus* shrubs (Fioretto et al., 2000) as well as among the *Q. ilex* litter incubated in the three woods (Papa, 2000).

Moreover, seasonality in enzyme activity during organic matter decay appeared only in Mediterranean area. In other areas, in fact, the activity of cellulase increased as decomposition progressed until a maximum and, then, rapidly declined (Linkins and Sinsabaugh 1990; Sinsabaugh et al., 1992).

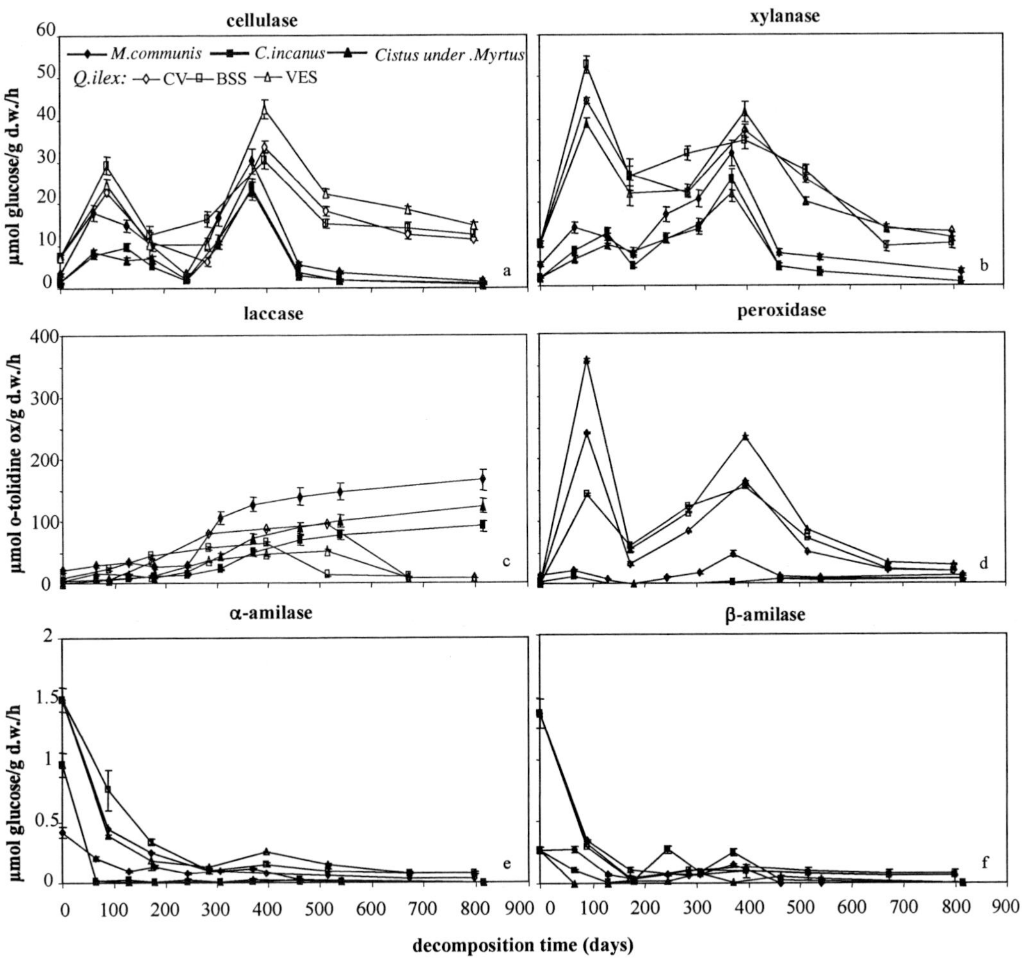

Figure 8. Enzyme activities (cellulase, xylanase, laccase, peroxidase, α-amylase, β-amylase) changes during leaf litter decomposition of *Cistus incanus* (in both incubation microsites, such as under *Cistus* and *Myrtus* shrubs) and *Myrtus communis* in a Mediterranean maquis and of *Quercus ilex* in three holm-oak woods of South Italy. Data from Fioretto et al., 2001; Papa, 2000.

The presence of cellulase and xylanase in the early stage of decomposition supports the view that cellulose and xylan are among the first decomposed compounds and their persistence also after about 2 years could be due to no exhausting decomposing substrates (Figure 4) (Fioretto et al. 2005 a). Even if some species of fungi synthesize the enzymatic complexes that hydrolysed, in particular, the cellulose and the hemicellulose (Joshi et al., 1993), no correlation was found among these activities and fungal biomass.

Among the lignolytic enzymes, laccase (EC 1.10.3.2) remained at low activity during the first 6-8 months for *Quercus* and *Cistus* and *Myrtus*, respectively (Figure 8). Subsequently, it increased rapidly in correspondence of the increase of fungal biomass and a strong correlation was found between these parameters ($P<0.001$). Such activity was specific for each litter type but no significant differences were found among incubation sites for *Quercus* litter or between the microsites for *Cistus* litter. Therefore a qualitative change of the microbial community occurred when the litters colonized by laccase producing fungi that take over when the opportunist's microorganisms start to decrease.

Other enzymes like α- and β-amylase, degrading the starch, were controlled by the substrate availability. They were high at the start of incubation and declined quickly as decomposition progressed in all incubation sites (Figure 8). The starch, in fact, degrades easily with respect to the other cellular components. The amylolytic microorganisms, r-strategists, colonize quickly and easily their substrate, and quickly disappear in lack of it.

It is to emphasize that after 2 years of decomposition the cellulolytic and ligninolitic activities were persistent because their substrate were not exhausted.

By comparison of the litters (Figure 8), *Quercus ilex* showed a higher activities than *Cistus* and *Myrtus* (Fioretto et al., 2000; Papa, 2000), in spite of its lower decomposition rate. This discrepancy could derive by the fact that the activities are assayed in vitro in saturating substrate where the effect of possible inhibitor compounds of the litter and/or produced by the colonizing microorganisms disappears. Peroxidase activity, nevertheless, in *Cistus* was below the detection limit of the assay for about a year of decay and a low activity was measured after.

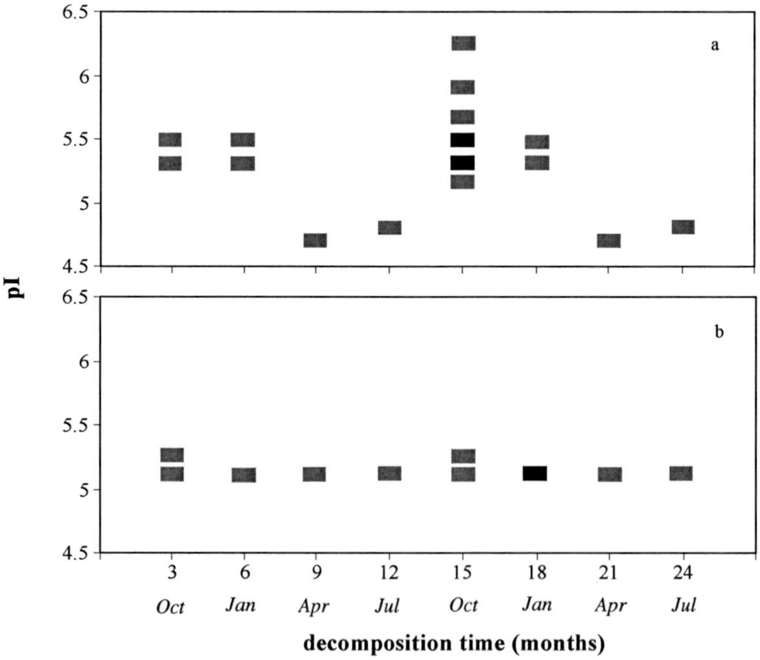

Figure 9. Laccase (A) and peroxidase (B) isoforms during *Quercus ilex* leaf litter decomposition in the holm-oak wood of WWF Oasis "Bosco di San Silvestro". Data from Di Nardo et al., 2004.

These data evidence changes in microbial succession during decomposition. It was specifically supported by electrophoretic analyses of extracts from *Q. ilex* leaf litter (Di Nardo et al., 2004). They were used to characterize laccase and peroxidase isoenzyme patterns. It shows that the decomposer community can involve different microorganisms with similar functional activity but expressing different isoenzymes (Sinsabaugh et al., 2002).

The main laccase and peroxidase isoenzymes, on decomposing *Q. ilex* leaves, occurred during the wet autumn and winter (Figure 9) (Di Nardo et al., 2004). Others isoenzymes were present in the dry spring and summer period when the lowest enzyme activity was recorded, suggesting that such conditions had severely restricted microbial growth (Fioretto et al., 2000) but enhanced diversity. However, the appearance of a new laccase isoforms in the second year of decomposition suggested the blooming of microorganisms the litter, probably favoured by the variations in physical and chemical composition of the litter. Changes in laccase isoforms during decomposition of *Q. ilex* leaf litter have also been reported in other Mediterranean areas (Criquet et al., 2000).

Fungal Succession

During the decomposition of plant material the composition of fungal community changes, evidencing a microbial succession (Frankland, 1998). This succession can be viewed as changes in taxonomic diversity and, if the role of the fungal population is know, also as functional diversity.

Fungal species distribution and successional changes occurring during the decomposition process have been extensively investigated on several litter types of the European forests (Dilly and Imler, 1998, Kjøller and Struwe, 1987, 1990, 2002; Rosenbrock et al., 1995) and of Mediterranean ecosystems (Pasqualetti et al., 1999; Tempesta et al., 2003, 2005; Sadaka and Ponge, 2003; Pasqualetti et al., 2006). Nevertheless, some of these studies only consider the presence and the frequency of species, but did not try to relate the results to microbial activity and litter quality changes.

Fungal colonization on needles litter of *P. pinea*, *P. laricio* and *P. sylvestris*, incubated in the *P. pinea* wood on Terzigno (Vesuvius), and of *A. alba*, incubated in silver fir wood on Monte Taburno, showed very similar pattern in all litter species (Virzo De Santo et al., 2002).

The invading fungi just after incubation (1-2 months) were the dematiaceous hyphomycetes, with *Cladosporium* and *Altemaria* sp. pl. dominants in *P. pinea* and *P. laricio.* In such litters the dematiaceous fungi were accompanied by the ascomycete *Lophodermium*, a typical primary colonizer. *Cladosporium* and *Altemaria* may be considered chiefly external, ubiquitous colonizers, able to utilize pectin, while their cellulolytic activity depend on species and conditions (Domsch et al., 1993). In silver fir litter the dominant dematiaceous hyphomycete was *Thysanophora penicilloides. Thysanophora* as well as *Lophodermium* are internal colonizers restricted to specific hosts. The litter of *P. sylvestris*, was instead colonized mainly by coelomycetes.

The frequency of the early stage dematiaceous hyphomycetes dropped significantly after the four months of incubation, and in the later stages these fungi disappeared. From the sixth month of incubation the frequency of other dematiaceous hyphomycetes, e.g., *Polyscitalum* sp. pl., increased and became dominant, together with tuberculariaceous hyphomycetes.

Table 8. The number of isolates of the four most frequent fungi during Q. ilex leaf litter decomposition expressed as a percentage of the total number of isolates in each sample. Data from Di Nardo, 2004

Genera	Month	Decompositon time (month)	Frequency (%)
Tricoderma	Oct	3	18
Penicillium			16
Alternaria			14
Mucor			11
Mucor	Jan	6	37
Tricoderma			23
Penicillium			12
Mortierella			5
Penicillium	Apr	9	31
Tricoderma			22
Mortierella			7
Beltrania			3
Penicillium	Jul	12	34
Tricoderma			27
Mucor			6
Mortierella			6
Penicillium	Oct	15	30
Tricoderma			22
Mortierella			19
Cladosporium			6
Tricoderma	Jan	18	24
Penicillium			24
Mucor			12
Mortierella			4
Penicillium	Apr	21	32
Tricoderma			30
Mortierella			6
Alternaria			4
Penicillium	Jul	24	38
Tricoderma			28
Mortierella			11
Mucor			5

On pine needles the basidiomycetes, the most significant group of secondary saprophytes able to carry out lignin degradation, became dominant about a year later, even if they were detected already in the first month of incubation of the *P. pinea* litter, which had a high water holding capacity (Virzo De Santo et al., 2002). Moreover, the occurrence of basidiomycetes appeared to show a seasonal pattern. Basidiomycetes were absent in the N-rich silver fir needle litter incubated at Monte Taburno, which was instead colonized by *Polyscitalum fecundissimum*, *Chalara*, and *Endophragmia*.

In the later phase of pine litter decomposition (after about 2 years) the frequency of ascomycetes with shield-like ascocarps (thyriothecia), became significantly high. Ascomycetes were also frequent on silver fir needles.

Microfungi succession was also studied during leaf litter decomposition of *Q. ilex* in the WWF Oasis "San Silvestro" (Table 8) (Di Nardo, 2004). *Trichoderma* and *Penicillium* were isolated at each sampling. They had cellulolytic and pectinolytic activity, respectively (Kjøller and Struwe, 1990), and was important in controlling decomposition rate (Figure 10). They, in fact, release low-molecular sugars for the accompanying Mucorales, whose frequency was higher in the winter months. After 6 months of incubation, appeared in the litter *Morteriella* spp., with chitinolytic activity. Nevertheless, their highest frequency was observed when a significant decrease of litter pH occurred.

Microfungi of the genera *Alternaria*, *Cladosporium* and *Beltrania* were also found (Di Nardo, 2004). The first two, as said before, are ubiquitous colonizers generally able to utilize pectine. *Beltrania*, typical of Italian microflora, instead, has a unknown function. However *Beltrania* isolates from our *Q. ilex* litter did not evidenced cellulolytic activity when plated on cellulose agar (Di Nardo, 2004). By considering the functional role of the isolated fungi, it appears that about 40% were able to degrade pectin and occurred in the whole study period (24 months) supporting the role of pectin degradation in making available other litter compounds, like cellulose; about 30% were cellulolytic and occurred also during the whole decomposition process.

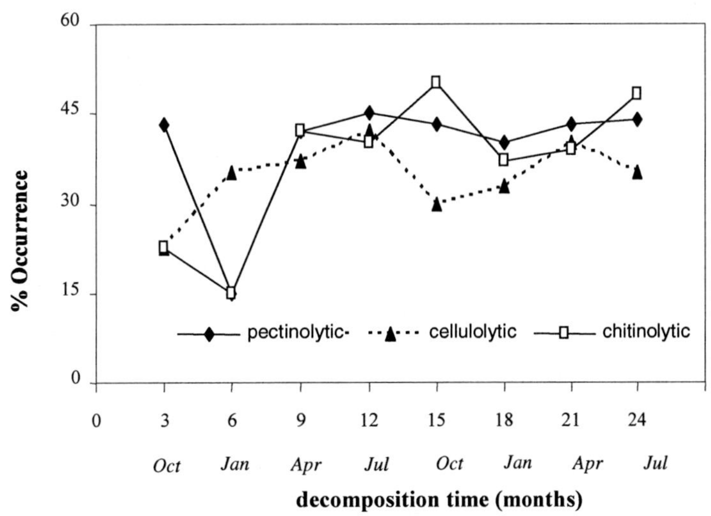

Figure 10. The occurrence of the four most frequent fungi able to utilize pectin, cellulose and chitin during *Quercus ilex* leaf litter decomposition in the holm-oak wood of WWF Oasis "Bosco di San Silvestro". Data from Di Nardo, 2004.

Chitinolytic fungi were approximately 20% in the first 6 months of decomposition and after raised to 40%, staying at such value in the whole period with a little increase at 15 months of incubation when *Mortierella* spp reached the highest frequency. This suggests a critical litter quality change at such a stage in which a relevant important fungal growth occurred. When fungal biomass significantly increase, the dead hyphae, rich in chitin, became a resource for chitinolytic fungi such as *Morteriella.* The role of chitin as a new substrate in the litter has been also evidenced in beech litter (Kjøller and Struwe, 1990).

CONCLUSION

In the Mediterranean area the interactions among litter quality, environmental factors, and microrganisms-fauna make more complex the decomposition process. However, its understanding moves in the general paradigma that consider: a) the initial concentration of the main components of litter, lignin, cellulose and hemicellulose, and nutrient content such as nitrogen N, the controls the decomposition rate determining not only the resource availability but also the palatability of organic matter, b) the degradation of lignin, cellulose and hemicellulose is related to climatic conditions, c) neither group of litter compounds was degraded independently, is a useful paradigma to understand such a process.

The dominant effect on litter decomposition in Mediterranean area was summer aridity because it highly reduces the activity and growth of microorganisms, affecting also the interactions with litter quality and soil organisms. The decomposition rate was positively related to the Lang aridity index for all the studied litter even if no statistical significance was evidenced.

The low moisture and the high temperature in the summer seasons strongly reduced microbial activity, as indicated by the high decrease of soil respiration and activity of cellulase and xylanase enzymes involved in the degradation of the main litter component. Seasonal fluctuations were evident also for lignin degrading enzymes like peroxidases (not laccases) in the studied litter, in agreement with the view that the overall microbial community was affected. The overall laccase activity, that did not show significant seasonal changes, was, however, dependent on multiple isoenzymes differently expressed along the year.

Litter quality affected decomposition. Two phase of litter decay were generally recognized in litter of tree species: the former in the first year occurred at high rate and the latter, subsequently, at low rate. However, It was true for tree leaf litter but not for shrub litter (*Myrtus* and *Cistus)* for which the decomposition rate was higher in the second phase than the first one. The plant habitus in the stand near the sea may account of significant changes of the microclimatic conditions (temperature and moisture of the litter) under the shrubs.

The effect of nutrient on decomposition rate was in line with that reported for temperate areas even if the influence of climatic conditions makes less evident the correlation. So nitrogen content was positively related to the rate of the first phase of decomposition, while initial lignin content was not positively related to the rate of second phase.

Most of the litter degraded lignin in the early stage of decomposition and appeared negatively related the initial nitrogen content differently from what found in other areas.

The positive relation with the Mn content still evidence the role of Mn dependent enzymes (Mn peroxidases) in such a degradation, but a soil site effect was also evidenced, because only in the original native stand the litter (*P. pinea*) was not affected by its low Mn initial content.

Fungal colonizers could be divided in species of ubiquitous genera (i.e. *Cladosporium, Alternaria*) and species of typical genera (i.e. *Beltrania*) of the Mediterranean area. They were related to initial litter characteristics and its changes during decomposition, evidencing typical microbial succession. Considering their functional role pectinolytic and cellulolytic fungi occurred during the whole decomposition period, while chitinolytic ones increased after six months from litter incubation. Ligninolytic fungi, even if isolated in the first month of decomposition became abundant later when lignin decay started.

REFERENCES

Aber J.D., Melillo J.M. 1991. *Terrestrial Ecosystems.* Saunders College Publishing, Philadelphia, P.A.

Anderson J.M. 1973. The breakdown and decomposition of sweet chestnut (Castanea sativa Mill.) and beech (Fagus sylvatica L. leaf litter in two deciduous woodland soils. *Oecologia* 12: 251-274.

Andrén O., Kätterer T., Hyvönen R. 2001. Projecting soil fauna influence on long-term soil carbon balances from faunal exclusion experiments. *Applied Soil Ecology* 18: 177-186.

Archibald F., Roy B. 1992. Production of manganic chelates by laccase from the lignin-degrading fungus *Trametes* (*Coriolus*). *Applied and Environmental Microbiology* 58: 1496-1499.

Arianoutsou M. 1993. Leaf litter decomposition and nutrient release in a maquis (evergreen sclerophyllous ecosystem of North-Eastern Greece. *Pedobiol.* 37: 65-71.

Berg B. 1986. Nutrient release from litter and humus in coniferous forest soils – a mini review. *Scandinavian Journal of Forest Research* 1; 359-369.

Berg B., Berg M.P., Bottner P., Box E., Breymeyer A., Calvo De Anta R., Couteaux M., Escudero A., Gallardo A., Kratz W., Madeira M., Mälkönen E., McClaugherty C., Meentemeyer V., Muñoz F., Piussi P., Remacle J. and Virzo De Santo A. 1993. Litter mass loss rate in pine forests of Europe and Eastern United States: some relationships with climate and litter quality. *Biogeochemistry* 20, 127-159.

Berg B., Ekbohm G. 1991. Litter mass-loss rates and decomposition patterns in some needle and leaf litter types. Longterm decomposition in Scots pine forest. VII. *Canadian Journal of Botany* 69: 1449-1456.

Berg B., Ekbohm G., Johansson M.B., McClaugherty C., Rutigliano F.A., Virzo De Santo A. 1996. Some foliar litter types have a maximum limit for decomposition – A synthesis of data from forest systems. *Canad. J. Bot.* 74: 659-672.

Berg B., Johansson M.B., Meentemeyer V., Kratz W. 1998. Decomposition of tree root litter in a climatic transect of coniferous forests in Northern Europe: a synthesis. *Scand. J. For. Res.* 13: 402-412.

Berg B., Laskowski R. 1997. Changes in nutrient concentrations and nutrient release in decomposing needle litter in monocultural systems of Pinus contorta and Pinus sylvestris – a comparison and synthesis. *Scand. J. For. Res.* 12: 113-121.

Berg B., Laskowski R. 2006. Litter decomposition: a guide to carbon and nutrient turnover. Advances in Ecological Research vol. 38, *Elsevier* Ltd.

Berg B., Lundmark J.E. 1987. Decomposition of needle litter in *Pinus contorta* and *Pinus sylvestris* monocoltures-a comparison. *Scand. J. For. Res.* 2: 3-12.

Berg B., McClaugherty C., Virzo De Santo A., Johansson M.B., Ekbohm G. 1995b. Decomposition of litter and soil organic matter – Can we distinguish a mechanism for soil organic matter buildup ?. *Scand. J. For. Res.* 10: 108-119, 1995.

Berg B., Staaf H., Wessen B. 1987. Decomposition and nutrient release in needle litter from nitrogen-fertilized Scots pine (*Pinus sylvestris*) stands. *Scand. J. For. Res.* 2: 399-415.

Berg B., Virzo De Santo A., Rutigliano F.A., Fierro A., Ekbohm G. 2003. Limit values for plant litter decomposing in two contrasting soils – influence of litter elemental composition. *Acta Oecologica* 24: 295-302.

Berg, B., Calvo de Anta, R., Escudero, A., Gärdenäs, A., Johansson, M.B., Laskowski, R., Madeira, M., Mälkönen, E., McClaugherty, C., Meentmeyer, V., Virzo De Santo, A., 1995a. The chemical composition of newly shed needle litter of Scots pine and some other pine species in a climatic transect. *Long term decomposition in a Scots pine forest.* X. Can. J. Bot. 73, 1423-1435.

Berg. B. 1998. Organic matter quality and C/N ratio as controlling factors of RSOM turnover. Mitteilunggen der Deutschen Bodenkundlichen Gesellschaft, 87: 79-91.

Berg. B. 2000. Litter decomposition and organic matter turnover in northern forest soils. For. *Ecol. and Management* 133: 13-22.

Blair J.M. 1988. Nitrogen, sulphur and phosphorus dynamics in decomposing deciduous leaf litter in the Southern Appalachians. *Soil Biol. Biochem.* 20 (5): 693-701.

Carelli M., Gnocchi S., Fancelli S., Mengoni A., Paffetti D., Scotti C. and Bazzicalupo M., 2000 - Genetic diversity and dynamics of *Sinorhizobium meliloti* populations nodulating different alfaalfa cultivars in Italian soils. *Appl. Environ. Microbiol.,* 66: 4785-4789.

Chadwick D.R., Ineson P., Woods C., Piearce T.G. 1998. Decomposition of *Pinus sylvestris* litter in litter bags: influence of underlying native litter layer. *Soil Biol. Biochem.* 30(1): 47-55.

Chernin L., Chet I. 2002. Microbial enzymes in the biocontrol of plant pathogens and pests. In: *"Enzymes in the Environment. Activity, Ecology and Applications*" (Burns R.C. and Dick R.P. Eds), Marcel Dekker, Inc. pp 171-226.

Cooke R.C., Whipps J.M. 1993. *Ecophysiology of fungi.* Blackwell, Oxford, U.K.

Cortez J. 1998. Field decomposition of leaf litters: relationships between decomposition rates and soil moisture, soil temperature and earthwortm activity. *Soil Biology and Biochemistry* 30: 783-793.

Cortez J., Bouché M.B. 1998. Field decomposition of leaf litters : earthworm-microorganism interactions – the ploughing-in effect. *Soil Biology and Biochemistry* 30: 795-804.

Cortez J., Demand J.M. Bottner P., Jocteur Monrozier L. 1996. Decomposition of Mediterranean leaf litter: a microcosm experiment investigating relationship between decomposition rates and litter quality. *Soil Biology and Biochemistry* 28: 443-452.

Coûteaux M.M., Bottner P., Berg B. 1995. *Litter decomposition, climate and litter quality. Tree* 10: 63-66.

Coûteaux M.M., McTiernan K., Berg B., Szuberla D., Dardenner P. 1998. Chemical composition and carbon mineralization potential of Scots pine needles at different stages of decomposition. *Soil Biology and Biochemistry* 30: 583-595.

Criquet S., Farnet A.M., Tagger S., Le Petit J. 2000. Annual variations of phenoloxidase activities in an evergreen oak litter : influence of certain biotic and abiotic factors. *Soil Biol. and Biochem.* 32: 1505-1513.

De Marco A., Forte A., Gentile E., Virzo De Santo A. 2004. Elemental composition and litter decomposition of *Phillyrea angustifolia* L. at burned and unburned sites. Proceedings 10th MEDECOS conference, Rhodes, Greece, Arianoutsou and Papanastasis Eds., 1-10.

Di Nardo C. 2004. Attività enzimatiche e successioni microbiche durante la decomposizione della lettiera di foglie di *Quercus ilex* L. in un ecosistema mediterraneo. PhD thesis.

Di Nardo C., Cinquegrana A., Papa S., Fuggi A., Fioretto A. 2004. Laccase and peroxidase isoenzymes durino leaf litter decomposition of Quercus ilex in a Mediterranean ecosystem. *Soil Biology and Biochemistry* 36: 1539-1544.

Dilly O., Bartsch S., Rosenbrock P., Buscot F., Munch J.C. 2001. Shift in physiological capabilities of the microbiota during the decomposition of leaf litter in a black alder (Alnus glutinosa (Gaertn.) L.) forest. *Soil Biol. Biochem.* 33: 921-930.

Dilly O., Imler U. 1998. Succession in the food web during the decomposition of leaf litter in a black alder (Alnus glutinosa (Gaertn.) L.) forest. *Pedobiologia* 42: 109-123.

Domsch K.H., Gams W., Anderson T.H. 1993. *Compendium of soil fungi.* Eching. IHW-Verlag.

Escudero A., Garrido M.V., Matias M.D. 1987. Decay curves of leaf litter from evergreen and deciduous tree species. Acta Oecologica, *Oecologia Plantarum* 8(1): 81-90.

Ferran A., Vallejo V.R. 1992. Litter dynamics in post-fire successional forest of Quercus ilex. *Vegetatio* 99-100: 239-246.

Fioretto A., Di Nardo C., Papa S., Fuggi A. 2005a. Lignin and cellulose degradation and nitrogen dynamics during decomposition of three leaf litter species in a Mediterranean ecosystem. *Soil Biology and Biochemistry*, 37: 1083-1091.

Fioretto A., Musacchio A., Andolfi G., Virzo De Santo A., 1998. Decomposition dynamics of litters of various pine species in a Corsican pine forest. *Soil Biol. Biochem.,* 30: 721-727.

Fioretto A., Musacchio A., Papa S., Andolfi G., Pellegrino G., Virzo De Santo A. 2001a. Nutrient dynamics in decomposing litter of three pine species in a Corsican Mediterranean pine forest. *J. of Mediterranean Ecology* 2: 63-73.

Fioretto A., Papa S., Curcio E., Sorrentino G., Fuggi A. 2000. Enzyme dynamics on decomposing leaf litter of *Cistus incanus* and *Myrtus communis* in a Mediterranean ecosystem. *Soil Biology and Biochemistry* 32: 1847-1855.

Fioretto A., Papa S., Pellegrino A. 2005b. Effects of fire on soil respiration, ATP content and enzyme activities in Mediterranean maquis. *Applied Vegetation Science* 8: 13-20.

Fioretto A., Papa S., Pellegrino A., Fuggi A. 2007. Decomposition dynamics of *Myrtus communis* and *Quercus ilex* leaf bitter: Mass loss, microbial activity and quality change. *Appl. Soil Ecol.* 36: 32-40.

Fioretto, A., Papa, S., Sorrentino, G., Fuggi, A., 2001b. Decomposition of *Cistus incanus* leaf litter in a Mediterranean maquis ecosystem: mass loss, microbial enzyme activities and nutrient changes. *Soil Biol. Biochem.* 33: 311-321.

Frankland J.C. 1998. Fungal succession-unravelling the unpredictable. *Mycol. Res.* 102: 1-15.

Gallardo J.F., Santa Regina I., Harrison A.F., Howard D.M. 1995. Organic matter and nutrient dynamics in three ecosystems of the "Sierra de Bejar" mountains (Salamanca Province, Spain). *Acta Oecol.* 16(4): 447-459.

Gelsomino A., Keijzer-Wolters A.C., Cacco G. and van Elsas J.D., 1999 - Assessment of bacterial community structure in soil by polymerase chain reaction and denaturing gradient gel electrophoresis. *J. Microbiol. Methods*, 38: 1-15.

Harmon M.E., Franklin J.K., Swanson F.J., Sollins P., Gregory S.V., Lattin J.D., Anderson N.H., Cline S.P. Aumen N.G., Sedell J.R., Lienkamper G.W., Cromack K., Cummins K.W. 1987. Ecology of coarse woody debris in temperate ecosystems. *Advances in Ecological Research* 15, 133-302.

Herlitzius H. 1983. Biological decomposition efficiency in different woodland soils. *Oecologia* 57: 78-97.

Hermann W., McGill A.W.B., Dormaar J.F. 1977. Effect of initial chemical composition on decomposition of roots of three grass species. *Can. J. Bot.* 55: 1632-1640.

Joshi S.R., Sharma G.D., Mishra R.R. 1993. Microbial enzyme activities related to litter decomposition near a highway in a sub-tropical forest of North East India. *Soil Biol. Biochem.* 25: 1763-1770.

Keyser P., Kirk T.K., Zeikus I.G. 1978. Ligninolytic enzymes of Phanerochaete chrysosporium: synthesized in the absence of lignin in response to nitrogen starvation. *J. Bacteriol.* 135: 790-797.

Kjøller A.H., Struwe S. 1982. Microfungi in ecosystems: fungal occurrence and activity in litter and soil. *Oikos,* 39: 389-422.

Kjøller A.H., Struwe S. 1987. Functional group of microfungi on decomposing ash litter. *Pedobiologia*, 30: 151-159.

Kjøller A.H., Struwe S. 1990. Decomposition of beech litter. A comparison of fungi isolated on nutrient rich and nutrient poor media. *Transactions of the Mycological Society of Japan*, 31:5-16.

Kjøller A.H., Struwe S. 2002. Fungal communities, succession, enzymes and decomposition. In: Enzymes in the Environmen: Activity, *Ecology and Applications* (Burns R.C. and Dick R.P. Eds), Marcel Dekker, Inc, New York, Basel: 266-284.

Kouki J., Hokkanen T. 1992. Long-term needle litterfall of Scots pine *Pinus sylvestris* stands: relation to temperature factors. *Oecologia* 89: 176-181.

Kurz C., Coûteaux M.M., Thiéry J.M. 2000. Residence time and decomposition rate of Pinus pinaster needles in a forest floor from direct field measurements under a Mediterranean climate. *Soil Biol. Biochem.* 32: 1197-1206.

Laskowski R., Niklińska, Maryański M. 1995. *The dynamics of chemical elements in forest litter. Ecology* 76: 1393-1406.

Linkins A.E., Sinsabaugh R.L. McClaugherty C.A., Melillo J.M. 1990. Cellulase activity on decomposing leaf litter in microcosms. *Plant and Soil* 123: 15-25.

Marschner P., Yang C.H., Lieberei R. and Crowley D.E., 2001 - Soil and plant specific effects on bacterial community composition in the rhizosphere. *Soil Biol. Biochem.*, 33: 1437-1445.

Marsden W.L., Gray P.P. 1986. Enzymatic hydrolisis of cellulose in lignocellulosic materials. CRC

McClaugherty C. and Berg B. 1987. Cellulose, lignin and nitrogen concentrations as rate regulating factors in late stages of forest litter decomposition. *Pedobiologia* 30, 101-112.

McTiernan, K.B., Coûteaux, M.M, Berg, B., Berg, M.P., Calvo de Anta, R., Gallardo, A., Kratz, W., Pissi, P., Remacle, J., Virzo De Santo A., 2003. Changes in chemical composition of *Pinus sylvestris* needle litter during decomposition along a European coniferous forest climatic transect. *Soil Biology and Biochemistry* 35, 801-812.

Meentemeyer V., Berg B. 1986. Regional variation in mass-loss of *P. sylvestris* needle litter in Swedish pine forests as influenced by climate and litter quality. *Scand. J. For. Res.* 1: 167-180.

Meentemeyer V., Box E.O., Thompson R. 1982. World patterns and amounts of terrestrial plant litter production. *BioScience 32*, 125-128.

Melillo J.M., Aber J.D., Linkins A.E., Ricca A., Fry B., Nadelhoffer K.J. 1989. Carbon and nitrogen dynamics along the decay continuum: plant litter to soil organic matter. *Plant Soil* 115: 180-198.

Merbach W., Mirus E., Knof G., Remus R., Ruppel S., Russow R., Gransee A. and Schulze J., 1999 - Release of carbon and nitrogen compounds by plant roots and their possible ecological importance. *J. Plant Nutrit. Soil Sci.*, 162: 373-383.

Miethling R., Wieland G., Backhaus H. and Tebbe C.C., 2000 - Variation of microbial rizhosphere communities in response to crop species, soil origin, and inoculation with *Sinorhizobium meliloti* L. 33. *Microbiol. Ecol.*, 40: 43-56.

Minderman G. 1968. Attrition, decomposition and accumulation of organic matter in forests. *Journal Ecology* 56:355-362.

Moore T.R., Trofymow J.A., Taylor B., Prescott C., Camiré C., Duschene L., Fyles J., Kozak L., Kranabetter M., Morrison I., Siltanen M., Smith S., Titus B., Visser S., Wein R., Zoltai S. 1999. Litter decomposition rates in canadian forests. *Global change biology* 5: 75-82.

Mulas B., Pasqualetti M., Rambelli A. 1995. Analysis of the litter microfungal communities in a Mediterranean maquis. Rendiconti di Scienze Fisiche e Naturali dell'Accademia dei Lincei, serie IX, vol. VI, fasc. 1: 65-86.

Nömmik H., Vahtras K. 1982. Retention and fixation of ammonium and ammonia in soils. In: Nitrogen in Agricultural Soils. Agronomy Monographs, no 22 (Stevenson F.J. Ed) pp. 123-171. *Agron. Soc. Am.* Madison, WI.

Odum E. 1983. *Basic Ecology.* CBS College Publishing, New York.

Olson J. 1963. Energy storage and the balance of producers and decomposers in ecological systems. *Ecology* 44: 322-331.

Osono T., Takeda H. 2001. Organic chemical and nutrient dynamics in decomposing beech leaf litter in relation to fungal ingrowth and succession during 3-years decomposition processes in a cool temperate deciduous forest in Japan. *Ecological Research,* 16: 649-670.

Papa S. 2000. Attività enzimatiche della microflora in relazione alla decomposizione della lettiera di foglie in area mediterranea. PhD Thesis

Papa S., Curcio E., Lombardi A., D'Oriano P., Fioretto A. 2002. Soil microbial activity in three evergreen oak (Quercus ilex) woods in a Mediterranean area. In: *Developments in Soil Science,* Violante A., Huang P.M., Bollag J.M., Gianfreda L. (Eds), *Elsevier Science* B.V., vol. 28B pp. 229-237.

Papa S., Pellegrino A., Ferrigno A. Fioretto A. 2007. Microbial activity of soil with different plant cover in Mediterranean area. Studi Trentini di Scienze Naturali, *Acta Biologica* 83 (in press).

Pasqualetti M., Mulas B., Zucconi L., Rambelli A., 1999. Succession of microfungal communities on Myrtus communis leaf litter in a Sardinian Mediterranean maquis ecosystem. *Mycological Researches,* 103: 724-728.

Pasqualetti M., Tempesta S., Rambelli A., Mulas B., 2006. Succession of microfungi during *Pistacia lentiscus* decomposition in a Sardinian Mediterranean maquis. *Plant Biosystems,* 140: 56-64.

Pausas J.G. 1997. Litter fall and litter decomposition in *Pinus sylvestris* forests of the eastern Pyrenees. *J. Veg. Sci.* 8, 643-650.

Perez J., Jeffrey T.W. 1992. Roles of manganese and organic acid chelators in regulating lignin degradation of biosynthesis of peroxidases by *Phanerochaete chrysosporium. Applied and Environmental Microbiology* 58: 2402-2409.

Pinzari F., Trinchera A., Benedetti A. and Sequi P., 1999 - Use of biochemical indices in the mediterranean environment: comparison among soils under different forest vegetation. *J. of Microbiol. Methods*, 36: 21-28.

Reid I.D. 1995. Biodegradation of lignin. *Canadian Journal of Botany* 73: 1011-1018.

Rosembrock P., Buscot F., Munch J.C. 1995. Fungal succession and changes in the fungal degradation potential during the initial stage of litter decomposition in a black alder (*Alnus glutinosa* (Gaertn.) L.). *European Journal of Soil Biology* 31: 1-11.

Rovira P., Vallejo V.R. 2002. Labile and recalcitrant pools of carbon and nitrogen in organic matter decomposing at different depths in soil: an acid hydrolysis approach. *Geoderma* 107: 109-141.

Rutigliano F.A., D'Ascoli R. and Virzo De Santo A., 2004 - Soil microbial metabolism and nutrient status in a Mediterranean area as affected by plant cover. *Soil Biol. Biochem.*, 36: 1719-1729.

Rutigliano, F.A., Virzo De Santo, A., Berg, B., Alfani, A., Fioretto, A., 1996. Lignin decomposition in decaying leaves of *Fagus sylvatica* L. and needles of *Abies alba* Mill. *Soil Biol. and Biochem.* 28, 101-106.

Sadaka N., Ponge J.F. 2003. Fungal colonization of phyllospere and litter of Quercus rotundifolia Lam. in a holm oak forest (High Atlas, Marocco*). Biology and Fertility of Soils*, 39: 30-36.

Saiya Cork, K.R., Sinsabaugh, R.L., Zak, D.R., 2002. The effects of long-term nitrogen deposition on extracellular enzyme activity in an *Acer saccharum* forest soil. *Soil Biology and Biochemistry* 34, 1309-1315.

Santa Regina I., Tarazona T., Calvo R. 1997. Aboveground biomass in a beech forest and a Scots pine plantation in the Sierra de la Demanda area of northen Spain, *Ann. Sci. For.*, 54: 261-269.

Sinsabaugh R.L. Antibus R.K. Linkins A.E. 1991. An enzymic approach to the analysis of microbial activity during plant litter decomposition. *Agriculture, Ecosystems and Environment* 34: 43-54.

Sinsabaugh R.L. Antibus R.K. Linkins A.E., McClaugherty C.A., Rayburn L. 1992. Wood decomposition over a first-order watershed: mass loss as a function of lignocellulase activity. *Soil Biol. Biochem.* 24: 743-749.

Sinsabaugh R.L., Carreiro M.M., Alvarez S. 2002. Enzyme and microbial dynamics of litter decomposition. In: Enzymes in the Environment: Activity, Ecology and Applications (Burns R.C. and Dick R.P. Eds), Marcel Dekker, Inc, New York, Basel: 249-266.

Sinsabaugh R.L., Moormead D.L. Linkins A.E. 1994. The enzymic basis of plant litter decomposition: emergence of an ecological process. *Applied Soil Ecology* 1: 97-111.

Stamou G.P., Pantis J.D., Sgardelis S.P. 1994. Comparative study of litter decomposition in two Greek ecosystems: a temperate forest and an asphodel semi-desert. *Eur. J. Soil Biol.* 30: 43-48.

Stevenson F.J. 1982. Humus chemistry: genesis, compositions, reactions. John Wiley and Sons, New York, USA.

Swift M.J., Heal O.W. and Anderson J.M. 1979. *Decomposition in terrestrial ecosystems.* Blakwell Scientific Publications. Oxford, London, Edinburg, Melbourne.

Taylor B.R., Parkinson D., Parsons W.J.F. 1989. Nitrogen and lignin content as predictors of litter decay rates: a microcosm test. *Ecology* 70: 97-104.

Taylor B.R., Prescott C.E., Parsons W.J.F., Parkinson D. 1991. Substrate control of litter decomposition in 4 Rocky-mountains coniferous forest. *Can. J. Bot.* 69: 2242-2250.

Tempesta S., Mulas B., Rambelli A., Pasqualetti M. 2005. Evolution of microfungal community on Chamaerops humilis leaf litter in a Sardinian Mediterranean maquis. Cryptogamie, *Mycologie,* 26: 15-26.

Tempesta S., Pasqualetti M., Fonck M., Mulas B., 2003. Succession of microfungi in Phyllirea angustifolia bitter in a Mediterranean maquis in Sardinia. *Plant Biosystems,* 137: 149-154.

Tian X., Takeda H., Azuma J. 2000. Dynamics of organic-chemical components in leaf litters during a 3.5-year decomposition. *European Journal of Soil Biology*, 36: 81-89.

Tuor U., Winterhalter K., Fiechter A. 1995. Enzymes of white-rot fungi involved in lignin degradation and ecological determinants for wood decay. *Journal of Biotechnology* 41: 1-17.

Vallejo V.R. 1993. Evaluation of C:N ratio as parameter of N mineralization. *Mitt. Oesterr. Bodenkd. Ges.* 47: 71-78.

Virzo De Santo A., Berg B., Rutigliano F.A., Alfani A., Fioretto A. 1993. Factors regulating early-stage decomposition of needle litters in five different coniferous forests. *Soil Biology and Biochemistry* 25: 1423-1433.

Virzo De Santo A., Fioretto A., Alfani A. 1985. Decomposizione della lettiera nelle faggete del Monte Taburno. *Società Italiana di Ecologia Atti* 5: 883-886.

Virzo De Santo A., Rutigliano F.A., Berg B., Fioretto A. Fierro A. 1998. Nitrogen dynamics of decomposing needle litters in three coniferous forests of the Mediterranean area. *Fresenius Envir. Bull.* 7: 510-517.

Virzo De Santo A., Rutigliano F.A., Berg B., Fioretto A., Puppi G., Alfani A., 2002. Fungal mycelium and decomposition of needle litter in three contrasting coniferous forests. *Acta Oecologica*, 23: 247-259.

Vogt K.A., Grier C.C., Vogt D.J. 1986. Production, Turnover and nutrient dynamics of above- and belowground detritus of world forests. *Advances in Ecological Research,* 15: 303-370.

Wise D.H., Schaefer M. 1994. Decomposition of leaf litter in a mull beech forest: comparison between canopy and herbaceous species. *Pedobiologia* 38: 269-288.

Witkamp M. 1966. Decomposition of leaf litter in regulation to environment, microflora and microbial respiration. *Ecology* 47: 194-201.

Witkamp M., Olson J.S. 1963. Breackdown of confined and non confined oak litter. *Oikos* 14: 138-147.

In: Soil Ecology Research Developments
Editors: Tian-Xiao Liu, pp. 63-87
ISBN: 978-1-60021-971-9

Chapter 2

RESPONSES OF SOIL MICROBIAL COMMUNITIES TO CLIMATIC AND HUMAN IMPACTS IN MEDITERRANEAN REGIONS

Efimia M. Papatheodorou[1]
Department of Ecology, School of Biology, Faculty of Sciences, Aristotle University,U.P.Box 119, 54 124 Thessaloniki, Greece

ABSTRACT

In Mediterranean regions climatic seasonality and exerted for millennia human impacts such as grazing and deliberate fire drive the structure and function of ecosystems. Recently, intensive agriculture, invasion of exotic species and atmospheric CO_2 enrichment are also involved. The function of soil subsystem is important for ecosystem maintenance providing nutrients released through microbially mediated decomposition. Specifically in shallow and nutrient-limited Mediterranean soils, the contribution of soil microbial communities to decomposition is even more important.

The aim of this paper is to examine the effect of climatic and human impacts on characteristics of the soil microbial communities; biomass, composition and activity. Estimates of biomass based on immobilized C and N, total PLFA or counts of bacteria and fungi. For describing structural diversity PLFA and FAME results are used, while for functional diversity BIOLOG data are considered. Results provided by molecular techniques support our discussion of genetic diversity. Finally, data on respiration and enzyme activities are used to assess microbial activity.

The survey shows that vegetation type exerts an influence on soil microbial features. However, in comparison with seasonality the effects of plant species are of lower importance. Grazing, by affecting vegetation patterning and soil compactness, is expected to exert a great influence on soil microbial characteristics. Nevertheless, when seasonality and grazing are compared variations due to season are greater than those induced by grazing. Similar results emerge when seasonality is compared with agricultural practices. The available information concerning the effect of fire and species' invasion on microbial communities is really sparse. Burning reduces microbial diversity and most enzyme

[1] e-mail: papatheo@bio.auth.gr.

activities while invasion by exotic species alters the rhizosphere microbial communities with the magnitude of influence to be increased with time after invasion. CO_2 enrichment, through its effect on plant productivity, affects significantly soil structure. However, no effects on microbial community structure and biomass are detected. Treatment with herbicides, compost or biosolids changes the activity, size and composition of microbes due to physicochemical characteristics of the amendments. Finally, transition from conventional to organic farming does not results in gradual changes in the characteristics of the microbial community.

INTRODUCTION

Apart from regions surrounding Mediterranean sea, areas characterized as Mediterranean or with Mediterranean type climate are those occupying mainly the western coasts of the five continents such as parts in south- western South Africa or in south-western Australia. These areas have a distinctive character that arises from both the physiographic conditions and the history of local societies' development (Brandt and Thornes 1996). The question is "which are the special features of these regions in terms of climate and in terms of their biocommunities' organization in time and space?"

Mediterranean climate is characterized by a hot summer with a high rate of evapotranspiration and a mild winter with rather high precipitation (Aschman 1973). The rainfall varies from about 1000 mm in the more northernly areas and areas above 800 m, to 250 mm in southern drylands. Mediterranean drylands, those with preciptation less than 500 mm per year, exhibit strong inter –annual variability (from 150 to 600 mm; Brandt and Thornes 1996). The principal limited factor for life in these areas is summer dryness.

Mediterranean-type ecosystems appear to be highly heterogeneous in time and space. Variations of temperature and humidity conditions in time exhibit interannual, seasonal and diurnal patterns (Stamou 1998). These variations are the more or less predictable components of the mediterranean climate while sudden catastrophic events such as heavy rainfalls (100 mm per hour or even more) could be of crucial importance (Lammott and Bladin 1989). Concerning space, a large-scale spatial heterogeneity is due to the inherent characteristics of the landscape enabling the coexistence of different biocommunities nearby each other, while a small-scale fragmentation of the biocommunity derived from human impacts exerted for millennia. Grazing, burning and cutting resulting to spatially fragmented habitats, are the main anthropogenic pressures exerted for at least the last 10000 yrs while intensive agriculture, invasion of exotic species and atmospheric CO_2 enrichment are factors recently involved. Although the influence of grazing and fire on the structure and function of Mediterranean regions has been extensively exploited (Perevolotsky and Seligman 1998, Henkin et al. 2006), the influence of the latter factors is still under investigation. Traditionally, the vast majority of studies conducted in Mediterranean areas focus on the structure and function of vegetation (Stamou and Pantis 1995, Kazanis and Arianoutsou 1996, Arianoutsou and Thanos, 1996, Puerto and Rico 1997, Kazanis and Arianoutsou 2004, Tzialla et al. 2006, Ovalle et al. 2006, Peco et al. 2006) while those related to soil sub-system are rather limited.

Likewise vegetation, the soils in Mediterranean regions are the product of interactions between the natural processes of pedogenesis and the activities of human societies. Brown soils (cambisols) are the dominant group in the Mediterranean basin. In general, they are fragile, shallow and with a distorted soil profile. In the north where climate is more humid, soils are generally richer in organic matter with relatively higher humidity, whereas in south there is an accelerated mineralization of soils. Subject to erosion soils are often "rejuvenated" and it results in soils with poor organic matter and low water-holding capacity (de Franchis and Ibanez 2003). The depth of the organic layers is usually thin (0-5 cm) while that of the mineral rocky layers ranges from 0.5 -1m. The humus layer is thin as well. The rocky part of the soil profiles exceeds 60%. Carbonate and iron oxides are abundant throughout or in parts of the soil profile (Papatheodorou and Stamou 2004). Mediterranean soils are rich in nitrogen and deprived in extractable phosphorus (Diamantopoulos et al. 1994, Papatheodorou 1996).

Studies referring to Mediterranean soils are dealing either with structure including soil physical characteristics (aggregation, infiltration, conductivity) which determine the resistance of soil to erosion (Lavee et al. 1998) or with function. Function was discussed in terms of rates of decomposition processes (Argyropoulou et al. 1993, Stamou et al. 1994), of the amounts of available nutrients supplied to plants (nitrogen, phosphorus, potassium etc) (Henkin et al. 1994, Willott 2000, Fillery 2001, Kabengi et al. 2003, Peco et al. 2006) or in terms of the community structure of soil biota with emphasis on arthropods (Sgardelis et al. 1981, Stamou and Asikidis 1992, Sgardelis and Margaris 1993, Stamou et al. 1993, Sgardelis et al. 1995, Stamou 1995, Stamou 1998). A highly important component of soil functioning that recently attracts substantial interest is the microbial community of soil. It relates to the development of appropriate techniques that offers the opportunity for greater insights into microbial community structure.

Microbes belonging to the lower trophic level are responsible for organic matter transformations. They are known as the early indicators of soil changes induced by modifications in climatic variables (Insam 1990) or in agricultural practices (Bending et al. 2004). Depending on the involved functions, microbes can be categorized into different functional groups such as ammonia oxidizers, denitrifiers, heterotrophs (Wertz et al. 2006) while depending on their relation to other organisms they are characterized parasites, pathogens and mutualistic symbionts (van der Putten 2007). Since in a bacterial world the species concept is obscure, the functional diversity is adopted as a way to overcome this difficulty. Moreover, crucial for the maintenance of soil system is not the species composition but the functional aspect of the microbial community (Øvreas 2000). Functional diversity can be defined as the numbers, types and rates at which an array of substrates is utilized by microbes (Zak et al. 1994). The substrates are provided in Biolog plates specified for bacteria or fungi respectively. This method is widely used since 1995 and although it is considerable criticized (Preston-Mafham 2002), it offers valuable information for the comparison of functional diversity from different sites provided that the same experimental protocol is used.

Apart from the functional diversity, structural diversity of microbial communities can be exploited by analysing the patterns of phospholipid fatty acids (PLFA; Tunlid and White 1992). Certain signature fatty acids in the overall PLFA profiles are specific for bacteria, fungi and actinomycetes respectively and can be used as indicators of stressful impacts on microbes (Guckert et al. 1986). There are standard methods by which phospholipids are converted to ester-bonded fatty acids (FAMEs). The advantage of this method is that it specifically describes the community of the living microbiota (Fernandes 2006). The total concentration of PLFA is also a measure of viable microbial biomass (Zelles et al. 1995). Moreover, by adding radioactive substrates, a measure of the active community could be obtained based on the organisms that are actively metabolizing the substrates (Roslev et al. 1998). Significant information about soil microbial community is also revealed by the consideration of genetic diversity. As significant components of living cells, DNA and RNA, are excellent signature molecules (Paul and Clark 1996). The DNA from environmental samples is obtained either by direct lysis of cells within the soil or by isolation of bacterial cells and then lysis of the bacterial fraction (Orveas 2000). From the different fractions of RNA, the most interest in microbial ecology is rRNA, and especially 5S and 16S rRNA. There is an array of methods used to analyse the microbial genetic diversity in environmental samples but I don't refer to it, since it is beyond the scope of this study.

For biomass assessment, the methods that are widely used are the fumigation-incubation method of Jenkinson and Powlson (1976) and the fumigation-extraction method (Vance et al. 1987). The exposure of soil organisms to chloroform vapor destroys their cells and allows cytoplasmic C, N, P, and S to leak into the soil. By the use of appropriate conversion factors these substances are converted to microbial biomass-C, N, P and S respectively. Alternative methods for biomass estimations are that of microscopy based on direct counts as well as that of substrate-induced respiration (SIR; Anderson and Domsch 1978). This latter can be used for assessing microbial activity called "basal respiration" as well. Direct counts provide limited information since only the culturable part of microbes is considered while the culturing environment differs from the soil environment. Soil microbial activity is estimated by the CO_2-release, when soil is incubated under aerobic conditions for a specific time period, or through the assessment of enzyme activities. In the case of enzymes, the activity is specific and refers to the degradation of known substrates.

It has been established that soil biota in Mediterranean regions are well adapted to face the adversity of environment induced by the oscillations of abiotic variables and the impacts of human interventions (Stamou 1998). In this context, the aim of this review is to explore the responses of soil microbial communities to the above mentioned influences. Therefore, the review is organized in a way that each section discusses the responses of microbial communities in terms of structure and function to grazing, fire, agriculture, invasion of exotic species, change of climatic variables. If relevant information is available, the responses of microbes to these interventions will be compared to those induced by the predicted oscillations of abiotic variables that constitute an inherent characteristic of mediterranean climate.

Microbial Communities in Agricultural Soils

The Effect of Management Practices

In studies focusing on the soil quality in agricultural systems, the community of soil microorganisms is examined in order to detect its ability to reflect changes induced by different management practices or by different cropping systems. Nowadays a substantial interest is devoted to the way organic and conventional cultivation affects soil properties. For most of these properties (e.g. soil organic matter) significant changes occur only after decades (Johnston 1986), while for microbiological properties few years are enough (Drinkwater et al. 1995).

In general, cultivations are verified as organic after 3 years of organic management. But what happens to the soil microbial community during this transition period? To what extent the characteristics of soil microbial communities in the newest organic cultivations are intermediate between those of conventional and the oldest organic ones? Can we talk for a gradual improvement of soil quality with the duration of organic cultivation? Although several studies examined differences in soil properties between transitional and established organic systems (Scow et al. 1994, Drinkwater et al. 1995), the existence of a number of confounding factors makes interpretations difficult. Some answers to these questions are given by the studies of Martini et al. (2004) and Monokrousos et al. (2006). Martini et al. (2004) examined the microbial community in established organic plots (> 5 yrs) and in transitional ones (< 1 yr). Conventional plots were also included. Based on FAME analysis, the authors showed differences between microbial communities from conventional and organic systems reflecting differences in communities composition or in abundances. Distinction was also revealed when the functional diversity was taken into account. The hypothesis that the microbial communities from transitional plots exhibit intermediate characteristics between conventional and organics systems was not supported by the data of Martini et al. (2004). In the study of Monokrousos et al. (2006) the duration of organic cultivation varied from 2 to 6 yrs (2, 3, 5 and 6 years). The composition of microbial community, in terms of bacterial and fungal contribution, was unaffected by the age of organic management but was differentiated between organic and conventional plots. The activities of almost all enzymes related to N and P-cycle were higher in plots with intermediate duration of organic cultivation and lowest in the conventional one. The data regarding microbial structure and activity supports the suggestion of Martini et al. (2004). Changes in microbial communities are not gradual from conventional to transitional and then to organic systems. This gradual response characterized only the communities of microbivores nematodes (bacterial and hyphal-feeding nematodes; Tsiafouli et al. in press). On the contrary, when the functional diversity was examined in phaseolus organic cultivations of different age (2, 5, 8, 10 and 14 yrs), the analysis revealed a gradual evolution from young organic plots with lower diversity and abundance to more diverse and with higher abundances older plots (unpublished data). In this study, the bacterial community in conventional plots seemed to be less functionally diversified and more stable over time. Evidently, in order to detect differences in soil microbial community between newest and older organic areas more than 6 yrs of cultivation are necessary.

Apart from organic farming, the agricultural soils of Mediterranean areas are under the influence of salinization. Soil salinization arises from the improper use of salt-rich waters in agriculture (Szabolcs 1998). The main influence of saline water in soil is the dispersion of clay particles with consequent changes in soil physicochemical properties. The modification of soil properties is expected to induce changes in bacterial communities. To test this hypothesis all aspects of bacterial diversity were measured and compared to those in an unaffected agriculture system; an organic one (Grecchio et al. 2004). Depression of microbial size and activity and modification of phenotypic and genetic diversity were recorded in salt-irrigated soils. It was concluded that different long-term irrigation systems exert a strong influence over the development of distinct microbial communities similar to the influence exerted by the plant type or the rhizosphere's products.

The effects of winter-cover cropping vs winter fallow practices were examined by Schutter et al. (2001). Relevant effects were compared to changes induced by season and soil type. For structure assessment microscopy and FAMEs analysis were employed. The results obtained by these two methods were not in accordance. The more sensitive variable to changes was the abundances of different FAMEs. When influences were ranked according to the magnitude of their effects on function and structure, soil type had the greater impact, while for the same soil type season became more important. A more general study referring to different types of land uses is that of Steenwerth et al. (2002) in California. Different grasssland and cultivated ecosystems provide a gradient of increasing intensity of soil disturbance. How the structure of microbial community changes in relation to the history of land use and the associated management inputs and practices? Do the communities clearly discriminate between the different land uses? In order to avoid bias introduced by differences in soil texture, in this study sites of similar soil texture were chosen. A sharp distinction was recorded between restored communities with longed lived buchgrasses, communities with formely–cultivated annual grasslands and cultivated ones. Remarkable was the finding that the structure of microbial communities of all annual grasslands was similar although the time since the last tillage event varied from 8 to 50 yrs. The explanation relies on the plant community composition which remains consistently similar in the same successional stage. On the contrary, the microbial communities of perennial grasslands varied widely. Concerning the extent of similarities among microbial communities, Buckley and Schmidt (2001) concluded that 7 yrs of abandonment after cultivation results in a microbial community more similar to that of cultivated areas than to never cultivated ones.

The Effects of Biosolids, Compost and Pesticide Application

Since biosolid wastes such as composts, sewage sludges or municipal solid wastes are recently produced in large amounts, there is an increasing concern about their recycling. Apart from recycling, the incorporation of these amendments to soil has a significant impact on soil quality. Biosolids' application could be used as a tool for restoring or improving the soil fertility in severely degraded areas. Since microbial community has a significant contribution to soil fertility, in order this latter to change, modifications in composition and function of soil microbial communities are expected. Most studies focusing on biosolids examined their effects on arid or semi-arid rangelands (Barbarick et al. 2004, Garcia-Gill et al. 2004). These ecosystems are considered vulnerable to degradation due to limited availability of water and the influence of grazing. While biosolids could affect plant productivity (improving the forage quality for livestock) and soil microbial communities, their effects depend on the amount, the chemical content and the frequency of application. The diversity of microbial community increased linearly with increases in the rate of application (Dennis and Fresquez 1989) while their influence on soil microbes continued to persist for years after application (Pascual et al. 1999, Barbarick et al. 2004, Sullivan et al. 2006). The rate of colour development in Biolog wells, reflecting microbial activity, was faster in biosolids-treated than in unamended soils and the rate of development increased as the dose of application increased (Sullivan et al. 2006). The effects on microbial biomass can be variable: decrease due to the biosolids toxic chemical concentrations, increase as a response to increasing plant productivity and to the added amounts of C, N and other nutrients or no effect. In the study of Sullivan et al. (2006) the lack of effect of biosolids on total microbial biomass was followed by a shift in the bacterial/fungal ratios in favor of bacteria.

Moreover, as biosolids application increased, the C/N ratio of plant material was reduced resulting in greater N mineralization activity.

Compost amendment is another way to improve soil conditions. It can stabilize soil structure (Barzegar et al. 2002), increase the amount of organic material inserted to soil (Filcheva and Tsadilas 2002), stimulate soil microflora (Ros et al. 2003) or induce antagonistic effects to native microflora. The impact of compost on microbial communities is ought to organic and inorganic compounds that they contain and to the interactions that compost endogenous microflora induced since it is a microbiologically active product (Schloss et al. 2003). Data concerning the influence of compost as microbial inoculum was presented in the study of Saison et al. (2006). The hypothesis that compost-borne pathogens, through their interactions with other soil microorganisms were responsible for the modifications of the native soil microflora community, was not supported by this study. They concluded that the changes in microbial characteristics were essentially due to the input of a compost matrix rich in organic matter. But which were the microbial parameters that were affected by compost amendment? Size, activity and structure. In all cases, the influences were dose-dependent. Changes in genetic diversity and activity were detected only when high amount of compost was added. Under these conditions, differences in genetic diversity were recorded only 4 days after the application, while the influences on microbial characteristics persisted for much longer period in comparison with effects induced by low doses.

Apart from materials incorporated into soils, pesticides used in liquid phase to control vegetation have a considerable influence on microbial communities. Pesticides are distinguished in herbicides, fungicides and insecticides. For the use of herbicides a lot of bare soil is needed, while for the others lower exposure of soil is enough. The disadvantages of pesticides' use, which increased considerably over the last 50 years, are related to the non-specific toxic influence on microorganisms. In some cases, the toxic effect was not due to the main constituent but to its secondary metabolites produced during its degradation (Bjørnlund 2000). The bioavailability of each pesticide is of crucial importance and is strongly related to soil type (Busse et al. 2001).

The effects of pesticides on microbes were examined in terms of overall microbial activity (Greaves 1982), but this cannot give any insight on what happens in the microbial communities. Some of the pesticide constituents could suppress the growth of specific microbes or enhance the proliferation of others, affecting the successional patterns in the microbial community (Johnsen et al. 2001). The effects of a herbicide (glyphosate) on soil microbial communities, was studied by Busse et al. (2001). Glyphosate is often the preferred herbicide in intensive forestry and is used mainly in California forests. There are controversial results concerning its influence on communities. In artificial media the effects on microbial growth were negative (Dick and Quinn 1995), while in field studies the effects were either absent or stimulating (Haney et al. 2000, Gianfreda et al. 1995), a fact that was associated with the soil characteristics. To test this hypothesis, glyphosate effects on biomass, activity, community size and metabolic diversity were examined in an array of soils differing in the amounts of clay, oxide and organic matter (Busse et al.2001). Soil respiration in glyphosate-treated soils was much higher than in the control implying that microorganisms can use glyphosate as energy source or for acquisition of nutrients. The variation in respiration between soils followed the soils differences in absorption potential which is related to their oxide and clay content. The other aspects of microbial community did not respond to glyphosate application. It is interesting to notice that the impact of time and site proved more significant than that of glyphosate. Very often the effects of pesticides are masked by variation in temperature, humidity, substrate availability and disturbance (Domsch et al. 1983). This becomes even more pronounced in mediterranean regions because the heterogeneity in time and space belongs to their inherent characteristics.

THE EFFECTS OF GRAZING

Millenia of human impacts have shaped the highly heterogenous "Mediterraenan mosaic landscape" (Naveh 1998). The major factor in the shaping of this landscape is the heterogeneity in grazing pressure by different species of domestic animals (Perevolotsky and Seligman 1998). For instance, in Israel cattle graze within fenced rangelands at a more or less constant stocking rate, while goats usually graze at unfenced areas during specific time periods (Henkin et al. 2006). At a plant community level, grazing could affect soil processes by altering plant species diversity and the plant dominance patterns (Ovalle et al. 2006, Peco et al. 2006). Positive relations between plant diversity and culturable soil bacterial activity and diversity were recorded by Stephan et al. (2000).

In addition, by grazing, a mosaic of microsites from sheltered areas under shrubs to exposed locations between shrubs is created. In each microsite different plant species can occur, producing litter of variable quantity and quality. Moreover, due to differences in cover because of the grazing management, soil micro-abiotic conditions vary from one site to another (Papatheodorou et al. 1998). All these influences are expected to affect significantly the soil microbial communities. Monokrousos et al. (2004) examined the extent to which soil chemical microenvironment - aspects of microbial community included - differed beneath various plant species. This study was conducted in a Greek grassland where grazing resulted on the coexistence of five shrub species interspaced by bare soil. Samples were taken from areas underneath each species in periods coinciding with seasonal changes in climatic variables. Variables describing the size and activity of microbial community (microbial biomass C and N, ergosterol as an index of fungal biomass and soil respiration) were affected significantly by sampling period and plant species. However, the variability in data induced by temporal variation was greater than that induced by spatial heterogeneity. Due to differences in microbial biomass, soils under evergreens were discriminated from soils under seasonal dimorphic species, indicating the importance of plant growth form in shaping microbial communities. Under the same experimental scheme but only for one sampling period the functional diversity of microbes was surveyed by Biolog GN plates. The produced pattern was different than the previously mentioned one. The microbial diversity in each microsite was completely distinctive from the others while the plant life form didn't seem to have any significant effect (unpublished data). The conclusion is that in this case the response of microbial functional diversity was indiosyncratic.

Apart from Monokrousos et al. (2004), the pronounced effect of season on microbial activity and diversity, compared to that of grazing, was also recorded in soil plots from Chihuahuan desert (Liu et al. 2000). Grazing exhibited insignificant independent effect but when it was combined with drought treatments, the effects on diversity became significant in specific seasons. According to the authors, the key factor for soil microorganisms in this ecosystem was the availability of litter substrate.

Grazing not only increases spatial heterogeneity but it could also affect soil processes in variable ways: by changing the amount and/or the chemical nature of produced litter, by creating "hotspots" of microbial activity due to animal dugs resulting in redistribution of organic material and nutrients (Haynes and Williams 1999) or finally by changing the soil physicochemical properties due to trampling. However, the available evidence in regard to this latter showed that although the soil physical properties changed significantly, the microbial communities (size, structure and activity) exhibited broad tolerance to compaction indicating poor link between physical and biological indices of soil health (Shestak and Busse 2005, Busse et al. 2006). When soil microbial biomass C and N were recorded underneath ungrazed, moderately grazed and overgrazed *Quercus coccifera* (L.) stands as well as in open areas between shrubs, similar values of microbial C were recorded all over the experimental sites (unpublished data). The lack of changes in the size of microbial biomass due to grazing, is recorded in the literature quite often (Li et al. 2005, Jackson et al. 2006). Non-significant effects of grazing on the structure of soil fungal communities were reported for grasslands soils in Spain (Maggi et al. 2005). On the contrary, microbial biomass N in *Q. coccifera* stands was significantly lower in soil under overgrazed shrubs, supporting the hypothesis that grazing increases mineralization of N instead of N immobilization (Holland and Detling 1990).

THE EFFECT OF FIRE

Fire occurred in the Mediterranean region for millennia and much of the natural vegetation is adapted either to irregular wildfires or to prescribed fires of low intensity (Fioretto et al. 2005). Alike the intensity of grazing fire frequency affects pattern and diversity of plant communities and landscapes. As mentioned by Traubaud and Galtie (1996), in some parts of the Massif Central (southern France) fires increased the homogeneity of the landscape, while in other parts they enhanced heterogeneity and diversity of plant communities.

In a special issue of the International Journal of Wildland Fire, Doerr and Cerda (2005) pointed out that although fire has an impact on entire ecosystems –flora, fauna, atmosphere and soil – special attention has been paid only on the first three ecosystem components. However, burning and the post-fire conditions can alter the physical soil properties (aggregation, pore size, water movement and runoff response), the chemical characteristics (pH, humus structure, availability of nutrients and C/N ratios) and the biological properties (structure of microbial community, biomass productivity e.t.c) as well. Burning of vegetation and litter causes a dual effect on soil. It induced a transient heat shock in the upper soil horizon and produces a layer of nutrient rich ash on the forest floor. Both phenomena being capable of affecting soil microbes (Pietikainen et al. 2000).

D'Ascoli et al. (2005) applied burning of low and high severity in an Italian maquis and studied immediate influences on microbial properties. Although, one week after the fire the functional diversity changed due to both burning per se as well as to the intensity of burning, a rapid recovery of diversity was recorded. On the contrary total microbial biomass was enhanced by fire. According to authors, the effect of fire on fungi was greater and persisted for a longer period compared to that on bacteria. Greater microbial biomass and respiration was also recorded in plots subjected to experimental fires (De Marco et al. 2005). Enhanced microbial biomass is an uncommon response to burning, because the soil structure is destroyed, the pH increases while the forms of N and C remaining after fire are more recalcitrant to microbial attack (Rutigliano et al. 1995, Diaz-Ravina et al. 2006). In this vein, Liu et al. (2000) mentioned decreasing of functional diversity and microbial activity due to burning, in spring and summer. The authors concluded that it was due to reduced carbon inputs into soil because of the delimitation of plants due to burning. It seems that most of the results reported in literature are quite controversial. One of the reasons might be related to the water status of the burned humus. In fact, when humus is dry an application of a prescribed burning of 150 °C could result in an immediate decrease of microbial biomass but when the humus is moist, the relevant effects could be different. Only after water evaporation – which can last for a certain time period- increasing temperature would result in decreasing microbial biomass (Pietikainen et al. 2000).

The work of Guerrero et al. (2005) related the response of microbes to the temperatures of applied fires. They studied the fate of organic C and microbial C in temperatures varying from 100 to 700 °C. For temperatures below 400 °C, the increase of extractable C enabled the recolonization of bacteria with consequent changes in respiration. In higher temperatures, heating caused a reduction in extractable C followed by negative impact on microbial biomass. For fungi, temperatures above 200 °C caused an almost 100% reduction in their biomass. The high sensitivity of fungi to fire is supported also by the findings of D' Ascoli et al. (2005). As shown, fires of relatively high temperatures (> 500 °C) have considerable impact on microbial population. In the same vein, Palese et al. (2003) found that the enzymatic activities decreased by fire, but a peak of activity was recorded in soil burned at 309 °C and it was related to higher amounts of soluble C. Another interesting finding is provided by the work of Fioretto et al. (2005) who examined the effects of prescribed fires on soil microbiota under three shrub species typical of mediterranean maquis (*Phyllirea angustifolia* L., *Cistus incanus* L. and *Myrtus communis* L.). They reported that although the microbial biomass and the activities of selected enzymes differed among soils collected underneath different plant species, in general the effects of fire were not dramatic. The impacts of fire on the size of microbial biomass exhibited less variation, while that on activity was largely dependent on vegetation mosaic and species. The lack of dramatic effects was attributed to the fact that temperature in soil surface did not increase extremely. Actually, with the application of fire, the surface temperature raised from 100 to 400 °C, for a few minutes, while during the next two hours dropped to 40-50 °C. In Mediterranean areas, soil biota are adapted to face temperatures in this range (40 – 50 °C) because such temperatures are frequently attained within the soil surface layers during the summer sunny days.

Diaz-Ravina et al. (2006) examined the effects of heating, fire retardant and soil texture on the composition of microbial communities. Retardants are usually polymers applied in order to combat wildfires and to control prescribed burnings. Among the examined parameters, heating at 350 °C provoked drastic changes in soil properties followed by decreasing microbial biomass and alteration of community structure. Although most of the microbes were killed by fire, Gram negative bacteria were able to use the high availability of substrate while they were also favored by pH increases. Bacteria recolonized rapidly burned areas, whereas the recovery of fungi was extremely slow. The effects of soil texture and retardant were of minor importance. The influence of retardant depended on dose and was differentiated between heated and unheated soils.

THE EFFECTS OF INVASION

Invasion has gained considerable attention only recently and therefore relevant literature is extremely limited. The invasion of exotic species is one of the most serious threats for local biodiversity and for ecosystem functioning (Mooney and Hobbs 2000). Due to plant-microbes interactions (symbiosis, herbivory or pathogenicity), colonization by invasive plant species could induce shifts in the structure of native soil microbial communities, ultimately resulting in changes to the appearance of the entire ecosystem. Apart from invasive plants there are also microbes invaders that could act as competitors to native microorganisms (Jules et al. 2002, Niwa et al. 2004, Wardle et al. 2004). Invasive microbes can change ecosystem processes (e.g. nitrogen fixation) by disturbing local patterns of symbiosis or pathogenicity or by disturbing local decomposition processes (van der Putten et al. 2007).

Among few studies focusing on how soil microbial communities respond to plant invasion and how then they affect further plant invasion is that of Batten et al. (2006). The aim of this study was to examine the impacts of invasive plants on the composition of soil community around rhizosphere and to further investigate how these impacts are related to the duration of invasion. Moreover, the spatial structure of invasion patches was taken into account. Actually, they studied how the rhizosphere microbial community was shaped in the center, the edge and outside the invasion patches. To cope with such a sudject, samples were taken from the center and the edge of the invasive patches as well as from the surrounding native plant community. Two invasive species (an annual forb and an annual grass named starthistle and goatgrass respectively) and five native annual forbs were studied.

To a large extent the invasive plant species spread within areas of low fertility soils such as serpentine grasslands in Northern California Coast Range. In such soils native species are unsuccessful occupants and therefore only refuges of these species exist. In these marginal areas, the overall microbial community composition in invasive edges was intermediate between that in the center and those in native plants. Invasion by goatgrasses seemed to change the community more rapidly than invasion by starthistle. In fact in patches of the former species microbial community in edges was very similar to that in the center. In starthistle sites, diversity was lower in the edges while the opposite occurred in goatgrass sites. Several years after invasion the rhizosphere microbial communities became more diversified in invasive patches than the native ones. In some cases, even in invaded areas the variation in microbial communities from year to year exceeded that between invasive and native plants.

The next question explored by Batten and his colleagues was how changes in microbial properties were related to soil structure (Batten et al. 2005), based on the finding that the rhizosphere of goatgrass consisted of fatty acids representative of arbuscular mycorrhizal fungi (AMF). By producing glomalin AMF could affect soil aggregation. Indeed, soil aggregation was enhanced in soil dominated by goatgrass compared to that dominated by native vegetation.

The Effects of Climate Changes and CO_2 Enrichment

There is a rather wide concensus within the scientific community regarding the climatic consequences of doubling CO_2 concentrations in the atmosphere and the potential negative impacts of climate changes on natural and human made systems. These impacts will exacerbate many of the existing problems (e.g. desertification, water loss etc) especially along the southern and eastern perimeters of the Mediterranean region (Giupponi and Shechter 2003). Changes in precipitation patterns and temperature regimes could introduce new threats to natural ecosystems. For Mediterranean areas, most models predict temperature increase and greater inter and intra-annual variability in rainfall distribution.

Few studies undertaken within the DEGREE project (Diversity Effects in Grasslands Ecosystems of Europe) examined the effects of climate changes on soil processes and diversity in European grasslands. Since the rest experimental sites are not identified as Mediterranean areas, data presented herewith concern only the soil function in a Greek grassland (Papatheodorou et al. 2004a). Soil temperature and humidity were artificially manipulated in the field in accordance to a climate change scenario proposed by Palutikof et al. (1996). According to this scenario, smoother temperature variations and more regular distribution of precipitation in time were predicted. The artificial modifications resulted in damped temperature fluctuations and in drier soil conditions. Due to their narrow range, these changes were considered as small-scale compared with large-scale ones. As large-scale changes were considered the seasonal fluctuations of climatic variables.

The size of microbial biomass was not affected by small-scale variations but activity did. This was related to the regular distribution of rainfall that induced episodes of enhanced activity during the drought period. All microbial parameters showed a significant response to large-scale variations, indicating that seasonal variations in temperature and humidity were more important than small–scale changes described by climate change scenarios. The functional diversity of GN bacteria together with evennenss and richness responded mainly to large-scale variations of temperature and humidity, reflecting seasonal differences in the functioning of the soil bacterial community (Papatheodorou et al. 2004b). The components of catabolic diversity decreased from summer to winter. Since loss in evenness relates to unstable or less resilient microbial function (Degens et al. 2001), a perturbation in winter can induce more dramatic changes in microbial community structure and function compared to changes caused by a summer disturbance.

The above studies concluded that small-scale variations in climatic variables had no significant effect on various aspects of soil microbial community. However, as Balser et al. (2002) mentioned the general idea of a rapid equilibrium of soil microbial communities with changing climate needs an exhaustive investigation. This is also supported by Balser and Firestone (2005) who examined the response of microbial composition to changing climatic factors by a transplanting experiment. They found that composition remained constant even after 2 yrs of exposure to new conditions.

The effects of drought on the microbial community structure involved in litter decomposition and specifically in Mediterranean pine needles' degradation were studied by Wilkinson et al. (2002). The microbial community colonizing litter during the initial stages of decomposition could be highly variable even between adjacent pine needles. This is due to biotic interactions e.g. grazing effects (Maraun et al. 1998) and to microclimatic conditions of the soil surface litter that in mediterranean areas can undergo frequent drying-rewetting stress. The experimental field treatments involved regular or irregular watering and a control receiving natural rainfall. Sampling was conducted in three occasions and data were compared to that referring to the decomposition of a temperate spruce litter in Germany. The shifts in microbial composition, during a period of 17 weeks were more pronounced in spruce than in pine litter. Pine litter seemed insensitive to drought treatments. It exhibited lower water content and was characterized by fungal dominance and low concentrations of bacterial PLFAs. Similar were the results concerning the effect of litterbags' mesh size on PLFA profiles. In pine litter, only microeukaryotic markers were affected by mesh size implying that microclimate appeared more important than biotic interactions in determining the structure of microbial communities.

Human activities result in considerable increases in the amounts of produced CO_2 and O_3. Under conditions of elevated CO_2, photosynthesis and plant productivity generally increase although a large variation among species and ecosystems is also reported (Poorter et al. 1996). The increase of plant productivity is followed by an increase in the amounts of C inputs via enhanced plant biomass turnover and/or root carbon losses. CO_2 enrichment is expected to affect the structure and function of soil microorganisms, since rhizodeposition plays a significant role in shaping soil microbial communities (De Lucia et al. 1999). On the contrary, O_3 is a phytotoxic pollutant with detrimental effects on productivity. Damage to the photosynthetic system in O_3-sensitive plant species could reduce productivity and soil C inputs, thus counteracting the effects of elevated CO_2 on microbial metabolism in soil. The response of soil microbial community beneath early- and late-successional plant species, under conditions of elevated CO_2 and O_3, were studied by Phillips et al. (2002). Specifically, soils were amended with two different substrates (cellobiose and N-acetylglucosamine) and the changes in microbial structure, biomass and activity were investigated. The hypothesis that increases of CO_2 affect positively microbial metabolism by affecting C inputs to soil was supported by this study. Respiration was 29% greater beneath plants growing under elevated CO_2. The amendment of the two different substrates affected differently the two groups of microbes; N-acetylglucosamine increased in a similar way the metabolism of bacteria and fungi, while that of cellobiose increased mainly the metabolism of fungi. Elevated CO_2 altered the composition of bacterial community; gram-negative bacteria became dominant over gram-positive while their total biomass remained unaffected. After 3 yrs of experiment, the authors concluded that microbial community developed under early-successional tree species seemed to be more sensitive than that developed under late-successional species.

Microbial Communities in Different Ecosystems

The responses of different groups of soil organisms to seasonal climatic variations in Mediterranean areas, were shown to be quite predictable (Stamou 1998). But what about the responses of soil microbial community? Does composition changes with season and are these changes predictable? Apart from climatic variables, the type of vegetation exerts also an important influence on microbial community composition. The presence of specific plant species induced heterogeneity in the availability of resources in time and space. In seasonally varying climate the effects of vegetation on microbial communities could be modulated or totally outweighted by the variation in climatic variables. To test these hypotheses, a long-term experiment was established, where the parameters of microbial community were examined in soil beneath an oak canopy and the nearest grassland, across seasons (Waldrop and Firestone 2006). Oak canopies and grassland areas represented two different habitats for soil microorganisms. The range of fluctuations of soil abiotic variables (temperature and humidity) differs between these two vegetation types. Because of the bare soil surface they are wider in grasslands and narrower in oaks. Due to the aforementioned differences, microbial respiration and biomass were higher under oak canopies, resulting in similar activity per unit of biomass in both vegetation types. The composition of microbial communities was affected by the type of vegetation, the season and the year. Biomarkers for Gram-positive, actinomycete and fungal groups characterized the oak samples, while biomarkers for Gram-negative characterized the grassland samples. The magnitude of the difference in microbial community composition attributed to vegetation type tended to be similar or even less than the attributable to seasonal cycles. Moreover, an intra-annual variability in microbial community composition was recorded in both plant communities. As Waldrop and Firestone (2006) concluded, microbial biomass should be mainly controlled by carbon inputs to soil while community composition by climatic variables.

And what about the relationship between composition and function? The functioning of soil microbial community displayed seasonal fluctuations which related to soil microclimate, but differences in function between oak and grassland soils were hardly detectable. Although the function of the overall soil microbial community showed no differences among vegetation types, functions mediated by specific groups of microbes respond differently (Waldrop and Firestone 2004). Such functions could be the decomposition of simple and recalcitrant compounds. Since the oak litter is a more complex C source in comparison with grassland litter, it is likely that soil microbes beneath oak are able to degrade more recalcitrant C compounds. Both in grassland and oak soils, the addition of substrates increased the response of Gram negative organisms. The incorporation of simple substrates (vanillin, starch and xylose) enhanced the activities of the same groups of microbes in both soils. When pine litter was added different microbial groups were activated showing that as the complexity of the substrate increases, the groups responsible for its degradation were different in oak from those in grassland soils. This supported the idea that the functional groups that degrade complex C sources are not functionally redundant across ecosystems. So, any changes in relation to these groups could have a great effect on ecosystem process. Similarly, the response of enzymatic activities to the addition of simple compounds was identical between soils, despite differences in the composition of microbial community. However, differences were recorded in enzymatic activities regarding the degradation of the pine litter and that of the soil organic matter.

The same line of experimentation followed the study of Fierer et al. (2003). They sought for the response of the bacterial genetic diversity in oak and grassland soils to drying-rewetting frequency. Soils of semiarid and arid Mediterranean-type ecosystems are particularly susceptible to drying-rewetting stresses, due to the infrequency of rainfall events and the warm-dry summer that enhance soil dryness. A rainfall event could induce lysis of microbial cells releasing 30-60% of microbial C (Kieft et al. 1987). Most of the existing information referred to enhanced rates of C and N mineralization as an immediate response to rewetting of dry soils, while remained unexploited the question of how these repeated drying-rewetting cycles affected the microbial community structure. Although a modification in structure is expected the direction of changes is unknown. There is an equal possibility for increases of fast growing microbes that could exploit the labile substrates produced by microbial death or for dominance of gram-positive bacteria and fungi with enhanced osmoregulatory capabilities, due to their cell walls characteristics. In accordance with Warldop and Firestone (2006), the bacterial communities in oak and grasslands soils, studied by Fierer et al. (2003), were distinctive in terms of genetic structure. Their response to frequency of drying-rewetting events differed between the vegetation types. In oak soils the higher the frequency of stress that soils were exposed, the more the bacterial communities differed from the unstressed control. On the contrary, the grass soil shows a high degree of community-level variability between samples and no apparent separation of bacterial communities on the basis of stress frequency was detected. The differences in the responses of the bacterial communities were well explained if their adaptation to field abiotic conditions was taken into account. For not preadapted to high variability in field soil moisture oak communities, the exposure to drying-rewetting cycles was more stressful. As Balser and Firestone (2005) mentioned the response of soil processes to environmental changes was also a function of the preexisting adaptation of the microbial community and its physiological plasticity.

The study of Balser and Firestone (2005) aimed at investigating the relation between composition and function. Specific functions (N-mineralization, C-mineralization, nitrification potential) and their relation to parameters of microbial communities were studied in an annual grassland and a mixed-conifer forest. Biomass and lipid diversity were unrelated to soil processes, while the opposite occurred with community structure, as this latter was depicted on the first principal component analysis of PLFAs. The production of N_2O, the gross N mineralization and the nitrification potential were related to composition of the community, the production of NO_3^- to structural and functional organization of the community, while the CO_2 production to temperature. It was revealed that community structure was strongly related to functions carried out by a narrow group of microbes while functions that are most common between soils such as CO_2 production, were rather controlled by abiotic variables than by the composition of the microbial community. These findings are in accordance with those of Warldrop and Firestone (2004) concerning the lack of redundancy of microbes mediating specific soil processes. The presence, absence or abundance of these microbes proved crucial for soil functioning. Cyclopropyl fatty acids, indicative of gram-negative bacteria were associated with N_2O production and nitrification potential while branched fatty acids (biomarkers for gram-positive bacteria) were associated with nitrate concentration. The study of Balser and Firestone (2005) illustrated the ability of fatty acids analysis to reveal the link between structure and function.

Another ecosystem in which the variability of soil microbial community parameters in time and space was studied, was a desert in Israel (Steinberger et al. 1999). The chosen sites represented a gradient from semi-arid to arid climate (rainfall varied from 650 to 110 mm and temperature from 13 to 20 ^{o}C). The main characteristic of desert ecosystems is the unpredictability of moisture availability that forces organisms to be active only for short time periods. The selective force of rainfall was proved by the positive relation between precipitation and microbial biomass (Steinberger et al. 1999). Apart from biomass, the number of fatty acids was also higher in sites with higher precipitation. Soil microbial communities in terms of their structural organization were distinctive between sites and among sampling periods. Changes in climatic variables induced shifts in the composition of the microbial communities. Increases in temperature are known to affect Gram-positive bacteria altering either the amount of branched relative to straight-chain fatty acids and/or the relative proportion of *iso* and *anteiso* branched fatty acids. Also, the amounts of cyclopropyl fatty acids increased with increasing temperature (Peterson and Klug 1994). Concluding Steinberger and his colleagues (1999) mentioned that microbes in desert ecosystems could exhibit one of the following responses: significant decreases of their biomass during the dry period which is attributable mainly to the decrease of Gram-negative bacteria, increases of biomass during the dry period due to the increase of Gram-positive and Gram-negative bacteria or no changes of biomass between rainy and dry periods. This latter response characterized microbes of the more arid zones.

CONCLUSION

The climatic variables, the plant rhizosphere products and the material used to improve soil fertility or to control vegetation are the selective forces influencing the size, activity and structure of soil microbial communities. However, in most instances soil type proved to exert the greatest influence on these communities. Differences in soil type could mask differences induced by climatic or human impacts. Therefore, only under similar soil types comparisons among microbial communities are feasible.

The contents of rhizosphere products are strongly correlated with the occurrence of specific plant species. Changes in plant diversity patterns are induced by grazing and fire regimes, by small or large-scale changes in climatic variables and by the increase of CO_2 in atmosphere. Moreover, plant dominance patterns are modified due to changes in land uses (e.g. from grasslands to cultivations). Areas occupied by plant communities belonging to similar successional stages are characterized by microbial communities of similar structural organization. Moreover, plant growth forms (evergreens or seasonal dimorphic) affect considerably the size of the microbial biomass while the microbial functional diversity responds idiosyncratically. It is worth noticing that irrigation with salt water exerts a negative influence on all aspects of microbial communities while the magnitude of influence is analogous to that exerted by plant type. The effects of fires on microbial communities could be dramatic or not because in different cases different driving factors such as the temperature of the applied fires, the duration of the burning or the water status of the humus are involved. However, especially for fungi, their responses to fire parameters are always reported negative.

Biosolids and composts are amended to soils for improving soil fertility and facilitating their recycling. The effects of former on microbial communities depend on the amount, the chemical content and the frequency of application. Microbial diversity increases linearly as the rate of biososlids' application increases. Also, due to their chemical content, biosolid effects on microbial populations vary considerablly from toxic to undetectable ones. Dose-dependent effects are also produced by compost amendment and pesticides application. Pesticides can alter the successional patterns of soil microbial communities whereas their influences are strongly related to soil type. Soil structure is decisive for their bioavailability. Compost influences strongly all aspects of microbial communities since it could act both as a microbial inoculum and at the same time as a source of organic matrix. The amount of organic material inserted into soil differentiated between cultivations with various durations of organic management. The gradual increase in the amount of supplied organic material, from the newly to oldest organic cultivations, does not induce gradual modifications in the structure or the function of the soil microbial communities. Probably, the fact that available evidence refers to a narrow range of durations of organic management (from 1 to 6 yrs) introduces bias in our conclusions. By contrary, characteristics of soil microbial communities are clearly different among conventional and the oldest organic cultivations. Similarly the structure of microbial communities of the oldest invaded, by plants, areas is much more distinctive from the communities of the native areas than from the newly invaded areas. Invasion resulted from the combined effects of human and climatic interventions is a threat for the function of entire communities. The effects of plants or microbes invasion on soil microbial populations are principally ought to the modifications of biotic interactions (plant-plant or plant-microbes interactions). Even in the case of plant invasion, invaders could be seen as microbial inoculums that induce an array of interactions. The study of invasion's influences is of special interest in Mediterranean areas since it is generally acknowledged that abiotic variables than biotic interactions is the principal force shaping communities in these areas.

Preadaptation of microbes to specific abiotic conditions and to specific substrates inserted into soil determines their response to newly created conditions. Soil microbial communities from different ecosystems exhibited characteristics that continue to persist for long after their establishment in new sites. So, the idea that microbes adapted quickly by altering their structure or function in response to changing conditions, because of their short life cycles, does not seem to occur in Mediterranean regions. In these areas where organisms are developed to cope with the inter- and intra- annual variability of abiotic variables, the changes in latter variables induced by human impacts are of lower magnitude. Unlike to non significant effects of small-scale variations described by climate change scenarios, large-scale (seasonal) variations of temperature and humidity affected significantly all microbial parameters. Compared to grazing, season exerted the most pronounced effect on microbial activity and diversity. Variations in microbial community composition between invasive and native areas were of less magnitude than variations in composition recorded from year to year. During the summer sunny days soil organisms experience high temperature regimes in Mediterranean regions. Thus, fires resulting in soil surface temperatures around 40-50 ^{o}C (soon after their outbreak) have no dramatic effects on soil microbial communities. Moreover, there are instances were the influence of time and site characteristics is more significant for microbes than the influence of pesticides.

By ranking the sensitivity of the three components of microbial communities, that is activity, structure and size, the former proved to be the most sensitive. Changes of respiration or of enzymatic activities in response to any disturbing factor either physical or anthropogenic were reported, since in most cases added materials were used as energy sources. On the contrary, changes in microbial biomass are rarely detected. Although, data on microbial structure are generally limited, there is evidence indicating shifts in the structure of microbial community in response to climatic or human impacts.

Finally, a finding with important consequences for the function of Mediterranean soil ecosystems refers to the correlation between microbial community structure and soil processes. Microbial groups that carry out more general processes such as CO_2 production are redundant between ecosystems, while those involved in specific processes (e.g nitrification) are not. In Mediterranean areas characterized as fine-grained mosaics due to enhanced spatial heterogeneity, redundancy in soil microbial communities could ensure the effective functioning of soil subsystem at least as regards processes of general concern.

REFERENCES

Anderson, J.P.E. and Domsch, K.H. (1978). A physiological method for quantitative measurement of microbial biomass in soils. *Soil Biology and Biochemistry*, 10, 215-221.

Aschmann, H. (1973). Distribution and peculiarity of Mediterranean ecosystems. In F. Di Castri and H.A. Mooney (Eds), *Mediterranean-type ecosystems: origin and structure* (pp. 11-19). Berlin, Heidelberg, New York, Springer.

Arianoutsou, M. and Thanos, C.A. (1996). Legumes in the fire-prone Mediterranean regions: An example from Greece. *International Journal of Wildland Fire,* 6 (2), 77-82.

Argyropoulou, M.D. and Asikidis, M.D and Iatrou, G.D. and Stamou, G.P. (1993). Colonization patterns of decomposing liter in a maquis ecosystem. *European Journal of Soil Biology,* 29, 183-191.

Balser, T.C. and Firestone, M.K. (2005). Linking microbial community composition and soil processes in a California annual grassland and mixed-conifer forest. *Biogeochemistry,* 73, 395-415.

Balser, T.C. and Kinzig, A. and Firestone, M.K. (2002). Linking soil microbial communities and ecosystem functioning. In: A. Kinzig and. C. Pacala and D. Tilman (Eds), *The Functional Consequences of Biodiversity: Empirical Progress and Theoretical Extensions* (pp.265-293). Princeton, Princeton University Press.

Barbarick, K.A. and Doxtader, K.G. and Redente, E.F. and Brobst, R.B. (2004). Biosolids effects on microbial activity in shrubland and grassland soils. *Soil Science*, 169, 176-187.

Barzegar, A.R. and Yousefi, A. and Daryashenas, A. (002). The effect of addition of different amounts and types of organic materials on soil physical properties and yield of wheat. *Plant and Soil*, 247, 295-301.

Batten, K.M. and Six, J. and Scow, K.M. and Rillig, M.C. (2005). Plant invasion of native grassland on serpentine soils has no major effects upon selected physical and biological properties. *Soil Biology and Biochemistry, 37*, 2277-2282.

Batten, K.M. and Scow, K.M. and Davies, K.F. and Harrison, S.P. (2006). Two invasive plants alter soil microbial community composition in serpentine grasslands. *Biological Invasions, 8*, 217-230.

Bending, G.D. and Turner, M.K. and Rayns, F. and Marx, M.C. and Wood, M. (2004). Microbial and biochemical soil quality indicators and their potential for differentiating areas under contrasting agricultural management regimes. *Soil Biology and Biochemistry*, 36, 1785-1792.

Bjørnlund, L. and Ekelund, F. and Christensen, S. and Jacobsen, C.S. and Kroh, P.H. and Johnsen, K. (2000). Interactions between saprotrophic fungi, bacteria ans protozoa on decomposing wheat roots in soil influenced by the fungicide fenpropimorph [Corbel (R)]: a field study. *Soil Biology and Biochemistry*, 32, 967-975.

Buckley, D.H. and Schmidt, T.M. (2001). The structure of microbial communities in soil and the lasting impact of cultivation. *Microbial Ecology*, 42, 11-21.

Busse, M.D. and Ratcliff, A.W. and Shestak, C.J. and Powers, R.F. (2001). Glyphosate toxicity and the effects of long-term vegetation control on soil microbial communities. *Biology and Biochemistry, 33*, 1777-1789.

Busse, M.D. and Beattie, S.E. and Powers, R.F. and Sanchez, F.G. and Tiarks, A.E. (2006). Microbial community responses in forest mineral soil to compaction, organic matter removal and vegetation control. *Canadian Journal of Forest Research*, 36(3), 577-588.

D'Ascoli, R. and Rutigliano, F.A. and De Pascale, R.A. and Gentile, A. and Virzo De Santo, A. (2005). Functional diversity of the microbial community in Mediterranean maquis soils as affected by fires. *International Journal of Wildland Fire,* 14(4), 355-363.

Degens, B.P. and Schipper, L.A. and Sparling, G.P. and Duncan, L.C. (2001). Is the microbial community in a soil with reduced catabolic diversity less resistant to stress or disturbance? *Soil Biology and Biochemistry*, 33, 1143-1153.

De Franchis, L. and Ibanez, F. (2003). *Threats in soils in Mediterranean countries. Document review*. Sophia-Antipolis, Plan Bleu.

De Lucia, E.H. and Hamilton, J.G. and Naidu, S.L. and Thomas, R.B. and Andrews, J.A. and Finzi, A. and Lavine, M. and Matamala, R. and Mohan, J.E. and Henrey, J.R. and Schlesinger, W.H. (1999). Net primary production of a forest ecosystem with experimental CO_2 enrichment. *Science*, 289, 1177-1179.

De Marco, A. and Gentile, A.E. and Arena, G. and Virzo De Santo, A. (2005). Organic matter, nutrient content and biological activity in burned and unburned soils of a Mediterranean maquis area of souethern Italy. *International Journal of Wildland Fire,* 14(4), 365-377.

Dennis, G.L. and Fresquez, P.R. (1989). The soil microbial community in a sewage-sludge-amended semi-arid grassland. *Biology and Fertility of Soils*, 7, 310-317.

Diamantopoulos, J. and Pirintsos, S.A. and Margaris, N.S. and Stamou, G.P. (1994). Variation in Greek phrygana vegetation in relation to soil and climate. *Journal of Vegetation Science*, 5, 355-360.

Díaz-Raviña, M. and Bååth, E. and Martin, A. and Carballas, T. (2006). Microbial community structure in forest soils treated with a fire retardant. *Biology and Fertility of Soils, 42*, 465-471.

Dick, R.E. and Quinn, J.P. (1995). Glyphosate-degrading isolates from environmental samples: occurrence and pathways of degradation. *Applied Microbiology and Biotechnology*, 43, 545-550.

Doerr, S.H. and Cerdá, A. (2005). Fire effects on soil system functioning: new insights and future challenges. *International Journal of Wildland Fire*, 14(4), 339-342.

Domsch, K.H. and Jagnow, G. and Anderson, T.H. (1983). An ecological concept for the assessment of side effects of agrochemicals on soil microorganisms. *Residue Reviews*, 86, 65-105.

Drinkwater, L.E. and Letourneau, D.K. and Shennan, C. (1995). Fundamental differences between conventional and organic tomato agroecosysems in California. *Ecological Applications*, 5, 1098-1112.

Fernandes, M.F. (2006). Fatty acid profiling of soil microbial communities: a comparison of extraction methods and temporal dynamics in plant residue amended soils. Dissertation, Oregon State University.

Fierer, N. and Schimel, J.P. and Holden, P.A. (2003). Influence of drying-rewetting frequency on soil bacterial community structure. *Microbial Ecology, 45*, 63-71.

Filcheva, E.G. and Tsadilas, C.D. (2002). Influence of clinoptilotite and compost on soil properties. *Communications in Soil Science and Plant Analysis,* 33, 595-607.

Fillery, I.R.P. (2001). The fate of biologically fixed nitrogen in legume-based dryland farming systems: a review. *Australian Journal of Experimental Agriculture,* 41(3), 361-381.

Fioretto, A. and Papa, S. and Pellegrino, A. (2005). Effects of fire on soil respiration, ATP content and enzyme activities in Mediterranean maquis. *Applied Vegetation Science*, 8, 13-20

Garcia-Gill, J.C. and Plaza, C. and Senesi, N. and Brunetti, G. (2004). Effects of sewage sludge amendment on humic acids and microbiological properties of a semi-arid Mediterranean soil. *Biology and Fertility of Soils*, 39, 320-328

Geaves, M.P. (1982). Effects of pesticides on soil microorganisms. In R.G. Burns and J.H. Slater (Eds), *Experimental microbial ecology* (pp.613-630). Oxford, Blackwell.

Gianfreda, L. and Sannino, F. and Violane, A. (1995). Pesticide effects on the activity of free, immobilized and soil invertase. *Soil Biology and Biochemistry*, 27, 1201-1208.

Grecchio, C. and Gelsomino, A. and Ambrosoli, R. and Minati, J.L. and Ruggiero, P. (2004). Functional and molecular responses of soil microbial communities under differing soil management practices. *Soil Biology and Biochemistry, 36*, 1873-1883.

Guckert, J.B. and Antworth, C.P. and Nichols, P.D. and White, D.C. (1985). Phospholipid, ester-linked fatty acid profiles as resproducible assays for changes in prokaryotic community structure of estuarine sediments. *FEMS Microbial Ecology*, 31, 147-158.

Guerrero, C. and Mataix-Solera, J. and Gómez, I. and Garcia-Orenes, F. and Jordán, M.M. (2005). Microbial recolonization and chemical changes in a soil heated at different temperatures. *International Journal of Wildland Fire,* 14(4), 385-400..

Giupponi, C. and Shechter, M. (2003).*Climate Change in the Mediterranean. Socio-economic perspectives of impacts on vulnerability and adaptation.* Northampton, Edward Elgar Publishing

Haney, R.L. and Senseman, S.A. and Hons, F.M. and Zuberer, D.A. (2000). Effect of glyphosate on soil microbial activity and biomass. *Weed Science*, 48, 89-93.

Haynes, R.J. and Williams, P.H. (1999). Influence of stock camping behaviour on the soil microbiological and biochemical properties of grazed pastoral soils. *Biology and Fertility of Soils,* 28, 253-258.

Henkin, Z. and Noumeir, I. and Kafkafi, U. and Seligman, N. and Gutman, M. (1994). Soil-phosphorus in a managed mediterranean woodland. Ecosystem-herbage response and cattl grazing effects. *Agriculture Ecosystems and Environment*, 47(4), 299-311

Henkin, Z. and Hadar, L. and Noy-Meir, I. (2006). Human-scale natural heterogeneity induced by grazing in a Mediterranean woodland landscape. *Landscape Ecology*, 22(4), 577-587.

.Holland, J.N. and Detling, J.K. (1990). Plant response to herbivory and belowground nitrogen cycling. *Ecology*, 71(3), 1040-1049.

Insam, H. (1990). Are the soil microbial biomass and basal respiration governed by the climate regime? *Soil Biology and Biochemistry*, 22, 525-532.

Jackson, R.D. and Allen-Diaz, B. and Oates, L.G. and Tate, K.W. (2006). Spring-water nitrate increased with removal of livestock grazing in a california oak savanna. *Ecosystems*, 9(2), 254-267

Jenkinson, D.S. and Powlson, D.S. (1976). The effect of biocidal treatments on metabolism in soil. V. A method for measuring soil biomass. *Soil Biology and Biochemistry*, 8, 209-213.

Johnsen, K. and Jacobsen, C.S. and Torsvik, V. and Sørensen, J. (2001). Pesticide effects on bacterial diversity in agricultural soils – a review. *Biology and Fertility of Soils, 33*, 443-453.

Johnston, A.E. (1986). Soil organic matter, effects on soils and crops. *Soil Use Management*, 2, 97-105.

Jules, E.S. and Kauffman, M.J. and Ritts, W.D. and Carroll, A.L. (2002). Spread of an invasive pathogen over a variable landscape: a nonnative root rot on Port Orford cedar. *Ecology* 83: 3167–3181.

Kabengi, N.J. and Zyrayk, R.A. and Baalbaki, R.Z. and Ryan, J. (2003). Phosphorus dynamics and characterization under long-term rotation trial. *Communications in Soil Science and Plant Analysis*, 34 (3-4), 375-392.

Kazanis, D. and Arianoutsou, M. (2004). Long-term post-fire vegetation dynamics in *Pinus halepensis* forests of Central Greece: A functional group approach. *Plant Ecology,* 171, (1-2), 101-121.

Kazanis, D. and Arianoutsou, M. (1996). Vegetation composition in a post-fire successional gradient of *Pinus halepensis* forests in Attica, Greece. *International Journal of Wildland Fire,* 6 (2), 83-91.

Kieft, T. and Soroker, E. and Firestone, M. (1987). Microbial biomass response to a rapid increase in water potential when dry soil is wetted. *Soil Biology and Biochemistry*, 19, 119-126.

Lamotte, M. and Bladin, P. (1989). Originalité et diversité des écosystèmes méditerranéens terrestres. *Biologia Gallo-Hellenica*, 16, 5-36.

Lavee, H. and Imeson, A.C. and Sarah, P. (1998). The impact of climate change on geomorphology and desertification along a Mediterranean-arid transect. *Land Degradation and Development*, 9(5), 407-422.

Li, Q. and Mayzlish, E. and Shamir, I. and Pen-Mouratov, S. and Sternberg, M. and Steinberger, Y. (2005). Impact of grazing on soil biota in a Mediterranean grassland. *Land Degradation and Development*, 16(6), 581-592.

Liu, X. and Lindemann, W.C. and Whitford, W.G.and Steiner, R.L. (2000). Microbial diversity and activity of distrurbed soil in the northern Chihuahuan Desert. *Biology and Fertility of Soils*, 32, 243-249.

Maggi, O. and Persiani, A.M. and Casado, M.A. and Pineda, F.D. (2005). Effects of elevation, slope position and livestock exclusion on microfungi isolated from soils of Mediterranean grasslands. *Mycologia*, 97(5), 984-995.

Maraun, M. and Visser, S. and Scheu, S. (1998). Oribatid mites enhance the recovery of the microbial community after a strong disturbance. *Applied Soil Ecology*, 9, 175-181.

Martini, E.A. and Buyer, J.S. and Bryant, D.C. and Hartz, T.K and Denison R.F. (2004). Yield increases during the organic transition: improving soil quality or increasing experience? *Field Crops Research, 86*, 255-266.

Monokrousos, N. and Papatheodorou, E.M. and Diamatopoulos, J.D. and Stamou, G.P. (2006). *Soil Biology and Biochemistry, 38*, 1282-1289.

Monokrousos, N. and Papatheodorou, E.M. and Diamantopoulos, J.D. and Stamou, G.P. (2004). Temporal and spatial variability of soil chemical and biological variables in a Mediterrannean shrubland. *Forest Ecology and Management, 202*, 83-91.

Mooney, H.A, Hobbs, R. (2000). Invasive *Species in a Changing World.* Washington, Island Press.

Naveh, Z. (1998). Culture and landscape conservation: a landscape-ecological perspective. In B. Gopal, and P.S. Pathak, and K.G. Saxena (Eds), *Ecology today: an anthology of contemporary ecological research* (pp. 19-48). New Delhi, International Scientific Publications.

Niwa, S. and Iwano, H. and Asada, S. and Matsuura, M. and Goka, K. (2004). A microsporidian pathogen isolated from a colony of the European bumblebee, *Bombus terrestris*, and infectivity on Japanese bumblebee. *Japanese Journal of Applied Entomology and Zoology*, 48: 60–64.

Ovalle, C. and Del Pozo, A. and Casado, M.A. and Acosta, B. and de Miguel, J.M. (2006). Consequences of landscape heterogeneity on grassland diversity and productivity in the espinal agroforestry system of central Chile. *Landscape Ecology*, 21 (4), 585-594.

Øvreas, L. (2000). Population and community level approaches for analyzing microbial diversity in natural environments. *Ecology Letters, 3*, 236-251.

Palese, A.M. and Giovannini, G. and Lucchesi, S. and Dumontet, S. and Perruci, P. (2004). Effects of fire on soil C, N and microbial biomass. *Agronomie,* 24, 47-53.

Palutikof, J.P. and Conte, M. and Casimiro, M.J. and Goodess, C.M. and Espiito, S.F. (19960. Climate and climatic change: In C.J. Brandt, and J.B. Thornes (Eds), *Mediterranean Desertification and Land Use* (pp.43-86). Chichester, Wiley.

Papatheodorou, E. (1996). Vegetation dynamics and nutrient cycling (N and P) in a degraded evergreen sclerophyllous ecosystem at Mt. Hortiatis. Ph.D. Thesis, Aristotle Univ. Thessaloniki (in Greek with an English summary).

Papatheodorou, E.M. and Pantis, J.D. and Stamou G.P (1998). The effect of grazing on phenology and biomass allocation in *Quercus coccifera* (L.). *Acta Oecologica*, 19, 339-347.

Papatheodorou, E.M. and Stamou G.P. (2004). Nutrient attributes of tissues in relation to grazing in an evergreen sclerophyllous shrub (*Quercus coccifera* L.) dominating in Mediterranean-type ecosystems. *Journal of Arid Environments*, 59, 217-227.

Papatheodorou, E.M. and Stamou, G.P. and Giannotaki, A. (2004a). Response of soil chemical and biological variables to small and large scale changes in climatic factors. *Pedobiologia, 48*, 329-338.

Papatheodorou, E.M. and Argyropoulou, M.D. and Stamou, G.P. (2004b). The effects of large- and small-scale differences in soil temperature and moisture on bacterial functional diversity and the community of bacterivorous nematodes. *Applied Soil Ecology, 25*, 37-49.

Pascual, J.A. and Garcia, C. and Hernandez, T. (1999). Lasting microbiologicaland biochemical effects of the addition of municipal solid waste to an arid soil. *Biology and Fertility of Soils*, 30, 1-6.

Paul, E.A. and Clark, F.E. (1996). Soil *Microbiology and Biochemistry* (Second Edition). San Diego, California, Academic Press.

Peco, B. and Sanchez, A.M. and Azcarate, F.M. (2006). Abandonment in grazing systems: Consequences for vegetation and soil. *Agriculture Ecosystems and Environment*, 113 (1-4), 284-294.

Perevolotsky, A. and Seligman, N.G. (1998). Grazing in Mediterranean ecosystems: inversion of a paradigm. *BioScience*, 48, 1007–1017.

Peterson, S.O. and Klug, M.J. (1994). Effects of sieving, storage, and incubation temperature on the phospholipids fatty acid profile of a soil microbial community. *Applied Environmental Microbiology*, 60, 2421-2430.

Phillips, R.L. and Zak, D.R. and Holmes, W.E. and While, D.C. (2002). Microbial community composition and function beneath temperate temperate trees exposed to elevated atmospheric carbon dioxide and ozone. *Oecologia, 131*, 236-244.

Pietikäinen, J. and Hiukka, R. and Fritze, H. (2000). Does short-term heating of forest humus change its properties as a substrate for microbes? *Soil Biology and Biochemistry*, 32, 277-288.

Poorter, H. and Roumet, C. and Campbell B.D. (1996). Interspecific variation in the growth response of plants to elevated CO_2 : A search for functional types. In C. Körner, and F.A. Bazaaz (Eds), *Carbon Dioxide, Populations and Communities* (pp.375-412). San Diego, Academic Press,.

Preston-Mafham, J. and Boddy, L. and Randerson, P.F. (2002). Analysis of microbial community functional diversity using sole-carbon source utilization profiles – a critique. *FEMS Microbial Ecology*, 42, 1-14.

Puerto, A. and Rico, M. (1997). Edaphic variability and floristic structure on Mediterranean grassland slopes. *Arid Soil Research and Rehabilitation*, 11 (1), 9-22.

Ros, M. and Hernandez, M.T. and Garcia, C. (2003) Soil microbial activity after restoration of a semiarid soil by organic amendments. *Soil Biology and Biochemistry*, 35, 463-469.

Roslev, P. and Iversen, N. and Henriksen, K. (1998). Direct fingerprinting of metabolically active bacteria in environmental samples by substrate specific radiolabelling and lipid analysis. *Journal of Microbiological Methods*, 31, 99-111.

Rutigliano, F.A. and Virzo De Santo, A. and Alfani, A. and Esposito, A. (1995). Attività metabolica di suoli campani in funzione degli degli incendi. *Società Italiana di Ecologia Atti*, 16, 535-538.

Saison, C. and Degrange, V. and Oliver, R. and Millard, P. and Commeaux, C. and Montange, D. and Le Roux, X. (2006). Alteration and resilience of the soil microbial community following compost amendment: effects of compost level and compost-borne microbial community. *Environmental Microbiology*, 8(2), 247-257.

Schloss, P.B. and Hay, A.G. and Wilson, D.B. and Walker, L.P. (2003). Tracking temporal changes of bacterial community fingerprints during the initial stages of composting. *FEMS Microbial Ecology*, 46, 1-9.

Schutter, M.E. and Sandeno, J.M. and Dick, R.P. (2001). Seasonal, soil type, and alternative management influences on microbial communities of vegetable cropping systems. *Biology and Fertility of Soils, 34*, 397-410.

Scow, K.M. and Somasco, O. and Gunapala, N. and Lau, S. and Venette, R. and Ferris, H. and Miller, R. and Shennan, C. (1994). Transition from conventional to low-input agriculture changes soil fertility and biology. *California Agriculture*, 48, 20-26.

Sgardelis, S.P. and Stamou, G. and Margaris, N.S. (1981). Structure and spatial distribution of soil arthropods in a phryganic (eastern Mediterranean) ecosystem. *Revue de Ecologie and Biologie du Sol*, 18, 221-230.

Sgardelis, S.P. and Margaris, N.S. (1993). Effects of fire on soil microarthropods of a phryganic ecosystem. *Pedobiologia*, 37, 83-94.

Sgardelis, S.P. and Pantis, J.D. and Argyropoulou, M.D. and Stamou, G.P. (1995). Effects of fire on soil macroinvertebrates in a Mediterranean phryganic ecosystem. *International Journal of Wildland Fire*, 5, 113-121.

Shestak, C.J. and Busse, M.D. (2005). Compaction Alters Physical but Not Biological of Soil Health. *Soil Science Society of America Journal,* 69, 236-246.

Stamou, G.P. and Asikidis, M.D. (1992). The effect of certain biotic factors on the demographic parameters od *Scheloribates* cf *latipes* (Acari: Oribatida). *Pedobiologia*, 36, 351-358.

Stamou, G.P. and Asikidis, M.D. and Argyropoulou, M.D. and Sgardelis, S.P. (1993). Ecological time versus standard clock time: the asymmetry of phonologies and the life history strategies of some soil arthropods from Mediterranean ecosystems. *Oikos*, 66, 37-35.

Stamou, G.P. and Pantis, J.D. and Sgardelis S.P. (1994). Comparative study of litter decomposition in two Greek ecosystems: a temperate forest and an asphodel semi-desert. *European Journal of Soil Biology*, 30, 43-48.

Stamou, G.P. (1995). Strategic responsis of oribatid mites to the severity of the Mediterranean environment. In D. Kropczynska, and J. Boczek, and A. Tomczyk (Eds.), *The Acari* (pp.295-304). Warszawa, Dabor.

Stamou, G.P. and Pantis, J.D. (1995). Responses of vegetation to the climate and anthropogenic severity of Greek Mediterranean ecosystems. In R. Fantechi, and D. Peter, and P. Balabanis, and J.L. Rubio (Eds.), *Desertification in a European context: physical and socio-economic aspects* (pp.321-338). Proc Eur Sch Climatol Nat Hazards. Luxembourg, Directorate-General XII.

Stamou, G.P. (1998). *Arthropods of Mediterranean-Type Ecosystems. Adaptations of Desert Organisms.* Berlin Heildelberg, Springer.

Steenwerth, K.L. and Jackson, L.E. and Calderón, F.J. and Stromberg, M.R. and Scow, K.M. (2002). Soil microbial community composition and land use history in cultivated and grassland ecosystems of coastal California. *Soil Biology and Biochemistry, 34*, 1599-1611.

Steinberger, Y. and Zelles, L. and Bai, Q.Y. and von Lützow, M. and Munch, J.C.(1999). Phospholipid fatty acid profiles as indicators for the microbial community structure in soils along a climatic transect in the Judean Desert. *Biology and Fertility of Soils,* 28, 292-300.

Stephan, A. and Meyer, A.H. and Schmid, B. (2000). Plant diversity affects culturable soil bacteria in experimental grassland communities. *Journal of Ecology, 88*, 988-998.

.Sullivan, T.S. and Stromberger, M.E. and Paschke, M.W. and Ippolito, J.A. (2006). Long-term impacts of infrequent biosolids applications on chemical and microbial properties of a semi-arid rangeland soil. *Biology and Fertility of Soils*, 42, 258-266.

Szabolcs, I. (1998). Concepts, assessment and control of soils affected by salinization. In Eger, H. and. Blume, H.-P and Hebel, A. and Steiner K.G. (Eds.), *Towards Sustainable Land Use: Furthering Cooperation Between People and Institutions* - Selected papers of the 9th Conference of the International Soil Conservation Organisation (ISCO) (pp. 469-476). Catena, Verlag.

Thornes, J.B. (1996). Introduction. In C.J. Brandt, and J.B. Thornes (Eds.), *Mediterranean Desertification and Land Use* (pp. 1-11). England, John Wiley and Sons Ltd.

Trabaud, L. and Gatlie, J-F. (1996). Effects of fire freqency on plant communities and landscape pattern in the Massif des Aspres (southern France). *Landscape Ecology*, 11 (4), 215-224.

Tsiafouli, M.A. and Argyropoulou, M.D. and Stamou, G.P. and Sgardelis, S. (in press). Duration of organic agriculture and nematodes. *Belgian Journal of Zoology*.

Tunlid, A. and White, D.C. (1992). Biochemical analysis of biomass, community structure, nutritional status, and metabolic activity of microbial communities in soil. In G. Stotzky, and J-M. Bollag (Eds.), *Soil Biochemistry* (Vol.7 pp. 229-262). New York, Dekker.

Tzialla, C.E. and Veresoglou, D.S. and Papakosta, D. and Mamolos, A.P. (2006). Changes in soil characteristics and plant species composition along a moisture gradient in a Mediterranean pasture. *Journal of Environmental Management*, 80 (1), 90-98.

Vance, E.D. and Brookes, P.C. and Jenkinson, D.S. (1987). An extraction method for measuring soil microbial biomass C. *Soil Biology and Biochemistry*, 19, 703-707.

Van der Putten, W.H. and Klironomos, J.N. and Wardle, D.A. (2007). Microbial ecology of biological invasions. *The International Society for Microbial Ecology Journal*, 1, 28-37.

Waldrop, M.P. and Firestone, M.K. (2004). Microbial community utilization of recalcitrant and simple carbon compounds: impact of oak-woodland plant communities. *Oecologia, 138*, 275-284.

Waldrop, M.P. and Firestone, M.K. (2006). Seasonal dynamics of microbial community composition and function in oak canopy and open grassland soils. *Microbial Ecology, 52*, 470-479.

Wardle, D.A. and Bardgett, R.D. and Klironomos, J.N. and Setala, H. and Van der Putten, W.H. and Wall, D.H. (2004). Ecological linkages between aboveground and belowground biota. *Science*, 304: 1629–1633.

Wertz, S. and Degrange, V. and Prosser, J.I. and Poly, F. and Commeaux, C. and . Guillaumaud, N. and Le Roux, X. (2006). Maintenance of soil functioning following erosion of microbial diversity. *Environmental Microbiology*, 8, 2162–2169.

Wilkinson, S.C. and Anderson, J.M. and Sgardelis, S.P. and Tsiafouli, M. and Taylor, A. and Wolters, V. (2002). PLFA profiles of microbial communities in decomposing conifer litters subject to moisture stress. *Soil Biology and Biochemistry*, 34, 189-200.

Willott, S.J. and Miller, A.J. and Incoll, L.D. and Compton, S.G. (2000). The contribution of rabbits (*Oryctolagus cuniculus* L.) to soil fertility in semi-arid Spain. *Biology and Fertility of Soils*, 31(5), 379-384.

Zak, J.C. and Willig, M.R. and Moorhead, D.L. and Wildman, H.G. (1994). Functional diersity of microbial communities: a quantitative approach. *Soil Biology and Biochemistry*, 26, 1101-1108.

Zelles, L. and Bai, Q.Y. and Rackwitz, R. and Chadwick, D. and Beese, F. (1995). Determination of phospholipid- and lipopolysaccharide-derived fatty acids as an estimate of microbial biomass and community structures in soils. *Biology and Fertility of Soils*, 19, 115-123.

In: Soil Ecology Research Developments
Editors: Tian-Xiao Liu, pp. 89-119
ISBN: 978-1-60021-971-9

Chapter 3

QUORUM-SENSING LANGUAGE IN BACTERIA AND ITS RELEVANCE IN THEIR INTERACTIONS WITH PLANTS

***B. Rodelas,* [1,2]*, R. Vílchez* [2,3]*, L. Mora* [1]*, A. Lemme* [3]*, C. Pozo* [1,2]*, M. Molina-Muñoz* [2]*, M.V. Martínez-Toledo* [1,2] *and J. González-López* [1,2,*]**

[1]Departamento de Microbiología, Universidad de Granada, Spain;
[2]Instituto del Agua, Universidad de Granada, Spain;
[3] Department of Cell and Immune Biology, Helmholtz Center for Infection Research (HZI), Braunschweig, Germany

ABSTRACT

The concept of bacteria living as single isolated cells has been replaced by the vision of these organisms as members of communities that mostly work coordinately, with the aid of communication systems of variable complexity that make possible for every single cell to perceive the rest of bacteria sharing an habitat and talk to them. The languages of bacteria rely on chemical signals, synthesized by the cells and sent out to the surrounding media, where their concentration rises correlated to the increase of cell numbers. When a certain threshold cell density (quorum) is reached, the accumulated chemical signals trigger a coordinated population response, often moved by the need of adaptation to a change in the environment. The term "quorum sensing" was coined to describe this ability of bacteria to monitor the size of their populations before taking the decision of expressing a certain genotype. There are many different molecules released as signals by bacteria, being the best characterized to date acyl-homoserine lactones (AHLs), autoinducing oligopeptides (AIPs), and autoinducer 2 (AI-2). Amongst the known phenotypes regulated by quorum sensing, there are a number of functions related to both beneficial and pathogenic interactions between bacteria and eukaryotic organisms. Bacteria that interact with plants are not an exception, and use sophisticated quorum sensing systems whose involvement on the regulation of important steps for survival and competition in the rhizosphere, root colonization, and establishment of symbiotic or

* Corresponding author (jgl@ugr.es).

pathogenic associations with plants, has been thoroughly investigated during the last 20 years, yet leaving many questions unanswered. This chapter reviews the basic concepts of the quorum sensing mechanism and its implications, with particular emphasis on the current knowledge of its importance in the interaction of legume plants and nodule-forming bacteria.

I. THE QUORUM-SENSING (QS) MECHANISM: AN OVERVIEW

1. Introduction

Bacteria are classically considered as strictly unicellular organisms. Even when grouped, the contact amongst cells is minimal, and their unions labile and temporal. Exceptions to the rule are the rudimentary multicellular organizations of fruiting bodies of myxobacteria, and of the filamentous cyanobacteria that develop specialized cells for nitrogen fixation (heterocysts) [Søgaard-Andersen *et al.*, 2003; Zhang *et al.*, 2006]. Nevertheless, this traditional conception of bacterial unicellular existence has been rewritten in the last two decades, when it has become apparent that homogeneous bacterial populations lacking cellular specialization also show a degree of multicellular organization. Individual bacterial cells are indeed able to differentiate simultaneously as part of a communal response, inducing changes of the morphology, physiology or behavior of the whole population. These changes aim for a particular common objective, and the factor that triggers the coordinate response is the cell density of the population. A certain numbers of individuals, or minimal unit (quorum), is thus required for the population to take a collective decision and synchronize its behavior [Fuqua *et al.*, 1994]. The quorum concept also implies that before any coordinate response of a bacterial population can take place, each individual cell requires an estimation of the size of the full population, and for this reason, bacteria have developed censing strategies. The mechanism is termed quorum-sensing (QS), and basically operates as a communal device of gene regulation relying on cell density [Fuqua *et al.*, 1994].

In order to survive in an open environment, bacteria need to adapt to the continuous changes of the milieu, and cooperation is required to attain these functions [Hense *et al.*, 2007]. The physiological roles known to be regulated by QS are hence widely diverse, as summarized in Table 1. However, most of these phenotypes share a common feature: they are only expressed when the bacterial population is able to grow to a certain cell density, which guarantees the success of its actions. In the case of bacteria pathogenic for plants and animals, the coordinate expression of virulence factors during the infection of the host organism is a crucial step for the success of the process. QS controls the expression of several bacterial functions which are relevant to humans, due to their importance in the fields of medicine, agriculture and industry. Some of these implications will be discussed in more detail later in this Chapter.

The quorum sensing language is composed of "chemical words": molecules synthesized by bacteria, also named cell density factors (CDF) [Loh *et al.*, 2002] or autoinducers [Hense *et al.*, 2007]. CDF are often low-molecular weight compounds, which are synthesized constitutively at basal levels by the individual cells. The signal molecules are mostly diffusible through the cell membrane and are sent out to the surrounding media, where they may accumulate if bacteria reach high cell densities or inhabit spatially-limited environments.

Table 1. Bacterial phenotypes regulated by quorum-sensing (de Kievit & Iglewsky, 2000; Whitehead *et al.*, 2001, Daniels *et al.*, 2004, Vendeville *et al.*, 2005, Waters & Bassler, 2005, González & Keshavan, 2006, Williams *et al.*, 2007)

Phenotype	Organism
Antibiotic production	*Chromobacterium violaceum, Erwinia carotovora, Pseudomonas aureofaciens, P. chlororaphis, P. fluorescens, Serratia marcescens, Serratia* spp., *Streptomyces* spp.
Aggregation	*Pseudomonas aureofaciens, Rhodobacter spharoides, Yersinia pseudotuberculosis*
Biofilm formation	*Aeromonas hydrophila , Burkholderia cenocepacia, Helicobacter pylori, P. aeruginosa, P. putida, Serratia liquefaciens, Streptococcus gordonii, S. mutans*
Bioluminescence	*Vibrio fischeri, V. harveyi*
Biosurfactant production	*Bacillus subtilis, Pseudomonas aeruginosa, Serratia liquefaciens, S. marcescens*
Cell division	*Escherichia coli*
Cyanide production	*Chromobacterium violaceum, Pseudomonas aeruginosa*
Exopolysaccharide production	*Pantoea stewartii, Pseudomonas syringae*
Exoproteases/exoenzymes production	*Aeromonas hydrophila, A. salmonicida, Burkholderia cenocepacia, B. pseudomallei, Chromobacterium violaceum, Erwinia carotovora, E. chrisantemii, Pseudomonas aeruginosa, P. aureofaciens, Serratia liquefaciens, S. proteomaculans, Staphylococcus aureus, Streptococcus pyogenes, Vibrio vulnificus*
Exotoxins	*Actinobacillus actinomycetemcomitans, Clostridium perfringens, Staphylococcus aureus, Vibrio harveyi*
Growth	*Rhizobium leguminosarum* bv. *viciae, R. etli*
Legume nodulation	*Rhizobium leguminosarum* bv. *viciae, R. etli, Sinorhizobum meliloti, Bradyrhizobium japonicum*
Nitrogen fixation	*Rhizobium etli*
Pigment production	*Chromobacterium violaceum, Serratia marcescens, Serratia* spp.
Plasmid transfer	*Agrobacterium tumefaciens, Rhizobium leguminosarum* bv. *viciae*
Secretion systems	*Pseudomonas aeruginosa, Escherichia coli*
Siderophore biosynthesis	*Burkholderia cenocepacia, Pseudomonas aeruginosa, Vibrio harveyi*
Stationary phase regulation	*Rhizobium leguminosarum* bv. *viciae, R. leguminosarum* bv. *phaseoli, Nitrosomonas europaea*
Swarming/motility	*Burkholderia cenocepacia, Campylobacter jejuni, Pseudomonas syringae, Rhizobium etli, Serratia liquefaciens, S. marcescens, Yersinia enterocolitica, Y. pseudotuberculosis*
Transposition frequency	*Serratia marcescens*
Virulence (other factors)	*Erwinia carotovora, Agrobacterium vitiae, Burkholderia* spp., *Ralstonia solanacearum, Serratia marcescens, Shigella flexneri, Staphylococcus aureus,V. harveyi, V. cholerae, V. vulnificus, Xanthomonas campestris, Xenorhabdus nematophilus*

Under these conditions, signal molecules rise in their concentrations (both intra- and extracellularly) until a critical threshold is reached, wich elicits their interaction with transcriptional regulatory proteins, inducing (or in some cases repressing) the expression of particular genes in a cell-density dependent manner. This way, communication and coordinated response can take place between bacterial cells over a distance, not requiring a direct contact between individuals.

More recently, interesting new concepts about second functions for QS mechanisms have been proposed. QS faces many problems in a complex environment, as concentrations of CDF are altered by many factors, mainly derived from spatial heterogeneity and biological diversity [Hense *et al.* 2007]. Winzer *et al.* (2006) proposed the concept of compartment sensing (CD) to take into consideration that, in order to achieve the accumulation of a QS signal, there is a need for a diffusion barrier, which ensures that more molecules are produced than lost from a given microhabitat. This way, the QS signal molecule estimates both the degree of compartmentalization and the means to distribute this information through the entire population. Redfield (2002) proposed another alternative explanation for QS, termed diffusion sensing (DS). This approach explains response to CDF as a way for bacteria to determine if secreted molecules are actually rapidly diffusing away from the cell. This will aid to monitor and control the secretion of effector molecules such as degradative exoenzymes, antibiotic, surfactants and siderophores, to minimize loses by extracellular diffusion [Redfield, 2002, González and Marketon, 2003]. This concept is independent of cell-density and spatial distribution, as DS will act as a mechanism to sense mass-transfer properties of the environment surrounding a focal cell [Hense *et al.*, 2007]. A more recent hypothesis is efficiency sensing (ES) [Hense *et al.*, 2007], which combines the ability of cells to sense population density, mass-transfer properties of the environment, and spatial cell distribution, in order to estimate the efficiency of producing extracellular diffusible effectors and to respond only when this is efficient.

2. Signaling Molecules in Bacterial QS

Currently, QS mechanisms have been described in over 50 bacterial species of Gram-negative and Gram-positive bacteria [de Kievit and Iglewsky, 2000; Whitehead *et al.*, 2001, Daniels *et al.*, 2004, Vendeville *et al.*, 2005, Waters and Bassler, 2005, González and Keshavan, 2006, Williams *et al.*, 2007], as well as in some eukaryotic microorganisms (yeasts) [Sprage and Winans, 2006]. There are many different molecules released as QS signals by bacteria (Figure 1), which are often mentioned as autoinducers to reflect the fact that the induction of gene expression is inflicted by self-produced signal molecules. *N*-acyl-homoserine lactones (AHLs) and autoinducer-2 (AI-2) are to date the best known chemical structures used as widespread language signals by bacteria. In Gram-negative bacteria, other molecules such as 4-quinolones, fatty acids and fatty-acid methyl esters have been reported as QS signals, often working in a same species in combination with signal molecules of a different type [Aendekerk *et al.*, 2005]. In Gram-positive bacteria, autoinducer oligopeptides (AIPs) and γ-butyrolactones have frequently been found involved in bacterial cell-to-cell communication. The only QS signal shared by both Gram-negative and Gram-positive bacteria is autoinducer-2 (AI-2), and there are evidences of its possible role as a universal signal for interspecies communication [Whitehead *et al.*, 2001, Xavier and Bassler, 2003, Sun *et al.*, 2004].

N-acyl-homoserine lactones 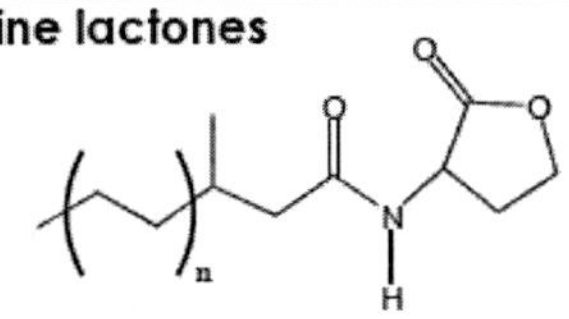 *Agrobacterium tumefaciens* *Burkholderia cepacia* *Erwinia carotovora* *Pseudomonas aeruginosa* *Rhizobium leguminosarum* bv. *viciae*, *R. etli* *Vibrio fischeri*	**Bradyoxetin** 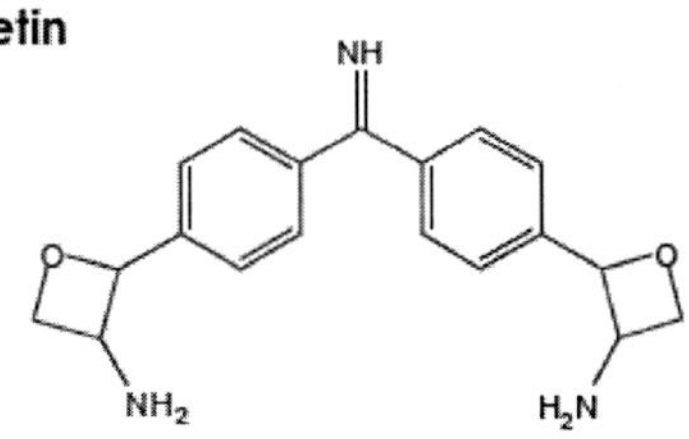 *Bradyrhizobium japonicum*
AI-2* (Furanosyl borate ester form) 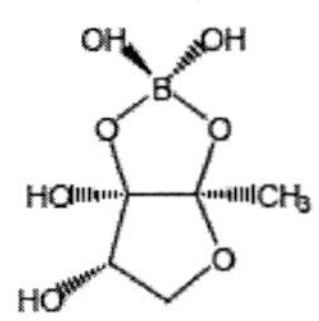 *Vibrio harveyi* (*hypothesized as universal signal in bacteria)	**2-heptyl-3-hydroxi-4(1*H*)-quinolone (*Pseudomonas* quinolone signal, PQS)** 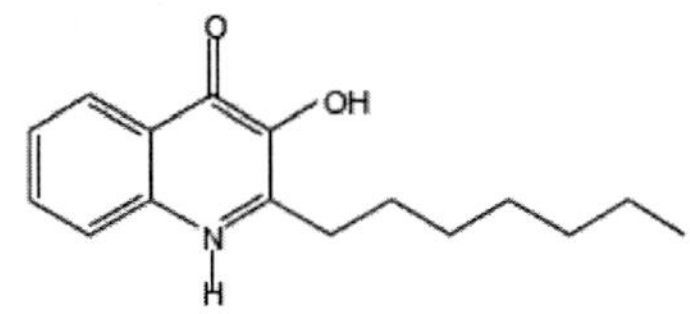 *Pseudomonas aeruginosa*
Hydroxyl-palmitic acid methyl ester (PAME) 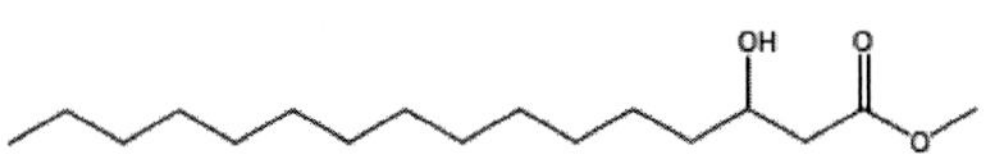 *Ralstonia solanacearum*	**Diketopiperazines** Cyclic (Δ-Ala-L-Val) 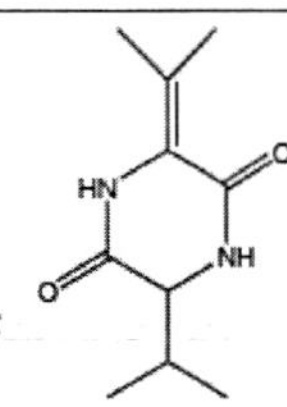 *Pseudomonas aeruginosa* *Enterobacter agglomerans* *Citrobacter freundii*
Methyl dodecenoic acid (Diffusible factor, DFS) 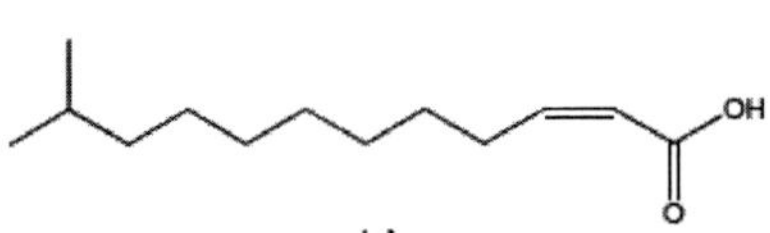 *Xanthomonas campestris*	**Farnesoic acid** 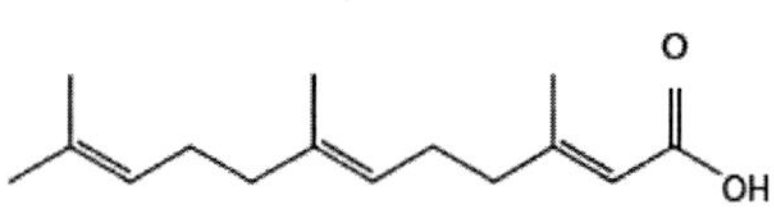 *Candida albicans*
Autoinducing oligopeptides (AIPs) Cyclic tiolactone (AIP I) 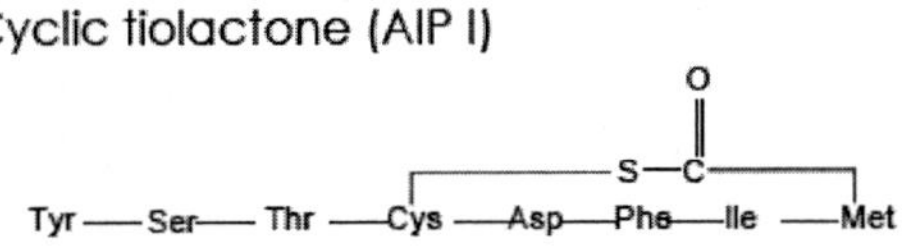 *Staphylococcus aureus* *Bacillus subtilis* *Lactobacillus plantarum* *Enterococcus faecalis*	**A-factor** 2-isocaproyl-3-hydroxymethil-γ-butyrolactone 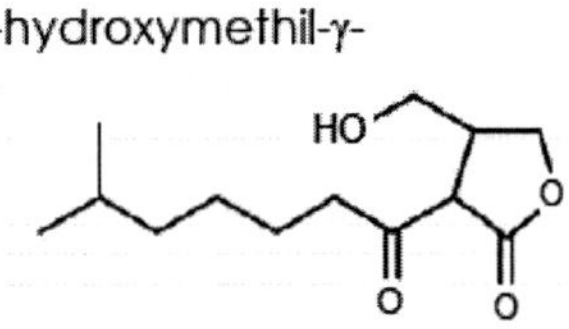 *Streptomyces* spp.

Figure 1. Structures of some representative QS-signal molecules (De Kievit & Iglewski, 2000, Holden & Swift, 2000, Zhang & Dong, 2004, Chhabra *et al.*, 2005, Williams *et al.*, 2007).

3. QS Mediated by *N*-Acyl-Homoserine Lactones (AHLs)

3.1. Structure and Synthesis of AHLs

The QS mechanism was first described in the symbiotic relationship formed by the bioluminescent bacteria *Vibrio fisheri* and the squid *Eupryma scolopes. E. scolopes* appears luminescent in dark marine environments due to the maintenance of high density *V. fisheri* populations in its specialized light organ [Whitehead *et al.*, 2001].

V. fisheri only initiates bioluminescence at high cell population densities due to the accumulation of an activator molecule, *N*-(3-oxohexanoyl)-homoserine lactone (3-oxo-C_6-HSL), first isolated and characterized by Eberhard *et al.* (1981). Further research carried out in the last two decades with several other bacterial groups demonstrated that AHL-based QS is widespread amongst Gram-negative bacteria. This mechanism has been studied in depth for many genera of Proteobacteria, and complex QS systems, involving hierarchical cascades of regulation and interconnection with other non-AHL based signaling strategies, have been thoroughly studied in *Pseudomonas aeruginosa* and *Vibrio harveyi* (for a review, see de Kievit and Iglewsky, 2000, Whitehead *et al.*, 2001, Waters and Bassler, 2005, Williams *et al.*, 2007). Gram-negative Proteobacteria that nodulate legume roots also communicate by AHLs, and the state of the art of QS involvement in these plant-bacterial symbiotic interactions will be further discussed at the end of this Chapter.

The basic structure of AHLs is a homoserine-lactone (HSL) ring, unsubstituted in the β- and γ-positions, which is *N*-acylated with a fatty acyl group at the α-position [Chhabra *et al.*, 2005]. The acyl side-chain is of variable nature in length, saturation level and oxidation state. Some examples of well-characterized AHLs structures are given in Figure 2. The HSL moiety originates from amino acid metabolism, being shown by *in vitro* studies using recombinant proteins that *S*-adenosyl methionine is the amino acid substrate for its synthesis [Whitehead *et al.*, 2001, Chhabra *et al.*, 2005]. The acyl side-chain is a product of fatty acid metabolism, mainly derived from cellular pools of the appropriate acyl-ACP (acyl carrier protein) rather than acyl-CoA [Hanzelka and Greenberg, 1996; Whitehead *et al.*, 2001]. The length of the acyl side-chain is commonly 4-12 C, and may be substituted in C-3 with a carbonyl or hydroxyl group [Hoang *et al.*, 2002, Watson *et al.*, 2002]. Some *Rhizobium* and *Sinorhizobium* species make AHLs with a side chain of up to 18 C atoms [Lithgow *et al.*, 2000, Marketon *et al.*, 2002], which may include unsaturated bonds.

It was firstly assumed that AHLs were able to freely diffuse through cellular membranes [Kaplan and Greenberg, 1985]. This hypothesis is supported by the fact that exogenously supplied AHL activate QS systems. However, more recent work demonstrated that long-chain AHLs ($>C_8$) may be actively transported. This is the case of an active efflux pumping system required for the effective translocation of AHLs produced by *Pseudomonas aeruginosa* [Pearson *et al.*, 1999]. The need of active transport for long-chain AHLs in other bacteria has not yet been reported though.

Most Gram-negative bacteria produce more than one type of AHL, but also different organisms can produce the same AHL [Hardman *et al.*, 1998]. These leads to the cross-talking phenomenon, whose implications which will be discussed in more detail later in this Chapter.

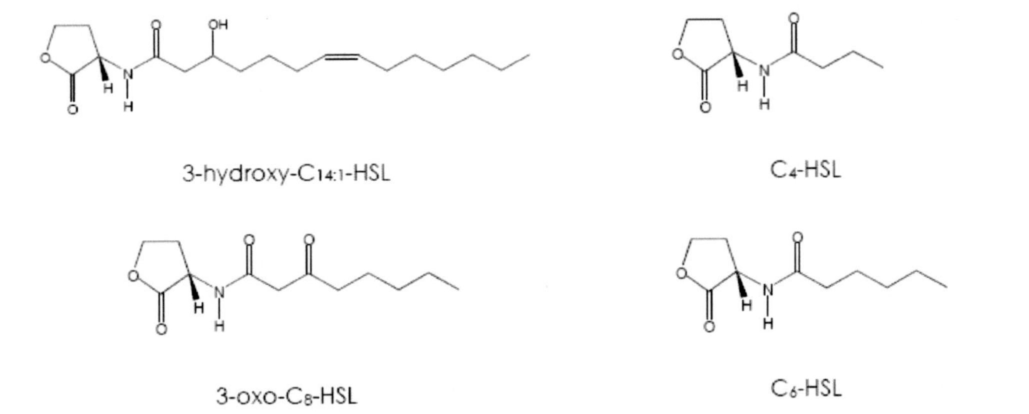

Figure 2. Chemical structure of some AHLs used as QS signals by Gram-negative bacteria (Dong *et al.* 2007; Williams *et al.*, 2007).

3.2. Functional Components of the AHL-Mediated QS Regulation

3.2.1. The LuxR and LuxI Protein Families

Bacteria commonly perceive changes in their surrounding environment by two-component systems, which typically consist of a sensor of the stimuli (which also works transmitting the signal), and a transcriptional regulator of the sensor's response to the signal. In the case of the regulation based on self-produced signals, a third component would be required, which generates the signals. However, the QS systems mediated by AHLs only have two main regulatory genes, which were firstly analyzed and sequenced in bioluminescent *Vibrio fisheri*, and named *luxR* and *luxI* [Engebrecht *et al.*, 1983]. Analysis of these genes in *V. fischeri* led to the basic model for AHL-dependent QS, which is now the archetype for other analogous cell-to-cell communication systems (Figure 3). By means of the AHL-based mechanism, Gram-negative bacteria can efficiently couple gene expression to fluctuations in cell-population density.

The *lux* genes are organized in two divergent transcriptional units, separated by ca. 155 bp. The first transcriptional unit carries seven genes in the *luxICDABEG* operon under the P_R promoter. The second unit consists of the regulatory *luxR* gene and the P_L promoter [Engebrecht and Silverman, 1984, 1987; Devine *et al.*, 1988]. LuxI is the autoinducer synthase that generates *N*-(3-oxohexanoyl)-L-homoserine lactone (3-oxo-C_6-HSL), and LuxR acts as a regulatory protein that binds the autoinducer molecule and activates the transcription of *luxICDABEG* by facilitating the binding of RNA polymerase to the target promoter [Stevens *et al.*, 1994]. Hence, LuxR covers both the roles of sensing/transducing the signal and the regulation of the response to it. LuxR binds to AHL through an amino-terminal domain, and the carboxy-terminal region of the protein carries an H-T-H motif [González and Marketon, 2002].

The *lux* genes are thus regulated by LuxR and the autoinducer product generated by LuxI. At low cell density, transcription of both *luxR* and *luxI* happens at a basal level, enough to allow a certain concentration of LuxR and LuxI to accumulate in the surround of cells.

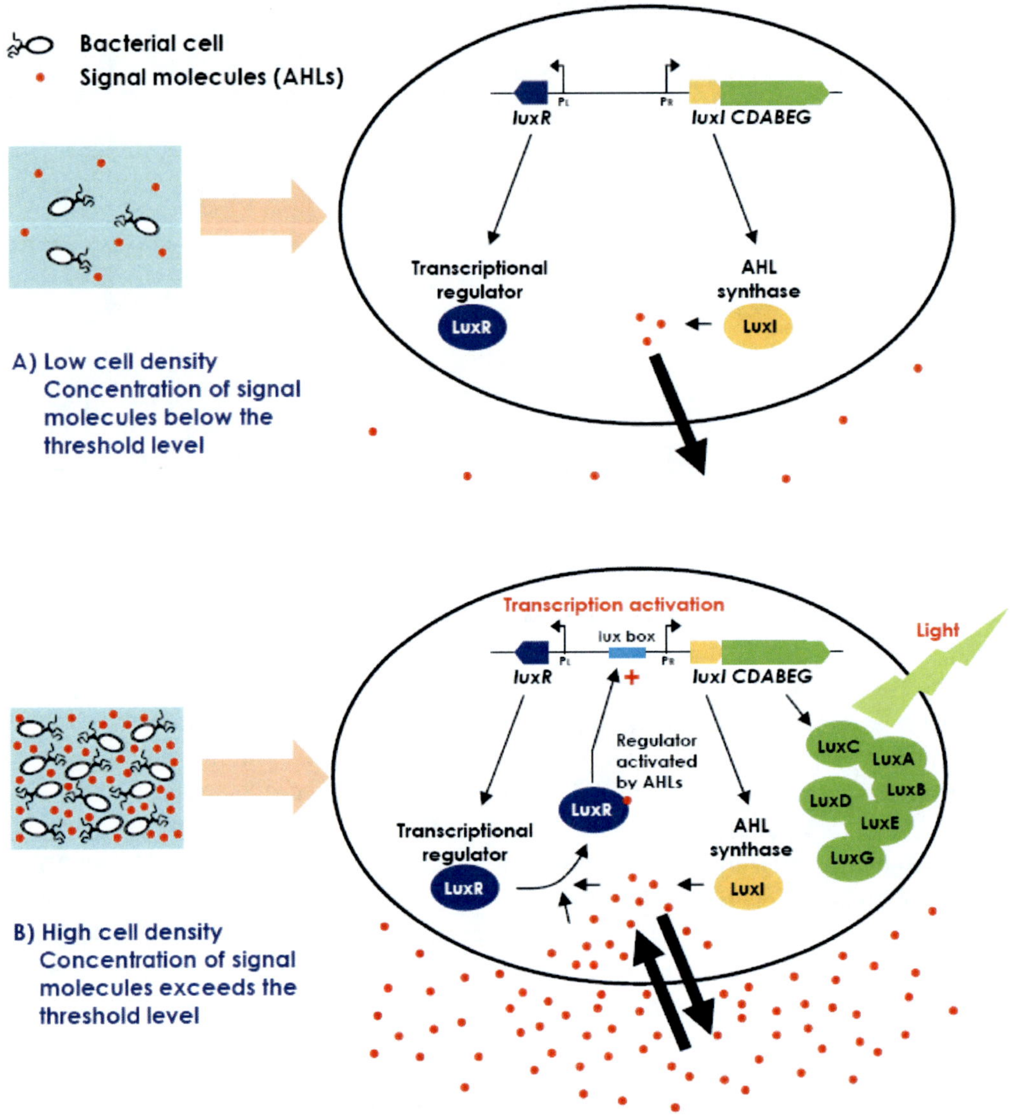

Figure 3. *Vibrio fisheri* model of QS mediated, cell-density dependent regulation of gene transcription. A) At low cell density, low levels of signal molecules (AHLs) are synthesized by LuxI and released, and transcription of the *luxCDABEG* genes is repressed B) When cell density rises, AHLs accumulate and exceed the threshold concentration. Signal molecules bind to the regulatory LuxR protein, that triggers transcription of the *luxCDABEG* genes and also of the *luxI* gene (autoinduction).

When cell density increases, concentration of the autoinducer molecules rises until a certain threshold level is reached, when signal molecules bind to the LuxR protein and activate it. AHL-bound LuxR changes its conformation, undergoes dimerization and binds to a specific *cis*-acting inverted repeat of 18-22 pb, named the **lux box** [Sitnikov *et al.*, 1995]. Subsequently, the transcription of the *luxICDABEG* operon is triggered, resulting in the increase of both emission of light and also 3-oxo-C_6-HSL synthesis. However, when 3-oxo-C_6-HSL concentration is abundant, the LuxR-autoinducer complex starts repressing transcription of *luxR,* self-limiting the autoinduction of bioluminescence [Whitehead *et al.*, 2001].

In most of the bacterial species analyzed to date, AHLs are synthesized by LuxI homologues and interact with LuxR-type transcriptional activators. Systems related to LuxR/LuxI have been identified as cell-density dependent regulators of many bacterial functions, often related to the relationships of bacteria with eukaryotic organisms, in both symbiotic and pathogenic interactions. LuxR and LuxI homologues constitute evolutively conserved families of regulatory proteins, although amino acid alignments of sequences disclose a low level of sequence similarity amongst members of each family of proteins [Whitehead *et al.,* 2001]. Phylogenetic analysis of the sequences of known LuxR and LuxI homologues reveals an early origin of these regulators during the evolution of Gram-negative bacteria, as well as the probable co-evolution of both types of genes as part of regulatory cassettes. In most cases, horizontal transfer of the *luxRI* genes is suggested, as several species carry multiple pairs of homologues of often poorly related sequences, which in addition are frequently harbored in conjugative plasmids [Gray and Garey, 2001; Wisniewski and Downie, 2002].

3.2.2. AHL Synthases Unrelated To LuxI: The LuxM/Ains and HDTS Families

LuxM/AinS family synthases have been only found to date in *Vibrio* species [Milton, 2006]. It was first revealed in *Vibrio harveyi* that the synthesis of 3-hydroxy-C_4-HSL was directed by LuxM, an autoinducer synthase lacking similarity to LuxI, although it catalyzes identical biochemical reactions to generate a specific AHL [Bassler *et al.*, 1993, Whitehead *et al.*, 2001]. Other synthases of the LuxM family have been found in *Vibrio fisheri* (AinS) and *V. anguillarum* (VanM) [Gilson *et al.*, 1995, Milton *et al.*, 2001]. A third type of AHL synthase named HdtS has been detected in *Pseudomonas fluorescens* [Laue *et al.*, 2000].

4. QS Mediated by Autoinducing Oligopeptides (AIPs)

In contrast to Gram-negatives, autoinducing oligopeptides (AIPs) are the principal QS signals known in Gram-positive genera such as *Lactococcus*, *Streptococcus*, *Staphylococcus* and *Bacillus.* AIPs range from 5-34 amino acids, usually arose after processing of a precursor peptide, and are often post-translationally modified [Camilli and Bassler, 2006, Williams *et al.*, 2007]. Mature peptides are secreted into the medium using one associated ATP-Binding Cassette (ABC) transporter complex [Donabedian, 2003]. At high cell densities, the accumulated peptides reach a critical threshold concentration and are recognized by a two-component signaling system. Signal transduction is relayed into the cell by phosporylation, resulting in an altered gene expression [Kleerebezem *et al.*, 1997]. AIPs were shown to be highly specific to their cognate sensor kinase, thus allowing only the specific recognition of one species [Havarstein *et al.*, 1997].

According to their structure, AIPs are divided in 3 families [Williams *et al.*, 2007]. One of the earliest known systems was that of *Streptococcus pneumoniae*. It was observed that at high cell densities *S. pneumoniae* was capable of competence [Tomasz and Hotchkiss, 1964, Tomasz, 1965, Tomasz and Mosser, 1966]. Further investigation elucidated an operon containing the *comCDE* genes, named after the observed phenomenon. The precursor of the autoinducing peptide is the product of the *comC* gene, whereas the two-component system is encoded by the *comD* and *comE* genes, respectively [Pestova *et al.*, 1996]. The autoinducing peptide, called competence stimulating peptide (CSP), is a 17-amino acid peptide including a double glycine motif in the leader sequence, which is cleaved by a proteolytic domain of the associated ABC transporter (ComA) [Hui and Morrison, 1991, Havarstein *et al.*, 1995, Zhou *et al.*, 1995]. Phosphorylation of ComE results in a positive feedback by inducing the *comCDE* as well as the *comAB* operon, respectively. Furthermore, the regulator *comX* is induced, which mediates expression of genes responsible for competence. Regulation of competence using the *comABCDE* system is known to be a general mechanism in streptococci. Besides competence regulation, there are more targets downstream of *comE* involved in environmental adaptations and production of virulence factors, i.e. biofilm formation or bacteriocin production, respectively [Cvitkovitch *et al.*, 2003, Peterson *et al.*, 2004, Kreth *et al.*, 2006, 2007].

The soil organism *Bacillus subtilis* also uses QS to control competence development, as well as for initiation of sporulation in a coupled regulation circuit [Grossmann, 1995, Lazazzera and Grossman, 1998]. Uptake of DNA is carried out during the transition from exponential growth phase to the stationary phase. Internalized DNA is assumed to serve as depository of genetic material used in chromosome repair. Limitation of nutrients in the environment leads to sporulation. Two autoinducing peptides are known to mediate these events. First is the processed ComX peptide, which acts similarly to the CSP peptide in streptococci by activating the *comP/comA* two component signaling system, resulting in the expression of the activator ComK. ComK promotes transcription of the genes that are required for competence development [Magnuson *et al.*, 1994, Solomon *et al.*, 1995, van Sinderen *et al.*, 1995, Turgay *et al.*, 1997, 1998]. A dual role in sporulation as well as in competence is connected to the second peptide, the CSF autoinducer (competence and sporulation factor). In contrast to other autoinducers, the CSF peptide posseses both extracellular and intracellular signaling roles, and is therefore imported into the cell [Solomon *et al.*, 1995, Lazazzera *et al.*, 1997]. Promotion of the competence development is mediated under low concentrations of the CSF peptide inside the cell by indirectly increasing the activity of the ComA response regulator. High intracellular concentrations prevent competence development and trigger sporulation by acting on a different two-component system [Perego *et al.*, 1994, Grossmann, 1995, Solomon *et al.*, 1995]. Thus, *B. subtilis* can react in an appropriate manner in response to environmental changes, due to the fluctuations in the internal CSF concentration.

Thiolactone peptides (Figure 1) constitute the best known class of AIPs, particularly those regulating the production of virulence factors in the pathogenic bacteria *Staphylococcus aureus* [Balaban and Novick, 1995, Ji *et al.*, 1995]. Staphylococcal AIPs vary in the primary amino acid sequence but contain a conserved cysteine at position 5 [Chhabra *et al.*, 2005]. In *S. aureus*, AIP-I production and regulation rely upon the *agrBDCA* genes. The autoinducing peptide is encoded by *agrD*, and a two-component sensor kinase-response regulator pair (AgrC and AgrA) acts as a receptor [Ji *et al.*, 1995]. When the extracellular concentration of the peptide signal accumulates to the minimal stimulatory level, the histidine sensor kinase protein detects it and subsequently, the response regulator is phosphorylated on a conserved aspartate residue. The phosphorylated response regulator activates the transcription of target genes [Miller and Bassler, 2001]. *S. aureus* strains are classified into four different groups according to the specificity of their AIPs [Schauder and Bassler, 2001].

5. QS Mediated By Autoinducer 2 (AI-2)

AI-2 was originally identified as a QS autoinducer in *Vibrio harveyi*, which controls bioluminescence in conjunction with an AHL-based mechanism [Henke and Bassler, 2004]. Interestingly, this regulation shares characteristics from QS-systems of both the Gram-negative and Gram-positive bacteria, as it combines AHLs and a two-component protein signal transduction system [Schauder and Bassler, 2001]. The gene responsible for ultimate production of AI-2 is named *luxS*, and seems widely distributed as it was found present in more than 60 bacterial species, including classes β, γ, δ and ε of the Proteobacteria, Spirochaetes, Firmicutes, Actinobacteria, and other genera [Vendeville *et al.*, 2005], leading to the suggestion that AI-2 can serve as one universal language [Sun *et al.*, 2004].

Chemically, AI-2 is a collection of interconvertible furanone-derived molecules [Chen *et al.*, 2002, Waters and Bassler, 2005, Chhabra *et al.*, 2005]. AI-2 emerges from DPD (4,5-dihydroxy-2,3-pentanedione), which is a byproduct of the activated methyl cycle (AMC), responsible for the generation of the methyl-donor *S*-adenosyl methionine (Figure 4) [Williams *et al.*, 2007]. DPD was isolated and purified by Cao and Meighen (1989), who also elucidated its structure. LuxS, a small metalloenzyme containing Fe^{2+}, is an *S*-ribosylhomocysteinase that cleavages *S*-ribosylhomocysteine to produce DPD [de Keersmaecker *et al.*, 2006].

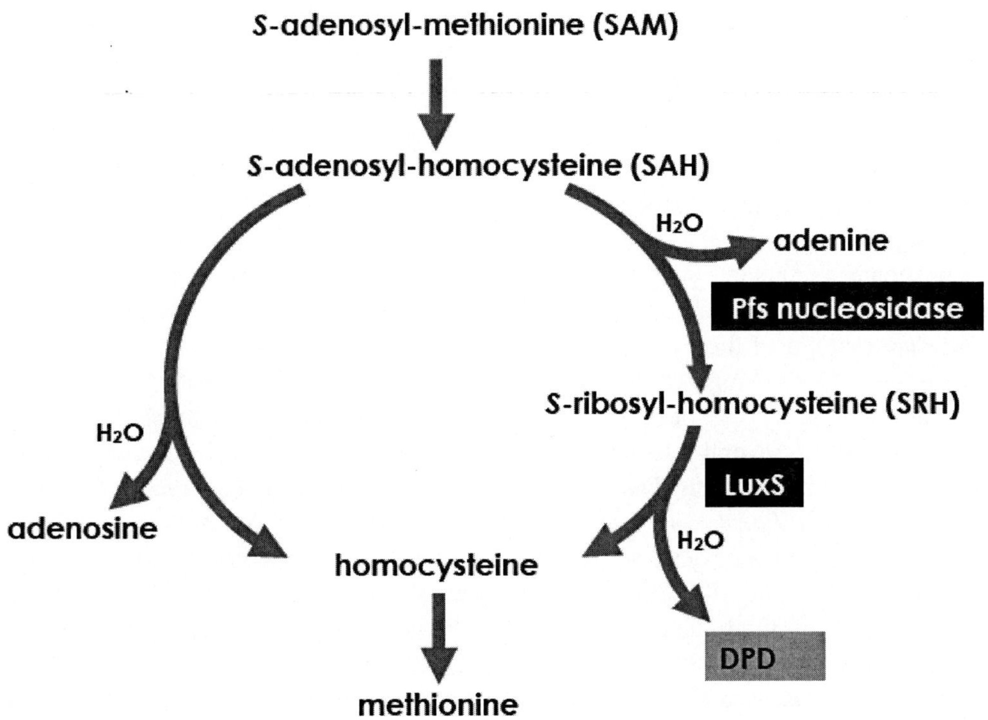

Figure 4. Pathways for detoxification of SAH and synthesis of DPD [Sun *et al.*, 2004].

AI-2 from *Vibrio harveyi* was chemically identified as a furanosyl borate diester [Chen *et al.*, 2002], and it was later demonstrated that DPD spontaneously cyclizes and forms AI-2 in the presence of borate [Semmelhack *et al.*, 2005]. However, other bacterial species produce AI-2 after different rearrangements of the DPD moieties [Camilli and Bassler, 2006]. To date, it is known that six compounds can arose from DPD. Therefore, all DPD-derived molecules that show QS-activity are named collectively autoinducer-2 [Vendeville *et al.*, 2005].

In addition to the production of the DPD precursor, the LuxS enzyme has an important metabolic function in the AMC, namely the recycling of the toxic metabolite *S*-adenosylhomocysteine (SAH) to homocysteine (Figure 4) [Williams *et al.*, 2007]. As in the case of AHLs, the amino acid precursor of AI-2 is *S*-adenosylmethionine (SAM), from which DPD is generated through three enzymatic steps [Schauder and Bassler, 2001]. Consumption of SAM as a methyl donor produces SAH, which is subsequently detoxified by the nucleosidase Pfs to yield adenine and *S*-ribosylhomocysteine (SRH). SRH is then converted to 4,5-dihydroxy-2,3-pentanedione (DPD) and homocysteine by LuxS. An alternative pathway does exist, via *S*-adenosylhomocysteine hydrolase (SAHh) [Sun *et al.*, 2004].

In *Vibrio harveyi*, the LuxS-generated signals are detected and transduced by the LuxP/LuxQ proteins. LuxP is a periplasmic binding protein which acts as receptor for AI-2, forming a complex that subsequently interacts with the inner membrane-bound LuxQ sensor histidine quinase, which transduces the AI-2 information into the cytoplasm and activates the response through a rather complex mechanism, involving a phosphorelay system and small regulatory RNAs [Vendeville *et al.*, 2005, Williams *et al.*, 2007]. However, the model of detection-transduction of the AI-2 signal is rather different in *Salmonella typhimurium*, *Escherichia coli* and other species [Vendeville *et al.*, 2005].

The function of AI-2 as a QS signal molecule in bacteria other than *Vibrio* spp. has been questioned, and this molecule has been suggested for most bacteria to be a metabolic side product in the AMC pathway [Winzer *et al.*, 2002, Chhabra *et a*l., 2005]. Analysis of 138 complete genomes revealed that even though the LuxS enzyme is widespread in bacteria, the periplasmic binding protein LuxP is only present in *Vibrio* strains [Sun *et al.*, 2004]. The open question is to elucidate whether other organisms use components different from the AI-2 signal transduction system of *Vibrio* strains to sense the signal of AI-2 (i.e. the *Salmonella typhimurium* case), or if they actually lack AI-2 based QS. Many studies evidence important implications of the AI-2 regulatory role in the expression of virulence factors by a number of pathogenic bacteria. It has been shown that by using chemically synthesized AI-2 it is possible to restore several phenotypes in *luxS*$^-$ mutants of seven different genera of pathogenic bacteria (*Clostridium perfringens*, enterohemorrhagic *Escherichia coli*, *Helycobacter pylori*, *Porphyromonas gingivalis*, *Shigella flexneri*, *Streptococcus mutans* and *Vibrio vulnificus*) (for a review, see Vendeville *et al.*, 2005). Studies based on microarray techniques revealed that AI-2 is involved in the regulation of over 400 genes in *Escherichia coli* strains, and of more than 300 genes in *Photorhabdus luminescens* [Gonzâlez and Keshavan, 2006, Krin *et al.*, 2006].

The use of AI-2 as a QS signal by both Gram negative and Gram positive bacteria suggest that is the earliest bacterial autoinducer and may have evolved before the divergence between Gram negative and Gram positive bacteria [Schauder and Bassler, 2001]. Consolidation of the idea of AI-2 acting as a universal bacterial language and its role in interspecies communication is an exciting challenge for future research on QS.

6. Eukaryotic Responses to Bacterial QS Systems

Recent work demonstrates that plants and algae are able to interfere with bacterial QS communication by secreting compounds that mimic the bacterial-made signals [Tepliski *et al.*, 2000, Bauer and Robinson, 2002, Mathesius *et al.*, 2003]. The best known mock AHLs are the structurally analogous halogenated furanones synthesized by the red algae *Delisea pulchra* (Figure 5), which act blocking communication by binding the bacterial AHL receptors [Givskov *et al.*, 1996]. Several higher plants (pea, alfalfa) produce compounds that either activate or inhibit the expression of bacterial genes regulated by QS [Teplitski *et al.*, 2000, Bauer and Mathesius, 2004, Waters and Bassler, 2005, González and Keshavan, 2006]. Bacterial phenotypes controlled by QS are frequently also regulated by additional environmental signals, ie. oxygen availability, nutrient starvation, iron limitation or catabolite repression [Newton and Fray, 2004]. Thus, it has been suggested that the plant-produced mimics are indirectly altering the bacterial QS response, rather than targeting it specifically. In either case, such compounds may determine the outcome of interactions between higher plants and a diversity of pathogenic and symbiotic bacteria [Newton and Fray, 2004].

Plants hereby seem able to listen to the bacterial language, and possibly also respond directly to AHLs-based signaling systems of either beneficial or harmful bacteria. This later hypothesis is also supported by data on the influence of physiologically relevant concentrations of AHLs on proteome profiles of *Medicago truncatula* roots, showing that AHLs induce the accumulation of root proteins involved in host defense responses [Mathesius *et al.* 2003]. Elicitation of defense responses by AHLs is also proven in tomato and bean plants [Bauer and Mathesius, 2004]. Finally, there is some evidence of plant produced compounds able to interact with bacterial communication mediated by AI-2 [Bauer and Mathesius, 2004].

Other responses of eukaryotic organisms to the AHL-based bacterial QS systems have been described. Motile zoospores of the sea green algae *Enteromorpha* preferentially attach to AHL-producing bacterial biofilms, due to AHLs acting as chemo-attractants [Joint *et al.*, 2002]. Bacterial AHLs elicit several responses in animals, such as immunomodulatory and haemodynamic effects in rats, inhibition of muscle contraction in pigs, and accelerated apoptosis in macrophages and neutrophils [Bauer and Mathesius, 2004].

Recent findings have given evidence of the ability of human cells to hamper bacterial QS by inactivation of the signal molecules. *Pseudomonas aeruginosa* is an opportunistic pathogen that causes infections mostly in immune compromised patients. Human respiratory epithelia have the capacity to inactivate the *P. aeruginosa* quorum-sensing signal 3-oxo-C_{12}-HSL, by means of a membrane-associated enzymatic mechanism [Chun *et al.*, 2004]. This capacity functions in some but not all mammalian cells [Hastings, 2004]. The ability of epithelial cells to inactivate 3-oxo-C_{12}-HSL may confer host protection against *P. aeruginosa* infections and could be exploited as a therapeutic target. Human hormones epinephrine and norepinephrine also interfere with the QS system of enterohemorrhagic *Escherichia coli* [Sperandio *et al.*, 2003]. The interference on bacterial QS mechanisms by plants and animals acting as hosts of bacterial pathogens is hypothesized as part of the multiple defense strategies developed by eukaryotes to avoid bacterial infections.

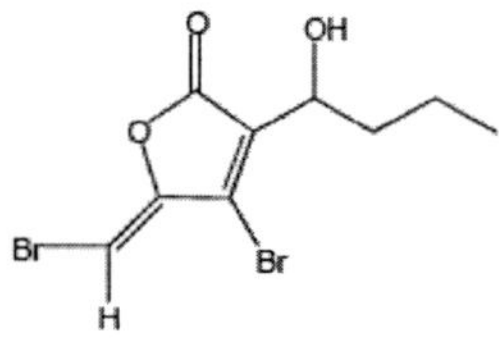

4-bromo-5-(bromomethylene)-3-(-1-hydroxybutyl)-2(*5H*)-furanone

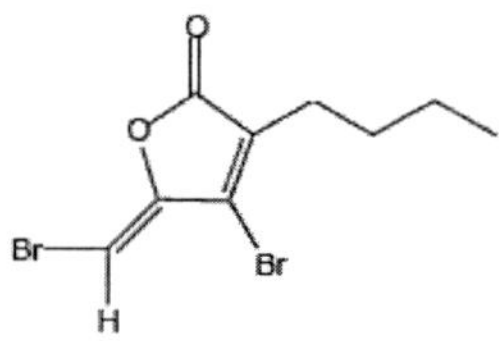

4-bromo-5-(bromomethylene)-3-butyl-2(*5H*)-furanone

Figure 5. Quorum sensing antagonistic furanones produced by red algae (*Delisea pulchra*).

7. Other Implications of the QS Mechanism

As it has been already stated along the Chapter, QS mechanisms play significant roles in microorganisms that live in symbiotic or pathogenic association with eukaryotic organisms. These facts highlight the potential implications of QS communication in the ecological interactions within microbial communities and the establishment of host-bacterial relationships. The advances in the knowledge and understanding of the bacterial QS mechanisms have also pointed out to the possibility of interfering with such communication systems to develop new approaches for the control of infectious diseases and the manipulation of biofilm formation [Zhang and Dong, 2004].

7.1. Cross Talking

QS signals generated by a particular microorganism often interfere effectively with the QS mechanisms of different organisms. This is partly due to the high structural similarities shared by the QS molecules of a same family (ie. AHLs). The QS-mediated interspecies communication is called cross-talking, and its foreseeable ecological significance has brought about new approaches to research in microbiology. In natural habitats, the ability of certain species of bacteria to alter or disrupt the communication of other organisms possibly confers a competitive advantage for the colonization of niches and uptake of available nutrients [Zhang and Dong, 2004]. Cross-talking has implications in many areas of microbiology, as bacteria in nature are regularly part of complex communities composed of a wide biodiversity of microorganisms.

7.2. QS Involvement in Biofilm Formation and Infectious Diseases

A common mechanism by which bacteria function as a multi-species community is to form a biofilm, defined as a community of bacterial cells that is adhered to a surface (biotic or abiotic) or air-liquid interface, encapsulated in a self-produced extracellular polymeric matrix [Costerton *et al.*, 1995]. Biofilms are architecturally complex structures with distinctive morphology and physiology, compared to planktonic cells. Biofilm bacteria predominate, numerically and metabolically, in virtually all nutrient-sufficient ecosystems. Bacteria in biofilms live in a microhabitat that has primitive homeostasis, a primitive circulatory system, and metabolic cooperativity, and which provides protection against changes in the surrounding milieu. Biofilms are hence paradigm examples of complex communities where QS and cross-talk can have a strong influence [Stanley and Lazazzera, 2004].

Biofilms are a major concern for human activities. Biofilm bacteria have detrimental effects on industrial installations, being involved in biofouling of water systems, microbially-induced corrosion of pipes and tanks, degradation of stored industrial products, spoilage of processed foods and pharmaceutical drugs [Pasmore and Costerton, 2003, Coetser and Cloete, 2005]. The pathogenesis of infections associated with many medical devices is also related to microorganisms growing in biofilms. Biomaterials that are most likely to be compromised by biofilm associated infections are urethral or intravascular catheters and implants (i.e. joint prostheses, intraocular lenses, artificial heart valves, stents) [Costerton *et al.*, 2005, Ha and Cho, 2006]. Biofilm formation in such devices greatly hampers eradication of the involved flora by antibiotics, as diffusion inside the biofilm structure is limited, making them highly resistant to treatment, which makes them a reservoir for bacterial contamination. Besides, within the high dense bacterial population, resistance and virulence genes are very efficiently transferred [Costerton *et al.*, 2005].

As already summarized in Table 1., the formation of bacterial biofilms is commonly regulated by QS mechanisms. The QS system of the opportunistic pathogen *Pseudomonas aeruginosa* has been thoroughly analyzed (reviewed by de Kievit and Iglewsky, 2000, Bjarnsholt and Givskov, 2007]. *P. aeruginosa* mutants defective in AHL production form biofilms with altered morphology and characteristics, compared to wild type strains. Biofilm formation by *P. aeruginosa*, involves the expression of a wide number of genes, and the whole process is regulated by the *lasR* and *lasI* genes by means of a single AHL (3-oxo-C_{12}-HSL). However, there are complex interactions of the 3-oxo-C_{12}-HSL-based QS system with other QS signals (a second AHL-producing locus, PQS, and diketopiperazins, see Figure 1) [de Kievit and Iglewsky, 2000].

Many other human pathogens causing bacteraemia, meningitis, and a variety of oral, gastrointestinal, and cutaneous infections communicate by QS mechanisms, either under the control of AHLs, AIPs or AI-2 based systems (reviewed by Donabedian, 2003, Vendeville *et al.*, 2005). Cell density not only influences biofilm formation on invaded tissues, but also regulates important events during the infection process, such as the synthesis and secretion of virulence factors (Table 1). Recent research has lead to the proposal that interference with this communication system offers potential targets for the design of novel antimicrobial drugs. In the future, treatments that inhibit the transcription of biofilm control genes or attenuate bacterial pathogenicity (rather than inhibit bacterial growth) might provide a successful strategy for the inhibition of infections [Pasmore and Costerton, 2003, Bjarnsholt and Givskov, 2007]. Drugs directed to blocking of QS will not exert selective pressure and hence resistances are less likely to arise [Vendeville *et al.*, 2005]. The introduction of QS signal blockers in anti-infectious therapy is a very attractive new approach, as it will provide a particularly useful tool against strains of bacteria exhibiting multi-resistance to conventional antibiotics.

7.3. Quorum Quenching

As already pointed out in Section 6., some organisms synthesize compounds which interfere with the bacterial QS communication systems. In the case of pathogenic bacterial-host interactions, the ability to interfere bacterial communication is crucial in the prevention of infection by pathogens that use QS to coordinate expression of virulence factors. These tactics of signal interference are termed quorum-quenching.

The quorum-quenching strategies take place at both the eukaryote-to-prokaryote and prokaryote-to-prokaryote levels [Waters and Bassler, 2005]. The main mechanisms of quorum quenching identified so far are: (i) inhibition of the biosynthesis of the signal molecule, (ii) premature degradation of the signal molecule or the LuxR homologue regulatory protein (in the case of AHLs), (iii) competitive antagonism due to the synthesis of a compound structurally similar to the signal molecule, and (iv) inhibition of sensor kinases. Several recent papers thoroughly review the details of these quenching mechanisms [Zhang and Dong, 2004, Waters and Bassler, 2005, Bjarnsholt *et al.*, 2007, Dong *et al.*, 2007].

A number of bacterial species are able to enzymatically inactivate QS signals. AHLs are degraded by AHL-lactonases and AHL-acylases. AHL-lactonases cleave the HSL lactone rings, while AHL-acylases hydrolyze the amide bond, releasing the corresponding fatty acid and inactive HSL [Zhang and Dong, 2004, Waters and Bassler, 2005, Dong *et al.*, 2007]. These enzymes are one of the most promising tools to develop new control strategies for infectious diseases.

Structural mimics of QS signals act outcompeting the legitimate signal for receptor binding [Dong *et al.*, 2007]. The best known structural analogues of AHLs are furanones (see section 5., Figure 4.), either natural or synthetically produced. Furanones also act accelerating the degradation of LuxR [Manefield *et al.*, 2002]. Synthetic analogues of naturally produced furanones have been developed which are potent inhibitors of QS in *P. aeruginosa* [Bjarnsholt and Givskov, 2007].

II. QS IN PLANT-BACTERIAL ASSOCIATIONS

QS is a common feature of bacteria (either beneficial or deleterious) that live in association with plants, regulating a wide range of phenotypes involved in the plant-bacteria interactions, such as plasmid transfer, virulence factors, competence, and production of antifungal compounds [Newton and Fray, 2004]. Production of AHLs has been reported for plant-associated strains of the genera *Agrobacterium* (*A. radiobacter, A. rhizogenes, A. tumefaciens, A. vitis*), *Azospirillum* (*A. lipoferum*), *Burkholderia* (*Burkholderia* sp.), *Erwinia* (*E. amylovora, E. carotovora, E. chrysantemi, E. herbicola*), *Pantoea* (*P. stewartii*), *Pseudomonas* (*P. aureofaciens, P. chloraphis, P. corrugata, P. fluorescens, P. putida, P. savastanoi, P. syringae*), *Ralstonia* (*R. solanacearum*), *Rhizobium* (*R. etli, R. fredii, R. leguminosarum* bv. *viciae*, bv. *phaseoli*, bv. *trifolii*), *Serratia* (*S. plymuthica*), *Sinorhizobium* (*S. meliloti*), and *Xanthomonas* (*X. campestris, X. oryzae*) [Cha *et al.*, 1998, Elasri *et al.*, 2001, Lithgow *et al.*, 2001, Vial *et al.*, 2006, Barnard *et al.*, 2007, Licciardello *et al.*, 2007, Liu *et al.*, 2007, Poonguzhali *et al.*, 2007].

Elasri *et al.* (2001) tested the ability to produce AHLs of a collection of 137 *Pseudomonas* spp. strains isolated form soil and rhizosphere of plants, and interestingly concluded that the percentage of AHL-producers was significantly correlated with the degree of relationship with the plant, decreasing from 49% among plant-pathogenic bacteria to 28% and 0% among nonpathogenic bacteria associated with the plant and soilborne bacteria, respectively. These data pointed out the possible roles of QS in rhizosphere colonization, competitive ability and plant invasion, by the regulation of relevant traits such as EPS production, root adhesion and biofilm formation [von Bodman *et al.*, 1998, Denny, 1999, de Kievit *et al.*, 2001, Marketon *et al.*, 2003].

AHLs activate the QS-dependent secretion of a number of virulence factors in well-known plant pathogenesis, such as the induction of crown gall tumors on susceptible hosts plants by *Agrobacterium tumefaciens,* (regulated by the TraI/TraR system), and the secretion of exoenzymes which macerate the plant cell walls and cause soft-rot in potato and other plant hosts by *Erwinia carotovora* (regulated by the ExpI/ExpR-CarI/CarR systems) [Barnard *et al.*, 2007, White and Winans, 2007]. QS regulation of virulence in *Xanthomonas campestris*, in this case mediated by DFS (methyl dodecenoic acid) as autoinducer, is also well characterized [Fouhy *et al.*, 2006].

8. QS in Symbiotic Plant-Bacterial Interactions

Rhizobium, Mesorhizobium, Sinorhizobium, Azorhizobium and *Bradyrhizobium* species are known for their ability to establish symbiotic interactions with leguminous plants by the formation and colonization of root nodules, where bacteria fix N_2 to ammonia and make it available for the plant [Gage, 2004]. This symbiosis is based on the specific recognition of signal molecules, which are produced by both the bacterial and plant partners. In response to flavonoids secreted by the plant, bacteria express the so-called nodulation genes (*nod*), which encode for the production of a lipochitin oligosaccharide signal for nodulation [Hirsch *et al.*, 2001]. The Nod proteins are recognized by the plant, inducing *de novo* organogenesis leading to nodule formation. The bacteria invade the root through root hairs by an infection thread and differentiate into bacteroids inside the nodule, where they fix N_2 while in return they are supplied by the plant with carbon substrates [Hirsch, 1992, Kinkema *et al.*, 2006].

Rhizobium leguminosarum bv. *viciae* strains enter into a symbiosis with legumes such as pea and vetch [van Rhijn and Vanderleyden, 1995]. QS signals are found in several species of legume-nodulating bacteria, being the cell-to-cell communication systems of *Rhizobium leguminosarum* bv. *viciae*, *R. etli*, *Sinorhizobium meliloti* and *Bradyrhizobium japonicum* particularly well known [González and Marketon, 2002, Wisniewski and Downie, 2002, González and Keshavan, 2006, Sánchez-Contreras *et al.*, 2007]. Most of the identified QS regulation systems in rhizobia use AHLs as autoinducers, while these organisms seem to lack AI-2 production [Sánchez-Contreras *et al.*, 2007].

8.1. QS in *Rhizobium* and *Sinorhizobium*

8.1.1. R. leguminosarum

The first AHL described in rhizobia was 3-hydroxy-$C_{14:1}$-HSL [*N*-(3*R*-hydroxy-7-*cis*-tetradecenoyl)-L-homoserine lactone] (Fig. 2), made by *Rhizobium leguminosarum* bv. *viciae.* Previously to its purification and identification, this compound was detected for its bacteriostatic bacteriocin-like activity, and it was first described as a highly diffusible antimicrobial compound, able to produce big sized zones of inhibition of growth of sensitive strains on agar plates, and apparently was produced by most field isolates of *Rhizobium leguminosarum* [Hirsch, 1979; Wijffelman *et al.*, 1983]. The compound was named *small* bacteriocin due to its low molecular weight, which allowed diffusion through dialysis membrane tubing. All *small* non-producing *R. leguminosarum* bv. *viciae* strains harbored highly self-transmissible plasmids that were able to repress *small* production and excretion, as a producing phenotype was observed in strain derivatives cured of these plasmids, which also became insensitive to the bacteriocin [Wijffelman *et al.*, 1983]. Therefore, production of *small*, and the presence of genes for repression and sensitivity in self-transmissible plasmids, were considered a characteristic of the *R. leguminosarum* species [Gray *et al.*, 1996]. More recently, the structure of *small* bacteriocin made by *R. leguminosarum* bv. *viciae* strain A34 was purified from chloroform extracts by HPLC and NMR spectra, revealing that in fact it was a QS signal of the AHL type [Schripsema *et al.*, 1996].

In strain A34, the production of 3-hydroxy-$C_{14:1}$-HSL is encoded by the *cinI* gene, and its expression is regulated by *cinR*, being the cluster of these two genes located chromosomically [Lithgow *et al.*, 2000]. The processes that are regulated by quorum sensing in rhizobia are still not known in full, but current knowledge indicates that it is a rather complex system involving several *luxRI* homologues. Six regulators of the LuxR class have been identified in strain A34, which regulate gene expression in response to AHLs [Sánchez-Contreras *et al.*, 2007]. These different regulatory systems operate as a network in which there is cross regulation mediated via different AHLs (Table 2, Figure 6.).

The *cinR* and *cinI* genes encode a LuxR-type regulator and an AHL synthase, respectively, and are common to all *R. leguminosarum* bv. *viciae* strains analyzed to date, being always found in the chromosome. In strain A34, CinI produces 3-hydroxy-$C_{14:1}$-HSL, which positively autoregulates *cinI* expression in a CinR-dependent manner [Lithgow *et al.*, 2000]. CinR and CinI are at the top of a hierarchical cascade that regulates the expression of at least other three AHL synthases, distributed between the symbiotic plasmid pLR1JI, and a non-symbiotic megaplasmid [Rodelas *et al.*, 1999, Lithgow *et al.*, 2000, Wilkinson *et al.*, 2002, Wisniewski-Dyé *et al.*, 2002]. One of these pairs of *luxRI* homologues, carried by pRL1JI, is the *rhiRI* gene cluster, responsible for regulation of expression of the *rhiABC* operon [Rodelas *et al.*, 1999, Lithgow *et al.*, 2000]. The *rhiABC* operon is also localized to plasmid pRL1JI, positioned adjacent to the nodulation and nitrogen fixation genes, and it is expressed in a cell-density dependent way in the rhizosphere. Although *rhiR* was one of the earliest QS-regulators sequenced in bacteria, the role of the cell-density regulated Rhi proteins remains unclear to date, due to little observable effects on mutants and the absence in the databases of similar gene products with recognized biochemical functions [Wisniewski and Downie, 2002]. Mutations of the *rhiA* gene affect nodulation in mutant strains already impaired for nodulation due to the deletion of the *nodFEL* genes [Cubo *et al.*, 1992]. RhiA and RhiB appear to be cytoplasmic proteins, while RhiC is a predicted periplasmic protein. Remarkably, RhiA is a protein specific of *R. leguminosarum* bv. *viciae* (not detectable in strains of the other two biovars) which is present in large amounts in the cytoplasm of cells in the rhizosphere, but not after bacteria differentiate to bacteroids inside the legume nodules [Dibb *et al.*, 1984]. A role of the *rhi* genes linked to the association of *R. leguminosarum* bv. *viciae* with the specific legume host is presumed.

Table 2. AHL production in *Rhizobium leguminosarum*

Genes for AHL production/regulation	Location	AHLs produced	Phenotype regulated	References
rhiR/rhiI	symbiotic plasmid	C_6-HSL C_7-HSL C_8-HSL	Nodulation efficiency	Cubo *et al.* (1992), Rodelas *et al.* (1999)
cinR/cinI	chromosome	3-hydroxy-$C_{14:1}$-HSL	Growth inhibition	Lithgow *et al.* (2000)
raiR/raiI	non-symbiotic plasmid	3-hydroxy-C_8-HSL C_6-HSL C_7-HSL C_8-HSL	Unknown	Wisniewski-Dyé *et al.* (2002)
traR/traI	symbiotic plasmid	3-oxo-C_8-HSL[a] C_8-HSL[a]	Transfer of the symbiotic plasmid	Wilkinson *et al.* (2002), Danino *et al.* (2003)
expR	chromosome	Unknown	Unknown	Sánchez-Contreras *et al.* (2007)

[a]: AHLs identified only on the basis of their migratory behaviour in thin layer chromatography (TLC).

The second QS system localized to pRL1JI is *traRI*, which regulates expression of the *tra-trb* operon, responsible for conjugal transfer of pRL1JI, in analogous ways to the *Agrobacterium tumefaciens* model of Ti-plasmid transfer [Wilkinson *et al.*, 2002, Sánchez-Contreras *et al.*, 2007]. In the *A. tumefaciens* model, the *traR* gene is transcribed in response to activation by opines made by the host plant, and then TraR binds to AHL made by TraI, inducing the plasmid transfer operon in a cell density dependent manner. A negative regulatory protein (TraM) inhibits premature induction of plasmid transfer when the TraR levels are basal [White and Winans, 2007].

Adjacent to the *tra-trb* operon in pRL1JI lies another *luxR*-type transcriptional regulator named *bisR*, which controls *traR* expression and also acts as a QS relay to regulate recipient induced transcription of pRL1JI [Danino *et al.*, 2003]. *traR* expression is activated by BisR specifically in response to 3-hydroxy-$C_{14:1}$-HSL. In pRL1JI-carrying cells, BisR represses *cinI* transcription (and hence production of 3-hydroxy-$C_{14:1}$-HSL), keeping levels of TraR low and transcription of *tra-trb* inhibited by TraM. But when pRL1JI-carrying cells are part of a mixed rhizobial population, BisR can respond to external 3-hydroxy-$C_{14:1}$-HSL made by potential plasmid recipient cells, then activating the genes for conjugal transfer. The BisR-mediated regulatory mechanism results in very high plasmid transfer rates [Danino *et al.*, 2003]. This recipient-dependent regulation of the transference of symbiotic plasmids in *R. leguminosarum* seems to date an exclusive feature of pRL1JI, as in other *R. leguminosarum* bv. *viciae* strains *traRI* are not found on the symbiotic plasmid [Sánchez-Contreras *et al.*, 2007].

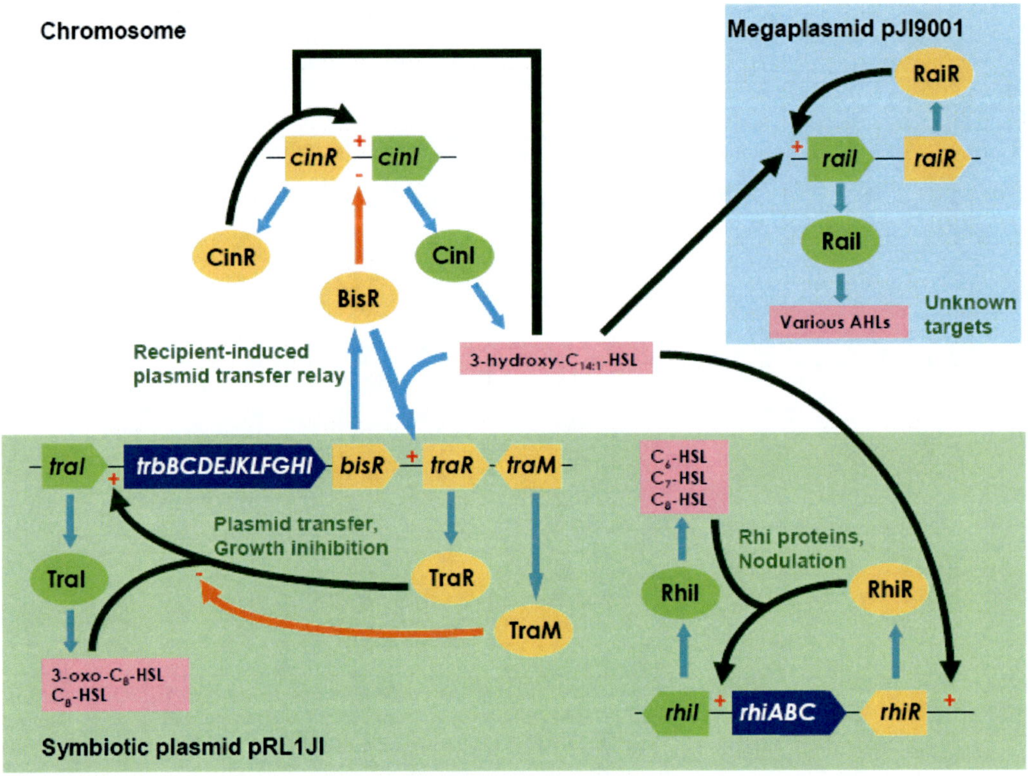

Figure 6. QS regulatory cascades in *R. leguminosarum* bv. *viciae* strain A34. Strain A34 has four well characterized QS systems. The *cinRI* system is chromosomal and responsible for the production of 3-hydroxy-$C_{14:1}$-HSL, which acts as a master inductor of the *tra, rhi* and *rai* systems, each of which has also its own autoinducing loop. The *tra* and *rhi* systems are localized to the symbiotic plasmid pRL1JI, together with genes that determine inhibition of growth under the regulation of 3-hydroxy-$C_{14:1}$-HSL. *tra-trb* genes are responsible for the conjugal transfer of pRL1JI, while the Rhi proteins play a role in nodulation. *raiRI* are localized to a megaplasmid (pIJ9001) and the genes regulated by this QS-system remain unknown. The BisR protein plays a double role, as it controls *traR* expression and hence induction of the *tra-trb* genes, but at the same time represses the expression of *cinI*, both in response to 3-hydroxy-$C_{14:1}$-HSL. Repression of *cinI* by BisR keeps low levels of 3-hydroxy-$C_{14:1}$-HSL production in strains already carrying pRL1JI (potential donors). Strains not carrying the pRL1JI plasmid (potential recipients) overproduce 3-hydroxy-$C_{14:1}$-HSL, which is detected by the pRL1JI-carrying strains, hereby inducing the expression of *tra-trb* genes and transfer of the plasmid to the recipient cells. See text for further details. The figure is based on data gathered from Cubo *et al.*, 1992, Rodelas *et al.*, 1999, Lithgow *et al.*, 2000, Wilkinson *et al.*, 2002, Wisniewski-Dyé *et al.*, 2002, Danino *et al.*, 2003.

Strain A34 carries a fourth QS-system (*raiIR*) under the control of 3-hydroxy-$C_{14:1}$-HSL, in a non-symbiotic megaplasmid [Wisniewsky-Dyé *et al.*, 2002]. Mutation of these genes does not show any effect on nodulation or symbiotic nitrogen fixation [Wisniewsky-Dyé *et al.*, 2002]. *raiIR* are absent from most analyzed *R. leguminosarum* bv. *viciae* strains [Sánchez-Contreras *et al.*, 2007] and its potentially regulated phenotypes are unknown [Wisniewski-Dyé *et al.*, 2002].

In brief, the quorum sensing regulatory cascade topped by the *cinRI* system in *R. leguminosarum* bv. *viciae* strain A34 has been shown related to the regulation of expression of the *rhi* genes and conjugative plasmid transfer, and appears to influence a variety of phenotypes including nodulation, expression of genes for adaptation to stress, entry into stationary phase, processing of exopolysaccharides, attachment, and biofilm production [Gray *et al.*, 1996, Rodelas *et al.*, 1999, Lithgow *et al.*, 2000, Wilkinson *et al.*, 2002, Wisniewski-Dyé and Downie, 2002, Danino *et al.*, 2003]. It was due to its ability to induce early entrance into stationary phase that 3-hydroxy-C14:1-HSL was mistaken for a bacteriostatic bacteriocin in early work. The role of AHLs in adaptation to stationary phase has also been reported for *R. leguminosarum* bv. *phaseoli* [Thorne and Williams, 1999]. Therefore, *small* is not a bacteriocin *per se*, but it acts antagonistically on sensitive strains inducing inhibitory functions, by mechanisms that are still under characterization. *cinR* and *cinI* mutants of strain A34 are unaffected in nodulation of pea and vetch [Lithgow *et al.*, 2000]

8.1.2. R. etli

The *cinRI* cluster was also isolated and sequenced in a *Rhizobium etli* strain (CNPAF512), and the autoinducer produced by the CinI protein is a saturated long chain AHL whose structure has not been totally elucidated, but it is seems to be different from *N*-(3*R*-hydroxy-7-*cis*-tetradecanoyl)-L-homoserine lactone, as it lacks at least the double bond [Daniels *et al.*, 2002]. This AHL also inhibits growth of other rhizobial strains in a bacteriocin-like way, and it has been demonstrated its implication in regulation of growth of *R. etli* strain CNPAF512, symbiosome development, and expression of nitrogen fixation in *Phaseolus vulgaris* nodules [Daniels *et al.*, 2002]. The autoinducer made by CinI also regulates swarming motility in strain CNPAF512 [Daniels *et al.*, 2004] and in addition to its signaling function, this long-chain AHL also has a direct role in surface movement of swarmer cells, as it has been demonstrated that these molecules possess biosurfactant activity [Daniels *et al.*, 2006]. Despite the high identity between the sequence of the CinRI proteins of *R. leguminosarum* and *R. etli* (over 95%), it seems clear that this QS system makes different signal compounds and regulates the expression of different phenotypes in each species.

Besides the long-chain AHL, two other autoinducers are synthesized by strain CNPA512 by a *railR* pair of genes, which have a high sequence similarity to those of *R. leguminosarum* bv. *viciae*. However, examination of different mutants of strain CNPAF512 for nodulation of beans showed an involvement of *raiI* in the restriction of nodule number, while nitrogenase activity remained unaffected. The nodulation suppression effect seems independent of RaiR [Daniels *et al.*, 2002].

In another strain of *R. etli* (CFN42), *cinRI* and *traRI* genes were also sequenced. Analysis of these QS systems does not show a direct involvement in the symbiotic process, although the *traRI* pair is involved in the regulation of the conjugative transfer of the symbiotic plasmid in strain CFN42 [Tun-Garrido *et al.*, 2003].

8.1.3. Sinorhizobium meliloti

Sinorhizobium meliloti strain Rm1021 carries at least two QS-systems [González and Marketon, 2002]. The *sinR/sinI* locus directs the production of several AHLs with a fatty acid side chain ranging from 12-18 C [Marketon *et al.*, 2002]. In Rm1021-derivative *sinI*⁻ mutants, short chain AHLs are produced, revealing the presence of at least a second QS system which has been named *mel* [Marketon *et al.*, 2002], but remains uncharacterized. Despite the genome of strain Rm1021 being fully sequenced, *in silico* searches failed to retrieve any *luxI* homologues. A gene homologous to HdtS is harbored, although its involvement in the synthesis of AHLs has not yet been assessed [González and Marketon, 2002]. When the *sinRI* genes are abolished, nodulation is delayed in *Medicago sativa*. In *S. meliloti* strains Rm41 and

Rm8530, *sinRI* genes are required for the synthesis of EPSs needed in the nodule infection steps [Marketon *et al.*, 2003].

S. meliloti strain Rm41 carries a third type of QS system, homologous to the *traRI* systems controlling plasmid transfer *in Agrobacterium* and *Rhizobium leguminosarum* bv. *viciae*. This system seems exclusive of strain Rm41 and is localized to a non-symbiotic plasmid [González and Marketon, 2002].

Most wild type strains of *S. meliloti* carry an *expR* gene, also encoding a LuxR-type regulator. *expR* defective mutants lack any nodulation phenotype, but interruption of *expR* influences EPS production and alters gene expression and protein levels in laboratory cultures [Marketon *et al.*, 2003, Sánchez-Contreras *et al.*, 2007].

8.2. QS in Bradyrhizobium japonicum

In contrast to *Rhizobium* and *Sinorhizobium*, no AHLs made by *Bradyrhizobium* have been purified nor characterized so far, although 22% of 142 *Bradyrhizobium* spp. isolates displayed AHLs-like activity when they were challenged in bioassays carried out with a QS-reporter strain [Pongsilp *et al.*, 2005]. *Bradyrhizobium japonicum* are the slow-growing bacterial symbionts of soy bean [Loh and Stacey, 2003]. In *B. japonicum* strain USDA110, it was demonstrated that the expression of the nodulation genes is repressed at high cell densities, evidencing the involvement of a QS mechanism in such regulation. A CDF named bradioxetin (Figure 1) has been characterized, which keeps structural similarities with the antibiotic oxetin and a siderophore (mugeneic acid) [Loh and Stacey, 2003]. Bradyoxetin synthesis is actually regulated by the available Fe^{+3} concentration, being maximally expressed under iron starvation conditions [Loh *et al.*, 2002].

In *B. japonicum*, expression of the nodulation genes in response to plant-made flavonoids is complexly regulated by several transcriptional activators [Loh and Stacey, 2003]. One of these proteins, NswB, is also involved in the QS regulated repression of the nodulation genes. NswB is the response regulator of a two-component system, which responds to cell density increases by detecting the rising of bradyoxetin concentrations. NswB response is the induction of *nolA*, which acts as a transcriptional activator of the *nodD2* gene [Loh and Stacey, 2002]. NodD2 is a regulatory protein required for the repression of nodulation genes. Thus, NwsB seems to work as a switch that results in either activation or repression of *nod* gene expression, depending on cell density [Loh and Stacey, 2003].

It is well established that the nodulation genes are not expressed in nodules, by a mechanism still unclear. Loh *et al.* (2002) hypothesized that the concentration of bacteroids inside the nodules reproduces a high cell density situation, thus leading to a QS-based repression of the *nod* genes. Conclusive evidence for the role of QS in bacteroids requests further studies using a QS-knockout mutant. To date, no mutants defective in bradyoxetin production are available, and the genes regulating its synthesis are yet unidentified. Many questions remain unanswered regarding the extent of QS implication in establishment and functioning of the *B. japonicum*-soy bean symbiosis.

III. CONCLUSION

The discovery of the QS mechanism as a widespread system for the control of synchronized gene expression in microorganisms in response to cell density has radically changed the conception of microbial communities, which are now understood as organized groups whose members are in constant communication and take global decisions, acting as multicellular organisms. The existence of many bacterial phenotypes of interest for humans that are regulated by intercellular communication provides a possibility to modify or modulate externally the bacterial responses, by interfering transmission and/or reception of their signal molecules. Advances in the knowledge and understanding of QS signaling systems, and identification of quorum quenching mechanisms in nature, will notably help to design new improved methods for the control of infectious diseases and biofilm formation. The study of the extent of the influence of QS systems in plant-bacterial interactions, and particularly on the establishment and performance of the rhizobia-legume symbiosis, is a task still in progress, as intercellular signaling is likely to play important roles for strain survival, colonization of plant roots, infection, and nodule development. Better understanding of these mechanisms may also allow the manipulation of QS systems for the improvement of the efficiency of inoculant strains.

IV. REFERENCES

Aendekerk, S., Diggle S.P., Song' Z., Høiby, N., Cornelis, P., Williams, P., and Cámara, M. (2005). The MexGHI-OpmD multidrug efflux pump controls growth, antibiotic susceptibility and virulence in *Pseudomonas aeruginosa* via 4-quinolone-dependent cell-to-cell communication. *Microbiology 151*,1113-1125.

Balaban, N., and Novick, R.P. (1995). Autocrine regulation of toxin synthesis by *Staphylococcus aureus. Proceedings of the National Academy of Sciences of the USA 92*,1619-1623.

Barnard, A. M. L., Bowden, S. D., Burr, T., Coulthurst, S. J., Monson, R. E., and Salmond, G. P. C. (2007). Quorum sensing, virulence and secondary metabolite production in plant soft-rotting bacteria. *Philosophical Transactions of the Royal Society B 362*, 1165–1183.

Bassler, B. L., Wright, M., Showalter, R. E., and Silverman, M. R. (1993). Intercellular signaling in *Vibrio harveyi* - sequence and function of genes regulating expression of luminescence. *Molecular Microbiology 9*, 773–786.

Bauer, W.D., and Robinson, J.B. (2002) Disruption of bacterial quorum sensing by other organisms. *Current Opinion in Biotechnology 13,* 234–237.

Bauer, W.D., and Mathesius, U. (2004). Plant responses to bacterial quorum sensing signals. *Current Opinion in Plant Biology 7*, 429-233.

Bjarnsholt, T., and Givskov, M. (2007) Quorum-sensing blockade as a strategy for enhancing host defences against bacterial pathogens. *Philosophical Transactions of the Royal Society B* 362, 1213-1222.

Camilli, A., and Bassler, B.L. (2006). Bacterial small-moleculre signaling pathways. *Science 311*, 1113-1116.

Cao, J.G., and Meigen, E.A. (1989). Purification and structural identification of an autoinducer for the luminescence system of *Vibrio harveyi. Journal of Biological Chemistry 264*, 21670-21676.

Cha, C., Gao, P., Chen, Y.C., Shaw, P.D., and Farrand, S.K. (1998). Production of acyl-homoserine lactone quorum-sensing signals by Gram-negative plant associated bacteria. *Molecular Plant-Microbe Interactions 11*, 1119-1129.

Chen, X., Schauder, S., Potier, N., van Dorsselaer, A., Pelczer, I., Bassler, B.L., and Hughson, F.M. (2002). Structural identification of a bacterial quorum-sensing signal containing boron. *Nature 415*, 545–49.

Chhabra, S.R., Philipp, B., Eberl, L., Givskov, M., Williams, P., and Cámara, M. (2005). Extracellular communication in bacteria. *Topics in Current Chemistry 240*, 279-315.

Chun, C.K., Ozer, E.A., Welsh, M.J., Zabner, J., and Greenberg, E.P. (2004) Inactivation of a *Pseudomonas aeruginosa* quorum-sensing signal by human airway epithelia. *Proceedings of the National Academy of Sciences of the USA 101*, 3587–3590.

Coetser, S.E., and Cloete, T.E. (2005). Biofouling and biocorrosion in industrial water systems. *Critical Reviews in Microbiology 31*, 213-32.

Costerton, J.W., Lewandowski, Z., Caldwell, D.E., Korber, D.R., and Lappin-Scott, H.M. (1995). Microbial biofilms. *Annual Review of Microbiology 49*, 711-745.

Costerton, J.W., Montanaro, L., and Arciola, C.R. (2005). Biofilm in implant infections: its production and regulation. *International Journal of Artificial Organs 28*,1062-1068

Cubo, M. T., Economou, A., Murphy, G., Johnston, A.W., and Downie, J. A. (1992). Molecular characterization and regulation of the rhizosphere-expressed genes *rhiABCR* that can influence nodulation by *Rhizobium leguminosarum* biovar *viciae. Journal of Bacteriology 174*, 4026–4035.

Cvitkovitch, D.G., Li, Y.H., and Ellen, R.P. (2003). Quorum sensing and biofilm formation in Streptococcal infections. *Journal of Clinical Investigation 112*, 1626–1632.

Daniels, R., De Vos, D.E., Desair, J., Raedschelders, G., Luyten, E., Rosemeyer, V., Verreth, C., Schoeters, E., Vanderleyden, J., and Michiels, J. (2002). The *cin* quorum sensing locus of *Rhizobium etli* CNPAF512 affects growth and symbiotic nitrogen fixation. *Journal of Biological Chemistry 277*, 462–468.

Daniels, R., Reynaert, S., Hoekstra, H., Verreth, C., Janssens, J., Braeken, K., Fauvart, M., Beullens, S., Heusdens, C., Lambrichts, I., De Vos, D.E., Vanderleyden, J., Vermant, J., and Michiels, J. (2006). Quorum signal molecules as biosurfactants affecting swarming in *Rhizobium etli. Proceedings of the National Academy of Sciences of the USA 103*, 14965-14970.

Daniels, R., Vanderleyden, J., and Michiels, J. (2004). Quorum sensing and swarming migration in bacteria. *FEMS Microbiology Reviews 28*, 261-289.

Danino, V. E., Wilkinson, A., Edwards, A., and Downie, J. A. (2003). Recipient-induced transfer of the symbiotic plasmid pRL1JI in *Rhizobium leguminosarum* bv. *viciae* is regulated by a quorum-sensing relay. *Molecular Microbiology 50*, 511–525.

de Keersmaecker, S.C.J., Sonck, K. and Vanderleyden, J. (2006). Let LuxS speak up in AI-2 signaling. *Trends in Microbiology 14*, 114-119

de Kievit, T.R., and Iglewsky, B. H. (2000). Bacterial quorum sensing in pathogenic relationships. *Infection and Immunity 68*, 4839-4849.

de Kievit, T.R., Gillis, R., Marx, S., Brown, C., and Iglewski, B.H. (2001). Quorum-sensing genes in *Pseudomonas aeruginosa* biofilms: their role and expression patterns. *Applied and Environmental Microbiology 67,* 1865–1873.

Denny, T.P. (1999). Autoregulator-dependent control of extracellularpolysaccharide production in phytopathogenic bacteria. *Europear Journal of Plant Pathology 105,* 417–430.

Devine, J.H., Countryman, C., and Baldwin, T.O. (1988). Nucleotide sequence of the *luxR* and *luxI* genes and structure of the primary regulatory region of the *lux* regulon of *Vibrio fisheri* ATCC 7744. *Biochemistry 27*, 837-842.

Dibb, N.J., Downie, J.A., and Brewin, N.J. (1984). Identification of a rhizosphere protein encoded by the symbiotic plasmid of *Rhizobium leguminosarum. Journal of Bacteriology 158*, 621-627.

Donabedian, H. (2003). Quorum sensing and its relevance to infectious diseases. *Journal of Infection 46*, 207-214.

Dong, Y.H., Wang, L.H., and Zhang, L.H. (2007). Quorum-quenching microbial infections: mechanisms and implications. *Philosophical Transactions of the Royal Society B, 362*, 1201-1211.

Eberhard, A., Burlingame, A.L., Eberhard, C., Kenyon, G.L., Nealson, K.H., and Oppenheimer, N.J. (1981). Structural identification of autoinducer of *Photobacterium fischeri* luciferase. *Biochemistry 20*, 2444-2449.

Elasri, M., Delorme, S., Lemanceau, P., Stewart, G., Laue, B., Glickmann, E., Oger, P. M. and Dessaux, Y. (2001). Acylhomoserine lactone production is more common among plant-associated *Pseudomonas* spp. than among soilborne *Pseudomonas* spp. *Applied and Environmental Microbiology 67*, 1198–1209.

Engebrecht, J., and Silverman, M. (1984). Identification of genes and gene products necessary for bacterial bioluminescence. *Proceedings of the National Academy of Sciences of the USA 81*, 4154–4158.

Engebrecht, J., and Silverman, M. (1987). Nucleotide sequence of the regulatory locus controlling the expression of bacterial genes for bioluminescence. *Nucleic Acids Research 15,* 10455-10467.

Engebrecht, J., Nealson, K., and Silverman, M. (1983). Bacterial bioluminescence: isolation and genetic analysis of functions from *Vibrio fischeri. Cell 32*, 773–81.

Fouhy, Y., Lucey, J.F., Ryan, R.P., and Dow, J.M. (2006). Cell-cell signaling, cyclic di-GMP turnover and regulation of virulence in *Xanthomonas campestris. Research Microbiology 157*, 899-904.

Fuqua, W.C., Winans, S.C., and Greenberg, E.P. (1994). Quorum sensing in bacteria: the LuxR-LuxI family of cell-density responsive transcriptional regulators. *Journal of Bacteriology 176,* 269-275.

Gage, D.J. (2004). Infection and invasion of roots by symbiotic, nitrogen-fixing rhizobia during nodulation of temperate legumes. *Microbiology and Molecular Biology Reviews 68,* 280-300.

Gilson., L., Kuo, A., and Dunlap, P.V. (1995). AinS, a new family of autoinducer synthesis proteins. *Journal of Bacteriology 177*, 6946-6951.

Givskov, M., Nys, R.D., Manefield, M., Gram, L., Maximilien, R., Eberl, L., Molin, S., Steinberg, P.D., and Kjelleberg. (1996). Eukaryotic interference with homoserine lactone-mediated prokaryotic signalling. *Journal of Bacteriology 178*, 6618-6622.

González, J.E., and Marketon, M.M. (2003). Quorum sensing in nitrogen fixing rhizobia. *Microbiology and Molecular Biology Reviews 67*, 574-592.

González, J.E., and Keshavan, N. (2006). Messing with bacterial quorum sensing. *Microbiology and Molecular Biology Reviews 70,* 859-875.

Gray, K.M., Pearson, J.P., Downie, J.A., Boboye, B.E.A., and Greenberg, E.P. (1996). Cell-to-cell signaling in the symbiotic nitrogen-fixing bacterium *Rhizobium leguminosarum*: autoinduction of a stationary phase and rhizosphere-expressed genes. *Journal of Bacteriology 178*, 372-376.

Gray, K.M., and Garey, J.R. (2001). The evolution of bacterial LuxI and LuxR quorum sensing regulators. *Microbiology 147*, 2379-2387.

Grossman, A.D. (1995). Genetic networks controlling the initiation of sporulation and the development of genetic competence in *Bacillus subtilis. Annual Review of Genetics 29*, 477-508.

Ha, U.S., and Cho, Y.H. (2006). Catheter-associated urinary tract infections: new aspects of novel urinary catheters. *International Journal of Antimicrobial Agents 28*, 485-490

Hanzelka, B. L., andGreenberg, E.P. (1996). Quorum sensing in *Vibrio fischeri*: evidence that *S*-adenosylmethionine is the amino acid substrate for autoinducer synthesis. *Journal of Bacteriology 178,* 5291–5294.

Hardman, A.M., Stewart, G.S., and Williams, P. (1998). Quorum sensing and the cell-cell communication dependent regulation of gene expression in pathogenic and non-pathogenic bacteria. *Antonie van Leeuwenhoek 74*, 199-210.

Hastings, J.W. (2004). Bacterial quorum-sensing signals are inactivated by mammalian cells. *Proceedings of the National Academy of Sciences of the USA 101*, 3993-3994.

Håvarstein, L.S., Diep, D.B., and Nes, I.F. (1995). A family of bacteriocin ABC transporters carry out proteolytic processing of their substrates concomitant with export. *Molecular Microbiology 16*, 229–240.

Håvarstein, L.S., Hakenbeck, R., and Gaustad, P. (1997). Natural competence in the genus *Streptococcus*: evidence that streptococci can change phenotype by interspecies recombinational exchanges. *Journal of Bacteriology 179*, 6589-94

Henke, J.M, and Bassler, B.L. (2004). Three parallel quorum-sensing systems regulate gene expression in *Vibrio harveyi. Journal of Bacteriology 186*, 6902-6914.

Hense, B.A., Kuttler, C., Müller, J., Rothballer, M., Hartman, A., and Kreft, J.U. (2007). Does efficiency sensing unify diffusion and quorum sensing? *Nature Reviews Microbiology 5*, 230-237.

Hirsch, A. M. (1992). Developmental biology of legume nodulation. *New Phytologist 122*, 211–237.

Hirsch, A.M., Lum, M.R., and Downie, J.A. (2001). What makes the Rhizobia-legume symbiosis so special? *Plant Physiology 127*, 1484-1492.

Hirsch, P.R. (1979). Plasmid-determined bacteriocin production by *Rhizobium leguminosarum. Journal of General Microbiology 113*, 219-228.

Hoang, T.T., Sullivan, S.A., Cusick, J.K., and Schweizer, H.P. (2002). Beta-ketoacyl acyl carrier protein reductase (FabG) activity of the fatty acid biosynthetic pathway is a determining factor of 3-oxo-homoserine lactone acyl chain lengths. *Microbiology 148,* 3849–3856.

Holden, M., Swift, S., and Williams, P. (2000). New signal molecules on the quorum-sensing block. *Trends in Microbiology 8*, 101-103.

Hui, F.M., and Morrison, D.A. (1991). Genetic transformation in *Streptococcus pneumoniae*: nucleotide sequence analysis shows *comA*, a gene required for competence induction, to be a member of the bacterial ATP-dependent transport protein family. *Journal of Bacteriology 173*, 372–381.

Ji, G., Beavis, R.C., and Novick, R.P. (1995). Cell density control of staphylococcal virulence mediated by an octapeptide pheromone. *Proceedings of the National Academy of Sciences of the USA 92*, 12055-12059

Kaplan, H.B., and Greemberg, E.P. (1985). Diffusion of autoinducer is involved in regulation of the *Vibrio fisheri* bioluminescence system. *Journal of Bacteriology 163*, 1210-1214.

Kinkema, M., Scott, P.T., and Gresshoff, P.M. (2006). Legume nodulation: successful symbiosis through short and long-distance signalling. *Functional Plant Biology 33*, 707–721.

Kleerebezem, M., Quadri, L.E., Kuipers, O.P., and de Vos W.M. (1997) Quorum sensing by peptide pheromones and two-component signal-transduction systems in Gram-positive bacteria. *Molecular Microbiology 24*, 895-904.

Kreth, J., Merritt, J., Zhu, L., Shi, W., and Qi, F. (2006). Cell density- and ComE-dependent expression of a group of mutacin and mutacin-like genes in *Streptococcus mutans FEMS Microbiology Letters 265*, 11-17.

Kreth, J., Hung, D.C., Merritt, J., Perry, J., Zhu, L., Goodman, S.D., Cvitkovitch, D.G., Shi, W., and Qi, F. (2007). The response regulator ComE in *Streptococcus mutans* functions both as a transcription activator of mutacin production and repressor of CSP biosynthesis *Microbiology 153*, 1799-17807

Krin, E., Chakroun, N., Turlin, E., Givaudan, A., Gaboriau, F., Bonne, I., Rousselle, J.C., Frangeul, L., Lacroix, C., Hullo, M.F., Marisa, L., Danchin, A., and Derzelle, S. (2006). Pleiotropic role of quorum-sensing autoinducer 2 in *Photorhabdus luminescens*. *Applied and Environmental Microbiology 72*, 6439-6451.

Kuo, A., Callahan, S.M., and Dunlap, P.V. (1996). Modulation of luminescence operon expression by *N*-octanoyl-L-homoserine lactone in *ainS* mutants of *Vibrio fischeri*. *Journal of Bacteriology 178*, 971-976.

Laue, B. E., Jiang, Y., Chhabra, S. R., Jacob, S., Stewart,G. S. A. B., Hardman, A., Downie, J. A., O'Gara, F., andWilliams, P. (2000). The biocontrol strain *Pseudomonas fluorescens* F113 produces the *Rhizobium* small bacteriocin *N*-(3-hydroxy-7-cis-tetradecenoyl) homoserine lactone via HdtS, a putative novel *N*-acylhomoserine lactone synthase. *Microbiology 146*, 2469–2480.

Lazazzera, B.A., and Grossman, A.D. (1998). The ins and outs of peptide signaling. *Trends in Microbiology 6*, 288-294.

Lazazzera, B.A., Solomon, J.M., and Grossman, A.D. (1997). An exported peptide functions intracellularly to contribute to cell density signaling in *B. subtilis*. *Cell 89*, 917-925.

Licciardello, G., Bertani, I., Steindler, L., Bella, P., Venturi, V., Catara, V. (2007). *Pseudomonas corrugata* contains a conserved *N*-acyl homoserine lactone quorum sensing system; its role in tomato pathogenicity and tobacco hypersensitivity response. *FEMS Microbiology Ecology 61*, 222-234.

Lithgow, J.K., Wilkinson, A., Hardman, A., Rodelas, B., Wisniewski-Dyé, F., Williams, P., and Downie., J.A. (2000). The regulatory locus *cinRI* in *Rhizobium leguminosarum* controls a network of quorum-sensing loci. *Molecular Microbiology 37*, 81-97.

Lithgow, J.K., Danino, V.E., Jones, J., and Downie, J.A. (2001). Analysis of *N*-acyl-homoserine lactone quorum-sensing molecules made by different strains and biovars of *Rhizobium leguminosarum* containing different symbiotic plasmids. *Plant and Soil 232*, 3-12.

Liu, X., Bimerew, M., Ma, Y., Müller, H., Ovadis, M., Eberl, L., Berg, G., and Chernin, L. (2007). Quorum-sensing signaling is required for production of the antibiotic pyrrolnitrin in a rhizospheric biocontrol strain of *Serratia plymuthica. FEMS Microbiology Letters 270*, 299-305.

Loh, J., and Stacey, G. (2003). Nodulation gene regulation in *Bradyrhizobium japonicum*: a unique integration of global regulatory circuits. *Applied and Environmental Microbiology 69*,10-17.

Loh, J., Carlson, R.W., York, W.S., and Stacey, G. (2002). Bradyoxetin, a unique chemical signal involved in symbiotic gene regulation. *Proceedings of the National Academy of Sciences of the USA 99*, 14446-14451.

Magnuson, R., Solomon, J., and Grossman, A.D. (1994). Biochemical and genetic characterization of a competence pheromone from *Bacillus subtilis. Cell 77*, 207-216.

Manefield, M., Rasmussen, T.B., Hentzer, M., Andersen, J.B., Steinberg, P., Kjelleberg, S., and Givskov, M. (2002). Halogenated furanones inhibit quorum sensing through accelerated LuxR turnover. *Microbiology 148*, 1119-1127.

Marketon, M. M., Gronquist, M. R., Eberhard, A. and Gonzalez, J. E. (2002). Characterization of the *Sinorhizobium meliloti sinR/sinI* locus and the production of novel *N*-acyl homoserine lactones. *Journal of Bacteriology 184*, 5686–5695.

Marketon, M. M., Glenn, S. A., Eberhard, A. and Gonzalez, J. E. (2003). Quorum sensing controls exopolysaccharide production in *Sinorhizobium meliloti. Journal of Bacteriology 185*, 325–331.

Mathesius, U., Mulders, S., Gao, M., Teplitski, M., Caetano-Anolles, G., Rolfe, B. G. and Bauer, W. D. (2003) Extensive and specific responses of a eukaryote to bacterial quorum sensing signals. *Proceedings of the National Academy of Sciences of the USA 100*, 1444–1449.

Miller, M.B., and Bassler, B.L. (2001). Quorum sensing in bacteria. *Annual Review of Microbiology 55*, 165-199

Milton, D. L., Chalker, V. J., Kirke, D., Hardman, A., Cámara, M., and Williams, P. (2001). The LuxM homologue VanM from *Vibrio anguillarum* directs the synthesis of *N*-(3-hydroxyhexanoyl) homoserine lactone and *N*-hexanoylhomoserine lactone. *Journal of Bacteriolgy 183*, 3537–3547.

Milton, D.L. (2006). Quorum sensing in vibrios: complexity for diversification. *International Journal of Medical Microbiology 296*, 61-71.

Newton, J.A., and Fray, R.G. (2004). Integration of environmental and host-derived signals with quorum sensing during plant-microbe interactions. *Cell Microbiology 6*, 213-214.

Pasmore, M, and Costerton, J.W. (2003). Biofilms, bacterial signalling, and their ties to marine biology. *Journal of Industrial Microbiology and Biotecnology 30*, 407-413.

Pearson, J.P., van Delden, C., and Iglewsky, B.H. (1999). Active efflux and diffusion are involved in transport of *Pseudomonas aeruginosa* cell-to-cell signals. *Journal of Bacteriology 181*,1203-1210.

Perego, M., Hanstein, C., Welsh, K.M., Djavakhishvili, T., Glaser, P., and Hoch, J.A. (1994). Multiple protein-aspartate phosphatases provide a mechanism for the integration of diverse signals in the control of development in *Bacillus subtilis. Cell 79*, 1047-1055.

Pestova, E.V., Havarstein, L.S., and Morrison, D.A. (1996). Regulation of competence for genetic transformation in *Streptococcus pneumoniae* by an auto-induced peptide pheromone and a two-component regulatory system. *Molecular Microbiology 21*, 853–862.

Peterson, S.N., Sung, C.K., Cline, R. Desai, B.V., Snesrud, E.C., Luo, P., Walling, J., Li, H., Mintz, M., Tsegaye, G., Burr, P.C., Do, Y., Ahn, S., Gilbert, J., Fleischmann, R.D., and Morrison, D.A. (2004). Identification of competence pheromone responsive genes in *Streptococcus pneumoniae* by use of DNA microarrays. *Molecular Microbiology 51*, 1051–1070.

Podbielski, A., and Kreikemeyer, B. (2004). Cell density-dependent regulation: basic principles and effects on the virulence of Gram-positive cocci. *International Journal of Infection Diseases 8,* 81-95.

Pongsilp, N., Triplett, E.W., and Sadowsly, M.J. (2005). Detection of homoserine lactone-like quorum sensing molecules in *Bradyrhizobium* strains. *Current Microbiology 51*, 250-254.

Poonguzhali, S., Madhaiyan, M., and Sa, T. (2007).Quorum-sensing signals produced by plant-growth promoting *Burkholderia* strains under *in vitro* and *in planta* conditions. *Research Microbiology 158*, 287-294.

Redfield, R.J. (2002). Is quorum sensing a side effect of diffusion sensing? *Trends in Microbiology 10*, 365-370.

Rodelas, B., Lithgow, J.K, Wisniewski-Dyé., F., Hardmann, A., Wilkinson, A., Economou, A., Williams, P., and Downie, J.A. (1999). Analysis of quorum-sensing dependent control of rhizophere-expressed (*rhi*) genes in *Rhizobium leguminosarum* bv. *viciae. Journal of Bacteriology 181*, 3816-3823.

Sánchez-Contreras, M. Bauer, W.D., Gao, M., Robinson, J.B., and Downie, J.A. (2007). Quorum-sensing regulation in rhizobia and its role in symbiotic interactions with legumes. *Philosophical Transactions of the Royal Society B, 362*, 1149-1163.

Schauder, S., and Bassler, B.L. (2001). The languages of bacteria. *Genes and Development 15*, 1468-1480

Schripsema, J., de Rudder, K. E. E., van Vliet, T. B., Lankhorst, P. P., de Vroom, E., Kijne, J. W., and van Brussel, A. A. N. (1996). Bacteriocin small of *Rhizobium leguminosarum* belongs to the class of *N*-acyl-L-homoserine lactone molecules, known as autoinducers and as quorum sensing co-transcription factors. *Journal of Bacteriology 178*, 366–371.

Semmelhack, M.F., Campagna, S.R., Federle, M.J., and Bassler, B.L. (2005). An expeditious synthesis of DPD and boron binding studies. *Organic Letters 7*, 569–572.

Sitnikov, D.M., Schineller, J.B., and Baldwin, T.O. (1995) Transcriptional regulation of bioluminesence genes from *Vibrio fischeri. Molecular Microbiology 17*, 801-812

Søgaard-Andersen, L., Overgaard, M., Lobedanz, S., Ellehauge, E., Jelsbak, L., and Rasmussen, A.A. (2003). Coupling gene expression and multicellular morphogenesis during fruiting body formation in *Myxococcus xanthus. Molecular Microbiology 48*,1-8.

Solomon, J.M., Magnuson, R., Srivastava, A., and Grossman, A.D. 1995. Convergent sensing pathways mediate response to two extracellular competence factors in *Bacillus subtilis. Genes and Development 9*, 547-558

Sperandio, V., Torres, A.G., Jarvis, B., Nataro, J.P., and Kaper, J.B. (2003). Bacteria-host communication: the language of hormones. *Proceedings of the National Academy of Sciences of the USA 100*, 8951-8956.

Sprague, G.F., and Winans, S.C. (2006). Eukaryotes learn how to count: quorum sensing by yeast. *Genes and Development 20*, 1045-1049.

Stanley, N.R., and Lazazzera, B.A. (2004). Environmental signals and regulatory pathways that influence biofilm formation. *Molecular Microbiology 53*, 917-924.

Stevens, A.M.N., Dolan, K.M., and Greenberg, E.P. (1994). Synergistic binding of the *Vibrio fisheri* LuxR transcriptional activator domain and RNA polymerase to the *lux* promoter region. *Proceedings of the National Academy of Sciences of the USA 91*, 12619-12623.

Sun, J., Daniel, R., Wagner-Döbler, I., and Zeng, A.P. (2004) Is autoinducer-2 a universal signal for interspecies communication: a comparative genomic and phylogenetic analysis of the synthesis and signal transduction pathways. *BCM Evolutionary Biology 4,* 36.

Teplitski, M., Robinson, J. B. and Bauer, W. D. (2000). Plants secrete substances that mimic bacterial *N*-acyl homoserine lactone signal activities and affect population density dependent behaviors in associated bacteria. *Molecular Plant Microbe Interactions 13*, 637–648.

The Quorum Sensing Site. http://www.nottingham.ac.uk/quorum

Thorne, S. H., and Williams, H. D. (1999). Cell density dependent starvation survival of *Rhizobium leguminosarum* bv. *phaseoli*: identification of the role of an *N*-acyl homoserine lactone in adaptation to stationary-phase survival. *Journal of Bacteriology 181*, 981–990.

Tomasz, A. and Mosser, J.L. (1966). On the nature of the pneumococcal activator substance. *Proceedings of the National Academy of Sciences of the USA 55*, 58–66.

Tomasz, A. (1965). Control of the competent state in *Pneumococcus* by a hormone-like cell product: an example for a new type of regulatory mechanism in bacteria. *Nature 208,* 155–159.

Tomasz, A., and Hotchkiss, R.D. (1964). Regulation of the transformability of pneumococcal cultures by macromolecular cell products. *Proceedings of the National Academy of Sciences of the USA 51*, 480–487.

Tun-Garrido, C., Bustos, P., González, V., and Brom, S. (2003). Conjugative transfer of p42a from *Rhizobium etli* CFN42, which is required for mobilization of the symbiotic plasmid, is regulated by quorum sensing. *Journal of Bacteriology 185*, 1681-1692.

Turgay, K., Hahn, J., Burghoorn, J., and Dubnau, D. (1998). Competence in *Bacillus subtilis* is controlled by regulated proteolysis of a transcription factor. *EMBO Journal 17*, 6730-6738.

Turgay, K., Hamoen, L.W., Venema, G., and Dubnau, D. (1997). Biochemical characterization of a molecular switch involving the heat shock protein ClpC, which controls the activity of ComK, the competence transcription factor of *Bacillus subtilis. Genes and Development 11*, 119-128.

van Rhijn, P., and Vanderleyden, J. (1995). The *Rhizobium*-plant symbiosis. *Microbiological Reviews 59*,124-142.

van Sinderen, D., Luttinger, A., Kong, L., Dubnau, D., Venema, G., and Hamoen, L. (1995). *comK* encodes the competence transcription factor, the key regulatory protein for competence development in *Bacillus subtilis. Molecular Microbiology 15*, 455-462

Vendeville, A., Winzer, K., Heurlier, K., Tang, C.M., Hardie, K.R. (2005). Making "sense" of metabolism: autoinducer-2, LuxS and pathogenic bacteria. *Nature Reviews Microbiology 3*, 383-396.

Vial, L., Cuny, C., Gluchoff-Fiasson, K., Comte, G., Oger, P.M., Faure, D., Dessaux, Y., Bally, R., Wisniewski-Dyé, F. (2006). *N*-acyl-homoserine lactone-mediated quorum-sensing in *Azospirillum*: an exception rather than a rule. *FEMS Microbiology Ecology 58*, 155-68.

von Bodman, S.B., Majerczak, D.R., and Coplin, D.L. (1998). A negative regulator mediates quorum-sensing control of exopolysaccharide production in *Pantoea stewartii* subsp. *stewartii*. *Proceedings of the National Academy of Sciences of the USA 95*, 7687–7692.

Waters, C.M, and Bassler, B.L. (2005). Quorum sensing: cell to cell communication in bacteria. *Annual Review of Cell and Developmental Biology 21*, 319-346.

Watson, W.T., Minogue, T.D., Val, D.L., von Bodman, S.B., and Churchill, M.E.A. (2002). Structural basis and specificityof acyl-homoserine lactone signal production in bacterial quorum sensing. *Molecular Cell 9*, 685–694.

Whitehead, N., Barnard, A.M.L., Slater, H., Simpson, N.J.L., and Salmond, G.P.C. (2001). Quorum sensing in Gram-negative bacteria. *FEMS Microbiology Reviews 25*, 365-404.

Wijffelman, C.A., Pees, E., van Brussel, A.A.N., and Hooykas, P.J.J. (1983). Repression of *small* bacteriocin excretion in *Rhizobium leguminosarum* and *Rhizobium trifolii* by transmissible plasmids. *Molecular and General Genetics 192*, 171-176.

Wilkinson, A., Danino, V., Wisniewski-Dye, F., Lithgow,J. K., and Downie, J. A. (2002) *N*-acyl-homoserinelactone inhibition of rhizobial growth is mediated by two quorum-sensing genes that regulate plasmid transfer. *Journal of Bacteriology 184*, 4510–4519.

Williams, P., Winzer, K., Chan, W.C., and Cámara, M. (2007). Look who's talking: communication and quorum sensing in the bacterial world. *Philosophical Transactions of the Royal Society B, 362*,1119-1134.

Winzer, K., Hardie, K.H., and Williams, P. (2002). Bacterial cell-to-cell communication: sorry, can't talk now - gone to lunch! *Current Opinion in Microbiology 5*, 216-222.

Wisniewsky-Dyé, F., and Downie, J.A. (2002). Quorum-sensing in *Rhizobium. Antonie van Leeuwenhoek 81*, 397-407.

Wisniewski-Dye, F., Jones, J., Chhabra, S. R., and Downie, J. A. (2002) railR Genes are part of a quorum-sensing network controlled by *cinI* and *cinR* in *Rhizobium leguminosarum. Journal of Bacteriology 184*, 1597–1606.

Xavier, K.B., and Bassler, B.L. (2003). LuxS quorum sensing: more than just a numbers game. *Current Opinion in Microbiology* 6, 191–217.

Zhang, C.C., Laurent, S., Sakr, S., Peng, L., and Bédu, S. (2006). Heterocyst differentiation and pattern formation in cyanobacteria: a chorus of signals. *Molecular Microbiology 59*, 367-75.

Zhang, L.H., and Dong, Y.H. (2004). Quorum sensing and signal interference: diverse implications. *Molecular Microbiology 53*,1563-1571. Balaban, N., and Novick, R.P. (1995). Autocrine regulation of toxin synthesis by *Staphylococcus aureus. Proceedings of the National Academy of Sciences of the USA 92*,1619-1623.

Zhou, L., Hui, F.M., and Morrison, D.A. (1995). Characterization of IS1167, a new insertion sequence in *Streptococcus pneumoniae*. *Plasmid 33*, 127–138.

In: Soil Ecology Research Developments
Editors: Tian-Xiao Liu, pp. 121-138
ISBN: 978-1-60021-971-9

Chapter 4

ECOLOGY AND FUNCTION OF CULTURABLE MICROBES IN SOIL AGGREGATION

T.C. Caesar-TonThat, A.J. Caesar, J.F. Gaskin, U.M. Sainju and W.B. Stevens
United States Department of Agriculture, Agricultural Research Service,
Sidney, MT 59270, USA

ABSTRACT

Soil structure plays a dominant role in the physical protection of soil organic matter by controlling microbial access to substrate, microbial turnover processes, and food web interactions. Good soil structure results in soil productivity, a cornerstone of agricultural sustainability. While there is a wealth of knowledge about soil aggregation, soil microbial biomass and microbial diversity of soil, there is little knowledge of the microbial community ecology of soil aggregates. This review intends to expand upon and examine the microbial nature of soil aggregation: species causal to aggregation and functional groups involved. One focal point will be to examine how a quest for identifying and characterizing key species associated with microaggregates can have implications for management practices to improve soil aggregation and ultimately soil structure. The review will attempt to identify some promising avenues for future research in this area of soil biology that is a central one to soil quality. Our goals are to catalyze rigorous, innovative research on current approaches and techniques on the microbial ecology of soil aggregation.

1. INTRODUCTION

Soil structure directly affects carbon sequestration, water holding capacity and aeration, each in turn affecting microbial activities controlling nutrient turnover and cycling processes. Conversely, the biotic basis of soil structure is described as due to microbial activity (fungi and bacterial activity with mycorrhizal fungi given primacy), root growth stimulating microbial activity leading to the production of both root mucigels and microbial exudates as "cementing agents", physical compaction by root growth, faunal activity and mineral interactions (Six et al., 2002). The prokaryotic basis of soil structure formation or aggregation is likely similar to the process described by Mueller (1996) with regards to flocculation in more aqueous conditions with aggregation occurring in steps: through microbiological degradation or modification of organic material, attachment of cells to soil particles, the production of extracellular polymeric substances (EPS) (through decomposition and utilization of organic matter or based on utilization of root exudates) and the concerted construction of biofilms by microbial communities, which can cause aggregation of soil particles. The production of EPS by bacteria are regarded as a means of stabilization against fluctuations in water potential, protection against antibiotics and regulation of the diffusion of nutrients and waste products and are thus necessary for survival. The formation of soil aggregates thus would not necessarily be an imperative for survival of the bacteria but a consequence of EPS and biofilm production that additionally may indirectly benefit the bacteria by providing the aeration and porosity for soil moisture, both required for growth of the microbes (Chenu and Strotzky, 2002). While there is a relative abundance of studies descriptive of soil aggregation, there is currently a paucity of research concerned with the species and proportional composition of bacteria involved in soil aggregation.

2. Possible New Approaches For Studying Soil Aggregation

Knowledge of the microbiology of soil aggregation can perhaps be informed by some major findings and studies in microbial adhesion and biofilm formation. Lectins are a prominent mode of bacterial adhesion and two similar lectins have been described from a soil-inhabiting basidiomycete species and a soil-inhabiting plant pathogenic bacterial species (Sudakevitz et al., 2002). Both lectins had affinities for plant hemicelluloses, essentially plant residues. Ofek et al. (2007) found that some bacterial lectins recognize glycolipids, terminal sugars and internal sugar sequences as well. Whether certain soil microbial species have lectins with higher affinities for binding to plant residues is in need of investigation. This may lead to a better understanding of the role individual species may play in soil aggregation. Perhaps species with such lectins might predominate in the densest part of the bacterial density and substrate gradients that mark the processes of residue degradation (Nunan et al., 2003) leading to EPS formation, and thus may be the most important contributors to soil aggregation.

Studies investigating structured networks or matrices such as "nanowires" and "honeycombs" (Schaudinn et al., 2007) in bacterial biofilms may provide the missing link as to the bacterial role in soil aggregation. Methodologies such as staining and thin sectioning of undisturbed soil (Nunan et al., 2003) to analyze spatial distribution of bacteria in soil matrices, and nano-scale secondary secondary ion mass spectrometry of microbes (Herrmann et al., 2007) that have assimilated isotopically labeled substrate in soil could lead to determination of whether biofilm networks occur in undisturbed soil. Such networks could provide the means for quorum sensing (cell-cell signaling) to occur.

Quorum sensing, the production of signaling molecules to sense bacterial population densities in the surrounding environment among members of the same and other species, has been described essentially as a mechanism that coordinates and reinforces community behaviors in many bacterial species (An et al., 2006). These signaling molecules, which include *N*-acyl homoserine lactone, peptides and a modified ribose, accumulate in the surrounding environment until a concentration threshold of such compounds is attained, which then instigates the expression of certain genes. A study involving spatial analysis of bacterial "patches" or population concentrations in the soil matrix concluded that the observed close proximity of cells to each other suggests concerted activity, that such clustering would better explain the reported time course of the nitrification process and further suggested the possibility of quorum sensing in such a case (Nunan et al., 2003). It has been increasingly recognized that quorum sensing is a far more complex process than solely for census taking (Dunn and Stabb, 2007). Other than its mention in the spatial analysis paper cited above, this aspect of bacterial ecology has not to our knowledge been considered in relation to soil aggregation. Many soil and rhizosphere bacteria have been shown to secrete quorum-sensing molecules. Among these are strains belonging to *Agrobacterium* (Fuqua and Winans, 1994; Zhang et al., 1993), *Burkholderia* (Lewenza et al., 1999), *Chromobacterium* (Chernin et al., 1998; McClean et al., 1997), *Pseudomonas* (Gray et al., 1994; Wood and Pierson III, 1996; Wood et al., 1997), *Ralstonia* (Flavier et al., 1997), *Rhizobium* species (and other related genera involved in nitrogen fixation symbioses with legumes (Pierson et al., 1998, Cha et al., 1998), *Rhodobacte*r (Schaefer et al., 2002) and *Serratia* species (Christensen et al., 2003). The behaviors controlled or affected by quorum sensing mechanisms include biofilm formation (Davies et al., 1998; Huber et al., 2001), expression of pathogenicity genes (Pierson et al., 1998; Andersson et al., 2000) and as in the original study on this phenomenon, bioluminescence (Fuqua et al., 1994). A newer alternative perspective postulates that individual cells practice quorum sensing to detect the degree of diffusion in their surroundings (Redfield, 2002) and not necessarily for any cooperative purposes. However, regardless of either interpretation, the net effect is that members of the same species appear to act as a single organism and threshold population levels must be reached (Bassler and Miller, 2006). This is interesting in light of the existence of structured biofilm networks described above. Such networks may be an answer to the question of how signaling is transmitted or signaling molecules propagated among communities of bacteria *in situ* (Battin et al., 2007). The theory that quorum sensing is actually done to assess the degree of diffusion in the surroundings of bacterial cells and the existence of structured biofilms would be compatible in the context of soil aggregate formation by bacteria. Biofilm formation, however structured, is affected both by quorum sensing and by the nutritional status of the bacteria (De Beer and Stoodley, 2006). Bacterial biofilm formation, which is preceded by EPS production (a stage of the process of soil aggregation involving the bacterial component), and shown to be under the control of cell-cell communication or quorum sensing, thus would require threshold population levels of the species involved.

Anaerobic bacteria have seldom to our knowledge been studied in relation to soil aggregation. Anaerobic microbial processes in soil have been discussed and the anaerobic conditions occurring in soil aggregates have been documented (Tiedje et al., 1984), but to our knowledge few studies have subsequently investigated the relationship between anaerobic bacteria and water stable aggregates (Andrade et al., 1998; Bethlenflavay et al., 1999). Interestingly, these two studies found correlations between population levels, culturable anaerobic bacteria, and water stable aggregates. Soil anaerobic conditions are supposed to be periodic or transient in occurrence and this along with over all low metabolic activity under such conditions would limit populations of anaerobes. However, in the limited number of recent studies, a range of aerotolerant anaerobes have been isolated and identified from diverse habitats (Kato et al., 2004; Matthies et al., 2004; Gossner et al., 2006; Kusel et al., 2006) and there is a documented ability of aerobic and anaerobic bacteria to stably coexist with aerobes in artificial communities and plant rhizospheres (Kato et al., 2005; Nikolausz, et al., 2006). This would perhaps indicate that less stringent and accessible conditions would be feasible for working with such anaerobes than the conditions prescribed for their study (Kaspar and Tiedje, 1994). Thus, studies of the role of anaerobes in soil aggregation need not be restricted to specialists.

3. The Need for Culture-Based Methods

Cause-and-effect investigation of the role of specific bacteria in the important issue of soil aggregation is highly constrained without cultural methods. Even a sophisticated study like that of Vaisenen et al. (2005) lacked direct evidence and identification of prominent or predominant players in the process of aggregation. Determination of key bacterial species in soil aggregation was to our knowledge last called for in a major work more than a decade previous (Tisdall, 1994) to the writing of the present paper. This need remained unmet to our knowledge until we initiated work to address it in a recent study (Caesar-TonThat et al., 2007). Progress in this area need not await the development of methods that overcome the admitted flaws of culture- (or molecularly-) based methods. We feel the most useful information currently would be to know what culturable organisms are predominant within soil aggregates and their likelihood to cause aggregation. This avenue has not been pursued to date. We are interested in culturable aggregators because it affords the ability to amend the soil with them in further work, and to develop means of assaying for their ability to form water-stable so aggregates. Much molecular method-based work in soil aggregation, while contributing in essential ways to the field, is not cause-and-effect in nature. Several practical handicaps of molecular methods presently preclude other than descriptive or correlative studies: the lack of a sufficient blast database of taxa with which to compare sequences for reliable identification, the up-front time and expense of methods such as real-time PCR (for which accurate primers must be developed and this would rely on identified, cultured organisms from which DNA would be extracted) that might allow quantification and the overall lack of any background on which to base studies that would compare those that could be identified using molecular methods with known aggregating bacteria. As described above, there is a documented relationship between rhizosphere-driven stimulation of bacterial populations, production of extracellular polysaccharides by such bacteria (and root mucigels, possibly partly in response to rhizosphere bacterial activity (Barber and Lynch, 1977) and resulting soil aggregation. We interpret this to be a relationship between an enrichment in the rhizosphere of larger populations of bacteria, the rhizosphere effect, (Katznelson 1965), the production of extracellular polysaccharides by fast-growing species such as *Pseudomonas* and *Serratia* spp. (Marilley and Aragno, 1999; Berg et al., 2005) and aggregation of soil. Thus, it is reasonable to hypothesize that population levels of bacteria that tend to be culturable would be determinative in soil aggregation. This is essentially an extension of the reasoning set forth by Ellis et al. (2003): that previous studies have shown that there are positive correlations between activity and cell size, cell size and culturability and activity and culturability. We believe based on this reasoning that isolating the bacteria occurring at the highest population levels in soil microaggregates is a necessary step to the elucidation of the key microbial effectors of soil aggregation. While it cannot be precluded that less abundant members of prokaryote communities associated with microaggregates might have some role in aggregation, it would be reasonable to conclude that the most abundant prokaryotic species found (by culturing) in microaggregates play the key roles in aggregation. Since the means for aggregating soil are regarded as essentially through binding with extracellular polysaccharides, one could conclude that the main players would tend to be culturable. And if they are to exert a major effect on their soil environment, they would need to be among the more numerous organisms in that environment.

While it has become almost axiomatic that the value of culture-based studies is highly compromised (such as to be negligible, judging from passing references made to, and quite unequivocally dismissive comments by the reviewers of a paper by the authors employing culture-based methods) the value of such studies has been alluded to quite often (Ellis et al., 2003; Janssen, 2006; Palleroni, 1997; Zinder and Salyers, 2001). This value deserves some emphasis here. The need for culture-based methods has been acknowledged by advocates of molecular methods for studying soil biology (Muyzer and Smalla, 1998), as providing "phylogenetic frameworks of protein-encoding gene sequences as well as for the classification and identification of closely related bacteria". The culture of even highly fastidious microbes is viewed as indispensable because pure cultures "allow a comprehensive understanding of the physiology, cell-cell interactions and permit access to metabolic pathways" of such microbes (Darby and Welburn, 2006). The latter authors further saw isolation in culture as "a limiting factor in the study of these organisms…", and the same applies to soil aggregation by microbes. The overall issue of the burgeoning of molecular methods in microbial ecology elicited this comment by others: "We need to guard against the research becoming too technology driven" (O'Donnell et al., 2001). A consensus has since developed that "new technologies will increasingly lead us down blind, non-generalist and expensive alleyways if studies are not directed and driven by theory" (Prosser et al., 2007). These workers further stressed the need to bridge the gap between descriptive studies of soil microbial diversity and soil function and between disciplines for microbial ecology to advance. It is also important to recall that knowledge of quorum sensing, discussed above, was initially derived from and has progressed chiefly through studies using culturable bacterial species, and perhaps this will be true of soil aggregation as well. Future studies suggested in the present review such as the role of bacterial lectins, quorum sensing, and biofilm networks or by others such as the need to determine which species are the most effective stabilizers (aggregators) of soil (Tisdall, 1994), will all rely on the need to culture the bacteria involved.

The isolation of the most abundant species of bacterial communities of microaggregates is thus supported by the several points outlined above: the correlation of culturability with activity, the need for culturable species to initiate ecological and physiological studies of the bacterial role in soil aggregation. For example, individual isolates found to be occurring at the highest dilution levels should be assessed for their ability to aggregate soil singly or in combination.

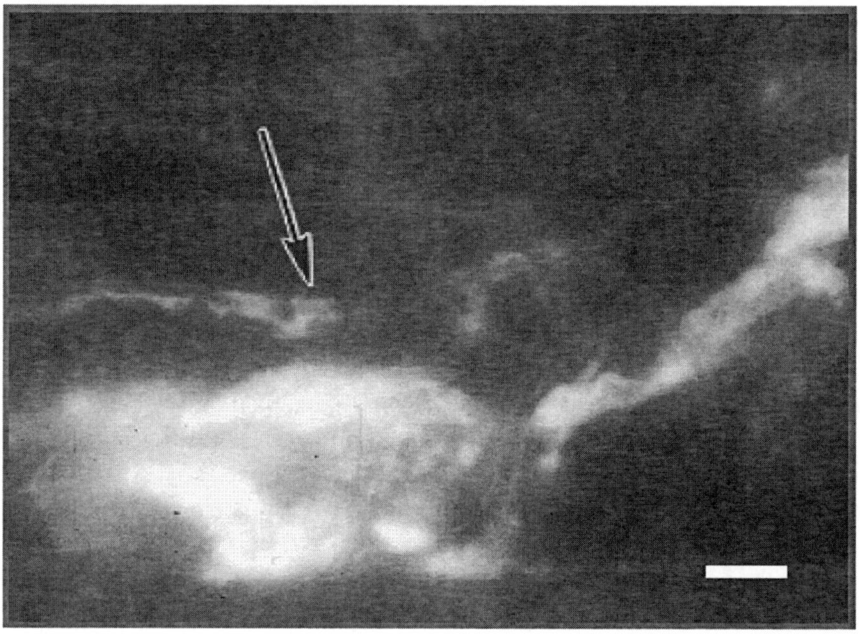

Figure 1. A saprophytic basidiomycete fungal isolate, identified as a member of the russoloid lineage closest to the genus *Peniophora* produces at its mycelial surface soil binding mucilage containing fucosyl sugar residues (arrow) as demonstrated by staining with FITC-UEA I lectin (Caesar-TonThat, 2002). Bar, 5 μm.

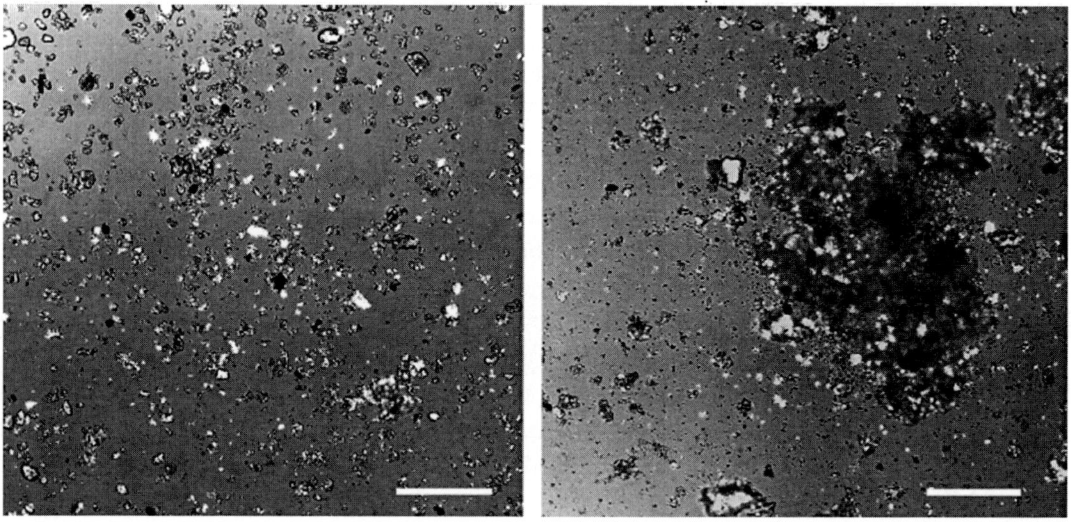

Figure 2. *Stenotrophomonas maltophilia*, a bacteria belonging to the class of γ-Proteobacteria and isolated from microaggregates of field cropped to barley under no-till and irrigation management, form aggregates with soil particles (< 0.050 mm size-fraction) in water. Left, soil particles without adding bacteria. Right. soil particles with bacteria added. Bar, 100 μm.

4. Fungi in Soil Aggregation

Perusal of the literature on the microbial role in soil aggregation leaves the impression that fungi have a prevalent role in soil aggregation compared to bacteria. While passing mention is made in the literature of bacterial involvement, the emphasis has consistently been on fungi. This is apparently based on such data as that indicating that fungal (biomass)-dominated soil communities provide increased carbon storage and that the fungal biomass consists of a higher C:N ratio (Six et al., 2006) than bacteria. This fungal dominance of microbial biomass in natural or uncultivated settings is furthermore linked to the "alteration of soil physical properties", apparently meaning soil aggregation (Tisdall and Oades, 1982). Fungi stabilize macroaggregates and bacteria help stabilize microaggregates (Tisdall, 1994). Endomycorrhizae have been described as the apparent driving force in soil aggregation, with their growth and production of the soil-binding glycoprotein glomalin (Wright and Upadhyaya, 1996) being stimulated by sloughed off root material and furthermore this combination stimulating increased bacterial populations (Dighton, 2003). These findings could have important future implications in the use of mycorrhizal fungi to promote the production of soil stable aggregates, improve water infiltration, and soil C sequestration in agricultural systems. However, there are some points to consider regarding the supposed primacy of such fungi in soil aggregation. A recent study has shown that bacteria have growth-yield efficiencies in soil that are equivalent to fungi, contrary to long-held assumptions (Thiet et al., 2006). This finding alone might properly call into question the larger role fungi are presumed to play in soil aggregation compared to bacteria.

Bacteria because of their smaller size compared to fungal hyphae have access to a greater variety of niches in the soil matrix than any other biotic contributor to microaggregation (Six et al., 2006) and presumably thereby, more substrate carbon. In addition, microaggregates, the formation of which are chiefly attributable to bacteria, have a greater capacity to protect carbon against decomposition than macroaggregates, which are formed chiefly by fungi (Tisdall, 1994; Oades andTisdall, 1982). A major caveat against microscopical methods of determining whether a particular soil is fungal-dominated is that much of the apparent fungal biomass may not be active (Foster, 1988, Six et al., 2006). Considering these findings collectively, it is difficult to conceive how on balance any particular ecological setting could be categorically "dominated" by either fungi or bacteria. Regarding the relatively more prominent role generally attributed to fungi, in particular to endomycorrhizae through production of glomalin, in soil aggregation, both in macroaggregates and microaggregates, perhaps recognition of these data about bacteria and aggregation might be cause for revising the perceived magnitude of that role to recognition of a greater role for bacteria. We agree with the interpretation that the formation and degradation of microaggregates may be more dynamic than generally predicted by carbon turnover times (Jastrow and Miller, 1998). Furthermore, our findings cause us to postulate that the microaggregates are dynamic to such an extent that bacterial species predominant in microaggregates differ in disturbed compared to non-disturbed conditions; we found gram-positive species in microaggregates from tilled soils with greater frequency than from no-tilled or uncultivated CRP settings (Caesar-TonThat et al., unpublished). Presumably, this is a result of the generally lower level of labile carbon and other nutrients in disturbed soil conditions, which would select for spore-forming species such as *Bacillus* or for *Arthrobacter* spp. or similar coryneform species, which can alter their morphology or physiology for dessication and starvation resistance (Lappin-Scott and Costerton, 1990).

Despite mention of a role for ectomycorrhizae (and thus by extension basidiomycetes) in soil aggregation in a relatively early assessment of the microbial ecology of soil aggregation (Tisdall, 1994), such fungi were subsequently largely unexplored for their role in soil aggregation until recently. A study by the senior author showed polysaccharide production as the basis of the ability of a basidiomycete fungus to form soil aggregates, (Caesar-TonThat and Cochran, 2000). In this study, pure cultures of the fungus and development of an assay for formation of water stable aggregates allowed the authors through experimentally derived data to contribute knowledge of the role of a basidiomycete to soil aggregation.

5. Microbial Diversity in Soil Aggregates

Inventories of species within any given ecological niche, thereby establishing the diversity index of that niche have been of enormous value in ecology. When this diversity is inventoried with molecular based methods, clearly the most comprehensive means to establish the magnitude of diversity, a picture emerges of a greater totality of species than any other means. Such measurements of total diversity move beyond the purely descriptive when they can be related to such questions as whether soil fertility or disease suppressiveness are affected or enhanced by microbial diversity per se or whether management methods or restoration efforts affect it and vice versa (Garbeva et al., 2006, Hartmann and Franco, 2006, Hiddink et al., 2005). The culturability/activity thesis described previously, unchallenged since its introduction, applies here. The ecological utility of assessing total bacterial diversity in water stable microaggregates with molecular methods compared to culturing the most abundant bacteria can be considered with an analogy drawn from this same authors work. The total microbial diversity in microaggregates can be likened to the seed-bank in plant ecology, which would represent "potential diversity", whereas the culturable species represent the plant community or "realized diversity" (Prosser et al., 2007). The uncultured species, as the analogous dormant members of a seed-bank, likely represent the adaptability to environmental change of either ecosystem rather than active members of the community. Thus, while great stress is placed on the low proportion of microbial species in soil that are typically culturable among the larger set of taxonomic units detected using molecular methods, whether the larger set represents other than inactive species, i.e., "potential diversity", remains to be shown. It remains true that most significant functions in soil such as nitrogen cycling, beneficial symbiotic associations between bacteria and plants, the mineralization of plant nutrients and suppressiveness to plant diseases are attributable (entirely or predominantly) to bacterial species that are culturable. In addition, a closer examination of the gap between species culturable and those not yet cultured indicates that the ability to culture representatives of the phylogenetic diversity in soil is better than the 1% often quoted (Janssen, 2006).

Typically in molecular studies the specification of the most abundant members of any measured community and their precise taxonomic identity have either not been directly addressed or remain elusive. These issues impede the practical application of the results for developing a fuller understanding of the microbial ecology and physiology of soil aggregation. For example, clarification is needed when the term "abundance" is used when describing the occurrence of microbial taxa in studies of diversity and community composition within microaggregates and soils in general. Also, the term "dominance" has been applied based on the frequency with which sequences of particular taxa are encountered from cloning procedures or sequencing of bands from denaturing or temperature gradient gel electrophoresis. Dominance and abundance applied this way may not be readily analogous to their relative abundance as cells or as a proportion of active cells in soil. This is because of the pitfalls affecting molecular-based, culture-independent methods, particularly when DNA is isolated from whole soils. These include differential lysis efficiencies of cells, the harshness of DNA extraction methods such as bead beating, purification steps leading to loss of DNA, differential target amplification, and intraspecies variation in rDNA repeated sequences (Witzengerode et al., 1997, Kirk et al., 2004, Green and Martin, 2006).

6. Techniques for Assessing Microbial Aggregative Function In Soil Aggregates and Biofilms

Various physical measurement techniques including resistance to pressure of artificial soil aggregates (Caesar-TonThat et al. 2007) or biofilms (Chen et al. 1998, Kreth et al. 2004; Xavier et al. 2005; Ahimou et al., 2007) amended with individual bacterial species have been used to pattern the degree of adhesiveness of the binding agents produced by microorganisms. Our research group has extended this work, using reflectance to measure the ability of individual species to sediment soil particles.

Electromagnetic spectrum is used for the measurement of spectral response of biomaterials (Alupoaei et al. 2004; Bohren and Huffman, 1983) and configured into three reactions: reflectance, absorptance, and transmittance. When the biomaterials receive incoming irradiation, a portion of the light is absorbed. When the incoming photons impinge on an object at a certain angle, a portion of the photons is reflected. Light neither absorbed nor reflected is transmitted. The portions of three spectral reactions vary in the electromagnetic wavelength of the illumination source and the physical properties of the target object. The measurement of the transmittance has been widely used to estimate the density of aggregate solutions (Ramos and McBride, 1996; Dontsova and Norton, 2002; Joyce and Smith, 2003; Kim et al., 2004), but requires special and expensive equipment combined with a time-consuming dilution procedure performed on one sample at a time and is often limited by a short measurement range (Gributs and Burns, 2004). Image-based reflectance measurement offers an advantage of quick and cost-effective measurement of the solution density by using a spectral camera capturing an image of the reflected light for multiple samples at a time. The captured image is calibrated by referencing white (255 in gray-scale value) and black (0 in gray-scale value) image spots as 100% and 0% reflectance, respectively, then the reflectance of each target solution image is converted to a gray-scale value that directly correlates to relative differences of the solution density.

We utilized the image-based reflectance method described above to measure the aggregative function of predominant culturable bacterial species isolated from microaggregates collected from a field study with different irrigation and tillage treatments (details of the field study are described in Sainju et al. (in press)). Bacteria from soils cropped to barley (*Hordeum vulgare* L.) under different irrigation (irrigation and no irrigation) and tillage (tillage or and no till) management were isolated and identified according to the procedure of Caesar-TonThat et al. (2007). The field study also included plots of grass and alfalfa that have been undisturbed and minimally managed for more than 20 years. Using the same sampling and analysis procedures as described above, bacteria were isolated from soil in these treatments (irrigated and non-irrigated) for the purpose of providing control comparisons. For the aggregation assays, suspensions consisting of soil particles (0.125g/ml, < 0.05 mm aggregate sieved-fraction) and bacteria from individual species were prepared in concentrations of 10^4, 10^5, and 10^6cells /ml in 10 ml of sterile deionized water. The soil/bacteria mixtures were consistently vortexed for 10 sec then soil particles were allowed to settle for 5 min. Images of the liquid suspension were immediately captured with a commercial digital camera using near infrared red settings (800-1000 nm), then images were converted into grey scale images using Adobe Photoshop software (version 7.0) in order to evaluate the reflectance of the controls which contain only soil particles without the bacteria. Isolates possessing the aggregative ability are indicated with shading in Table 1. Among all the identified species, it is obvious that some species aggregate soil better than others and some do not aggregate at all; more gram-negative species from α-Proteobacteria (*Brevundimonas, Sphingomonas, Sphingopyxis, Novosphingobium* spp.), β-Proteobacteria (*Burkholderia* spp.), γ-Proteobacteria (*Flavimonas, Pseudomonas, Stenotrophomonas* spp.), Flavobacteria (*Chryseobacterium, Flavobacterium* spp.), and Sphingobacteria classes (*Sphingobacterium* spp.) were soil aggregators compared to the gram-positive species which included mostly *Bacillus* and *Microbacterium* spp. Regardless of the irrigation practices, the percentage of isolates (regardless of species) with ability to aggregate soil were higher under no till than under till (under irrigation, 32.84 % vs. 21.13 %, respectively; under no-irrigation, 23.53% vs. 7.14 %, respectively) and a greater percentage of species among all the taxa identified from each treatment were soil aggregators under no-till compared to till (under irrigation 21.52 % vs. 13.92 %, respectively; under no irrigation 13.92 % vs. 5.06 %, respectively), suggesting that tillage management has an impact on the growth and survival of the dominant soil aggregating bacterial communities in the microaggregates. Furthermore, the total number of the soil aggregating isolates and species in microaggregates of irrigated treatments was higher than in non-irrigated treatments indicated that conditions under irrigation favor the presence of these beneficial bacteria.

A no-till system is characterized by sowing of crops directly into the crop residues that remain on the soil surface from the previous crop without soil preparation measures (e. g., plowing and disking), except for the narrow slit required for seed placement. Slow decomposition in no-till systems due to the partial contact between crop residues and soil results in higher organic content in the surface soil, promoting better structure, fertility, water infiltration and water–holding capacity. In contrast, conventional tillage mainly accomplished by a moldboard plow or disk, is characterized by a high degree of soil disturbance and loss of organic matter; soil layers are inverted, crop residues are completely incorporated into the plow layer leaving no organic residues on the surface. The long-term maintenance of a significant amount of organic matter on the soil surface is important for microbial growth and activity because it minimizes extreme temperature shifts, maintains adequate moisture and contributes to aeration. More accumulation of crop-derived C in free microaggregates has been observed in no tillage compared to conventional tillage (Six et al., 1999)

Table 1. Distribution of bacterial isolates of the predominant aerobic, heterotrophic, culturable species identified by fatty acid methyl ester (FAME) profiles from microaggregates of various tillage and irrigation systems collected at 0-5 cm depth before planting (spring) and assessment of their ability to aggregate soil using an *in vitro* soil sedimentation assay

	Management practices[a]						Reflectance[b] (%)		
GN	BNTnirr	BNTirr	BTnirr	BTirr	CRPNTnirr	CRPNTirr	10^6 c/ml[c]	10^5 c/ml	10^4 c/ml
α-Proteobacteria									
Brevundimonas diminuta (0.624)[d]	0	0	0	0	0	1	*78.28	55.93	54.96
Brevundimonas vesicularis (0.864)	1	0	0	1	0	1	*70.99	56.24	56.43
Sphingomonas echinoides (0.642)	0	0	0	0	0	1	*83.00	*71.34	*62.82
Sphingomonas sanguinis (0.714)	0	0	0	2	0	0	*92.15	*68.13	58.06
Sphingopyxis macrogoltabida (0.553)	0	1	0	0	0	0	*83.54	*90.01	*60.06
Novosphingobium aromaticivorans (0.401)	0	0	0	1	0	0	*60.07	*66.97	*59.74
Novosphingobium capsulatum (0.894)	1	1	0	1	2	1	56.12	53.73	*59.84
Phyllobacterium rubiacearum (0.527)	0	0	0	1	0	0	NA	NA	NA
Roseomonas fauriae (0.385)	0	1	0	0	0	0	56.53	48.83	51.41
β-Proteobacteria									
Burkholderia cepecia GC subgroup B (0.352)	0	1	0	0	0	0	*90.86	*67.20	53.92
Variovorax paradoxus GC subgroup A (0.575)	1						51.47	49.12	50.76
γ-Proteobacteria									
Flavimonas oryzihabitans (0.873)	2	0	0	0	0	0	*60.65	*61.68	*59.58
Pseudomonas amyloderamosa (0.358)	0	1	0	0	0	0	*76.78	*83.87	*61.89
Pseudomonas chlororaphis (0.333)	0	0	0	0	0	1	*60.93	*57.02	48.45
Pseudomonas fluorescens biotype G (0.937)	0	0	0	0	0	1	*75.85	51.35	44.29
Pseudomonas pertucinogena (0.287)	0	1	0	0	0	0	*87.55	*56.98	51.62
Pseudomonas putida biotype B (0.848)	0	0	0	0	0	3	59.55	43.64	46.28
Pseudomonas savastanoi (0.860)	0	1	0	0	0	0	*73.38	52.90	52.87
Pseudoxanthomonas spp. (0.528)	2	0	0	0	0	0	*63.24	*58.89	*60.76
Ralstonia pickettii (0.798)	0	0	0	0	1	0	*73.07	*59.77	58.43
Stenotrophomonas maltophilia (0.795)	1	3	0	0	1	3	*74.80	*60.48	55.51
Flavobacteria									
Chryseobacterium balustinum (0.849)	0	1	0	0	0	0	*71.24	*57.72	57.59
Chryseobacterium meningosepticum (0.527)	1	1	0	0	0	0	*72.44	*60.22	57.20
Flavobacterium johnsoniae (0. 551)	0	1	0	0	0	1	51.70	42.74	49.38
Flavobacterium hydatis (0.374)	0	0	0	0	1	0	*74.43	*57.17	54.31
Zobellia uliginosa (0.303)	0	1	0	0	0	0	37.89	40.26	*60.35
Sphingobacteria									
Sphingobacterium spiritivorum GC subgroup A (0.654)	0	0	0	1	0	0	*71.44	53.31	45.45
Sphingobacterium spiritivorum GC subgroup B (0.696)	0	1	0	1	0	0	*83.00	*70.92	55.00
GP									
Arthrobacter aurescens (0.743)	0	0	1	0	0	0	33.58	33.85	52.57
Arthrobacter globiformis GC subgroup A (0.918)	5	4	17	7	4	4	34.28	32.25	31.84
Arthrobacter mysorens (0.709)	1	0	2	2	1	0	58.51	51.20	51.07
Arthrobacter oxydans (0.902)	10	6	14	19	10	17	50.86	48.21	55.20
Arthrobacter pascens (0.915)	6	0	4	4	3	1	34.20	38.19	50.06
Arthrobacter ureafaciens (0.867)	1	1	1	0	0	0	59.01	55.21	52.57
Arthrobacter viscosus (0.742)	1	0	0	1	4	0	*75.15	46.95	42.69
Bacillus GC group 22 (0.346)	0	0	0	0	0	2	*75.59	*60.01	45.58
Bacillus alcalophilus (0.341)	0	0	0	0	0	1	59.08	51.50	48.16
Bacillus atrophaeus (0.775)	1	1	1	2	4	0	54.53	52.57	58.54
Bacillus circulans (0.830)	0	0	1	0	0	1	*79.77	*61.07	46.74
Bacillus coagulans (0.562)	0	1	0	0	0	0	NA	NA	NA
Bacillus flexus (0.605)	0	1	0	1	0	0	*77.85	*67.85	*61.11
Bacillus globisporus (0.346)	2	0	0	0	0	0	*62.54	*60.22	58.13
Bacillus megaterium GC subgroup A (0.969)	8	5	6	3	7	4	39.23	30.62	24.42
Bacillus megaterium GC subgroup B(0.711)	2	0	0	0	1	5	53.98	50.84	52.72
Bacillus psychrosaccharolyticus (0.791)	0	0	0	0	0	1	59.40	53.67	47.45
Bacillus pumilus GC subgroup B (0.861)	0	17	1	1	3	0	58.86	47.98	38.30
Bacillus sphaericus GC subgroup D (0.802)	0	1	1	0	1	0	*66.94	43.57	41.92
Bacillus subtilis (0.860)	0	0	0	1	4	1	31.90	35.32	35.98
Brevibacillus brevis (0.542)	0	0	0	0	1	0	*78.48	*75.43	49.25
Brevibacillus choshinensis (0.833)	3	0	5	4	4	2	56.21	46.62	50.42
Brevibacillus parabrevis GC subgroup A (0.769)	2	4	0	0	1	0	55.11	51.02	54.43
Brevibacillus parabrevis GC subgroup B (0.632)	0	0	0	1	1	1	53.82	35.13	32.35
Brevibacterium liquefaciens (0.462)	0	0	0	0	0	1	54.92	41.01	32.52
Brevibacillus reuszeri (0.531)	0	0	1	0	0	0	30.84	34.93	37.45
Cellulomonas fimi GC subgroup B(0.911)	2	1	1	0	2	0	43.30	34.69	43.67
Curtobacterium flaccumfaciens (0.845)	1	0	0	1	0	1	55.42	51.95	54.54
Curtobacterium pusillum (0.347)	1	0	0	0	0	0	59.39	56.37	55.46
Kocuria erythromyxa (0.684)	0	1	1	0	0	1	43.30	45.64	51.27
Kocuria kristinae (0.841)	2	2	0	2	0	2	*65.22	51.05	36.31
Kocuria rosea (0.705)	0	0	1	2	0	1	*69.33	*57.54	55.19
Kurthia sibirica (0.517)	0	0	0	1	2	0	*83.56	*68.67	46.12
Kytococcus sedentarius (0.315)	0	0	0	1	0	0	*73.00	*74.44	56.21
Lechevalieria flava (0.331)	1	0	0	0	0	0	NA	NA	NA
Microbacterium esteraromaticum (0.910)	2	2	2	0	0	1	*76.44	*58.68	47.54
Microbacterium flavescens (0.337)	1	0	0	0	0	0	44.74	41.44	48.17
Microbacterium hominis (0.541)	0	0	0	1	0	0	*82.41	*73.99	*75.11
Microbacterium lacticum GC subgroup A(0.670)	0	2	0	1	0	0	*67.80	53.11	56.03
Microbacterium laevaniformans (0.727)	0	1	0	1	0	1	*78.93	*59.33	49.22
Microbacterium liquefaciens (0.643)	0	0	0	0	0	1	59.38	56.89	55.84
Microbacterium saperdae (0.844)	1	0	0	0	0	0	*80.96	*57.04	52.03
Micrococcus luteus GC subgroup B(0.885)	3	2	1	1	0	2	58.47	53.62	53.54
Micrococcus lylae GC subgroup B (0.388)	0	0	1	0	0	0	38.86	37.91	54.43
Paenibacillus alginolyticus (0.750)	2	0	1	1	0	0	50.98	42.42	47.76
Paenibacillus alvei GC subgroup A (0.340)	0	0	0	0	1	0	50.40	49.99	45.50
Paenibacillus chondriotinus (0.674)	0	0	0	0	0	2	33.40	43.05	44.38
Paenibacillus lautus (0.629)	0	0	0	0	1	0	20.22	31.21	22.48
Paenibacillus pabuli (0.802)	0	0	0	1	1	0	42.28	42.65	39.34
Paenibacillus polymyxa (0.759)	0	0	4	2	4	2	53.18	33.25	38.99
Paenibacillus validus (0.581)	0	0	1	2	4	0	*93.60	*81.72	*62.21
Paracoccus denitrificans (0.569)	0	0	0	0	0	2	21.52	21.93	31.58
Rhodococcus erythropolis (0.932)	1	0	0	0	0	0	35.95	33.34	49.55
Vogesella indigofera (0.634)	0	0	1	0	0	0	53.15	48.22	43.82
Control[d]							44.99	44.99	44.99
Isolates with SIM < 0.300	2	1	1	0	2	2			
No match	2	4	2	1	1	0			
Total GP isolates analyzed	72	72	72	72	72	72			
Soil aggregating isolates (%)	23.53	32.84	7.14	21.13	24.63	24.29			
Soil aggregating species (%)	13.92	21.52	5.06	13.92	11.39	16.46			

[a] Management practices: barley no-till irrigated (BNTirr) and non-irrigated (BNTnirr); barley till-irrigated (BTirr) and non-irrigated (BTnirr); CRP no-till irrigated (CRPNTirr) and non-irrigated (CRPNTnirr).

[b] Bacterial and soil suspensions were mixed and reflected images were captured after 5 minutes. Images were calibrated by referring white (255 in gray-scale value) and black (0 in gray-scale value) image spots as 100% and 0% reflectance and converted to gray scale value. A set of controls was prepared in a similar manner but without the presence of bacteria.

*, numbers significantly higher than the control at $P < 0.001$ (Tuker-Kramer HSD test).

[c] Bacterial concentrations.

[d] Similarity index based on MIDI Aerobic Bacterial Library TSBA50.

and it has already been demonstrated that the production of EPS by the bacteria is dependent on their environment (White, 1995). The results of this study indicated that a combination of no till and irrigation management for barley cropping appears to provide a balance between the microorganisms and the physicochemical components, favoring growth of specific bacterial species with ability to aggregate soil, development and potentially the excretion of polymers, and aggregation.

Soil sedimentation by bacteria in liquid suspension measured by image-based reflectance is an improvement over the method using pressure to break artificial aggregates amended with bacteria. Soil aggregation assay based on the formation of artificial aggregates by adding bacteria to natural soil involves air-drying that induces intermolecular associations between organic molecules and mineral surfaces (Kemper and Rosenau, 1984) besides the "adhesive" properties of soil binding agents excreted by bacteria. These physical artefacts can be avoided with the soil sedimentation method.

Conclusion

The ultimate aim of research on soil aggregation, most if not all of it publicly funded, should be improved practices in soil health management. There are different pathways to reach this goal and we have sought to indicate some approaches in this important field of research that are justified but perhaps underemphasized or overlooked. Based both on precedent (discussed above) and on rationale drawn from a significant body of literature, we have concluded that use of cultural methods will enhance progress in the study of the microbial ecology of soil aggregation, thus allowing knowledge derived from such work to be integrated with molecular methods. This means studies with cultures detected at the highest levels of abundance in soil aggregates through isolation onto low nutrient or selective media, identification of these cultures and assessing them for traits possibly related to soil aggregation such as EPS production and further, the development and application of assays to test their ability to cause water stable soil aggregates. Also, if the goal of managing agroecosystems, in turn to manage microbial populations in soil and positively affect microbial functional groups which contribute to soil structure (Elliot and Coleman, 1988) then cause-and-effect studies can uniquely contribute to that end.

References

Ahimou, F., Semmens, M. J., Novak, P. J., and Haugstad, G. (2007). Biofilm cohesiveness measurement using a novel atomic force microscopy methodology. *Appl. Environ. Microbiol.*, 73, 2897-2904.

Alupoaei, C. E., Olivares, J. A., and García-Rubio, L. H. (2004). Quantitative spectroscopy analysis of prokaryotic cells: vegetative cells and spores. *Biosensors and Bioelectronics,* 19, 893-903.

An, D., Danhorn, T., Fuqua, C., and Parsek, M. R. (2006). Quorum sensing and motility mediate interactions between *Pseudomonas aeruginosa* and *Agrobacterium tumefaciens* in biofilm cocultures. *Proc. Natl Acad. Sci.* USA, 103, 3828–3833.

Andersson, R. A., Eriksson, A. R. B., Heikinheimo, R., Mäe, A., Pirhonen, M., Kõiv, V.; Hyytiäinen, H., Tuikkala, A., and Palva, E. T. (2000). Quorum sensing in the plant pathogen *Erwinia carotovora* subsp. *carotovora*: the role of *expR* (Ecc). Mol. Plant–Microbe Interact., 13, 384–393.

Andrade, G., Mihara, K. L., Linderman, R. G., and Bethlenfalvay, G. J. (1998). Soil aggregation status and rhizobacteria in the mycorrhizosphere. Plant Soil, 202, 89–96.

Barber, D. A. and Lynch, J. M. (1977). Microbial growth in the rhizosphere. *Soil Biol.* Biochem., 9, 305-308.

Bassler, B. L. and M. B. Miller. (2006). Quorum Sensing. In M. Dworkin, S. Falkow, E. Rosenberg, K.-H. Schleifer and E. Stackebrandt (Eds.), The Prokaryotes, Volume 2: *Ecophysiology and Biochemistry* (pp. 336–353). Springer-Verlag.

Battin, T. J., Sloan, W. T., Kjelleberg, S., Daims, H., Head, I. M., Curtis, T. P. and Eberl L. (2007). Microbial landscapes: new paths to biofilm research. *Nature Reviews Microbiology*, 5, 76-81.

Berg, G., Opelt, K., Zachow, C., Lottmann, J., Gotz, M., Costa, R., and Smalla, K. (2005). The rhizosphere effect on bacteria antagonistic towards the pathogenic fungus *Verticillium* differs depending on plant species and site. *FEMS Microbiol. Ecol.*, 56, 250–261.

Bethlenfalvay G. J., Cantrell, I. C., Mihara, K. L., and Schreiner, R. P. (1999). Relationships between soil aggregation and mycorrhizae as influenced by soil biota and nitrogen nutrition. *Biol. Fertil. Soils*, 28, 356–363.

Bohren, C. F. and Huffman, D. F. (1983). *Absorption and scattering of light by small particles*. New York, N.Y.: Wiley.

Boylen C.W. and Mulks M. H. (1978). The survival of coryneform bacteria during periods of prolonged nutrient starvation. *J. Gen. Microbiol.,* 105, 323–334.

Caesar-TonThat, T. C. and Cochran, V. L. (2000). Soil aggregate stabilization by a saprophytic lignin decomposing basidiomycete fungus. I. Microbiological aspects. *Biol. Fertil. Soils*, 32, 374-380.

Caesar-TonThat, T. C., Caesar, A. J., Gaskin, J. F., Sainju, U. M., and Busscher, W. J. (2007). Taxonomic diversity of predominant culturable bacteria associated with microaggregates from two different agroecosystems and their ability to aggregate soil in vitro. *Appl. Soil Ecol.*, 36, 10-21.

Cha, C., Gao, P., Chen, Y. C., Shaw, P. D., and Farrand. S. K. (1998). Production of acyl-homoserine lactone quorum-sensing signals by gram-negative plant associated bacteria. *Mol. Plant-Microbe Interact.,* 11, 1119–1129.

Chen, M. J., Zhang, Z., and T.R., Bott. (1998). Direct measurement of the adhesive strength of biofilms in pipes by micromanipulation. Biotechnology Techniques, 12, 875-880.

Chenu, C. and Strotzky, G. (2002). Interactions between microorganisms and soil particles: a review. In P. M. Huang, J.-M. Bollag, and N. Senesi (Eds.), *Interactions between Soil Particles and Microorganisms* (pp. 3-40). West Sussex, England: John Wiley and Sons, Ltd.

Chernin, L. S., Winson, M. K., Thompson, J. M., Haran, S., Bycroft, B. W., Chet, I., Williams, P., and Stewart, G. S. (1998). Chitinolytic activity in *Chromobacterium violaceum*: substrate analysis and regulation by quorum sensing. *J. Bacteriol.,* 180, 4435–4441.

Christensen, A. B., Riedel, K., Eberl, L., Flodgaard, L. R., Molin, S., Gram, L., and Givskov, M. (2003). Quorum-sensing-directed protein expression in *Serratia proteamaculans* B5a. *Microbiology,* 149, 471–483.

Darby, A. C. and Welburn, S. C. (2006). Symbiont culture. In K. Bourtzis and T. A. Miller (Eds.), *Insect Symbiosis* (vol. 1. p. 141-155). Boca Raton, FL: Taylor and Francis.

Davies, D. G., Parsek, M. R., Pearson, J. P., Iglewski, B. H., Costerton, J. W., and Greenberg, E. P. (1998). The involvement of cell-to-cell signals in the development of a bacterial biofilm. Science, 280, 295–298.

De Beer, D. and Stoodley, P. (2006). Microbial Biofilms. In M. Dworkin, S. Falkow, E. Rosenberg, K.-H. Schleifer and E. Stackebrandt (Eds.), The Prokaryotes, Volume 1: Symbiotic associations, Biotechnology, *Applied Microbiology* (pp. 904–937). Springer-Verlag.

Dighton, J. (2003). Fungi in Ecosystem Processes (pp. 432). New York, N.Y. : Marcel Dekker.

Dontsova, K. M. and Norton, I. D. (2002). Clay dispersion, infiltration, and erosion as influenced by exchangeable Ca and Mg. *Soil Science,* 167, 184-193.

Dunn, A. K. and Stabb, E. V. (2007). Beyond quorum sensing: the complexities of prokaryotic parliamentary procedures. *Anal. Bioanal. Chem.*, 387, 391-398.

Ellis, R. J., Morgan, P., Weightman, A. J., and Fry, J. C. (2003). Cultivation dependent and -independent approaches for determining bacterial diversity in heavy-metal contaminated soil. *Appl. Environ. Microbiol.*, 69, 3223–3230.

Elliott, E. T. and Coleman, D. C. (1988). Let the soil work for us. Ecol. Bull., 39, 23–32.

Ensign, J. C. (1970). Long-term starvation survival of rod and spherical cells of *Arthrobacter crystallopoietes. J. Bacteriol.*, 103, 569–577.

Flavier, A. B., Ganova-Raeva, L. M., Schell, M. A., and Denny, T. P. (1997).Hierarchical autoinduction in *Ralstonia solanacearum*: control of acyl-homoserine lactone production by a novel autoregulatory system responsive to 3-hydroxypalmitic acid methyl ester. *J. Bacteriol.,* 179, 7089–7097.

Foster, R. C. (1988). Microenvironments of soil microorganisms. *Biol. Fertil. Soils*, 6, 189-203.

Fuqua, W.C. (1994). Quorum sensing in bacteria: the LuxR–LuxI family of cell density responsive transcriptional regulators. *J. Bacteriol.*, 176, 269–275.

Fuqua, W. C. and S. C. Winans. (1994). A LuxR-LuxI type regulatory system activates *Agrobacterium* Ti plasmid conjugal transfer in the presence of a plant tumor metabolite. *J. Bacteriol.,* 176, 2796–2806.

Garbeva, P., Postma, J., van Veen, J. A., and van Elsas, J. D. (2006). Effect of above-ground plant species on soil microbial community structure and its impact on suppression of *Rhizoctonia solani* AG3. Environ. Microbiol., 8, 233–246.

Gossner, A. S., Kusel, K., Schulz, D., Trenz, S., Acker, G., Lovell, C. R., and Drake H. L. (2006). Trophic interaction of the aerotolerant anaerobe *Clostridium intestinale* and the acetogen *Sporomusa rhizae* sp. nov. isolated from roots of the black needlerush *Juncus roemerianus*. *Microbiology,* 152, 1209–1219.

Gray, K. M., Passador, L., Iglewski, B. H., and Greenberg, E. P. (1994). Interchangeability and specificity of components from the quorum-sensing regulatory systems of *Vibrio fischeri* and *Pseudomonas aeruginosa*. *J. Bacteriol.,* 176, 3076–3080.

Green, B. D. and Martin, K. (2006). Capturing the uncultivated majority. *Curr. Opin. Biotechnol.*, 17, 236–240.

Gributs, C. E. W. and Burns, D. H. (2004). Optical property determination of turbid media by time-resolved transmittance measurements and wavelet transform analysis. *Can. J. Anal. Sci. Spectrosc.*, 49, 193-201.

Hartmann, M. and Franco, W. (2006). Community structure analyses are more sensitive to differences in soil bacterial communities than anonymous diversity indices. *Appl. Environ. Microbiol.*, 72, 7804–7812.

Herrmann, A. M., Clode, P. L., Fletcher, I. R., Nunan, N., Stockdale, E. A., O'Donnel, A. G., and Murphy, D. V. (2007). A novel method for the study of the biophysical interface in soils using nano-scale secondary ion mass spectrometry. *Rapid Commun. Mass Spectrom.*, 21, 29-34.

Hiddink, G. A., Termorshuizen, A. J., Raaijmakers, J. M., and van Bruggen, A. H. C. (2005). Effect of mixed and single crops on disease suppressiveness of soils. *Phytopathology*, 95, 1325-1332.

Huber, B., Riedel, K., Hentzer, M., Heydorn, A., Gotschlich, A., Givskov, M., Molin, S., and Eberl, L. (2001) The *cep* quorum-sensing system of *Burkholderia cepacia* H111 controls biofilm formation and swarming motility. *Microbiology*, 147, 2517–2528

Janssen, P. H. (2006). Identifying the dominant soil bacterial taxa in libraires of 16S rRNA and 16S rRNA genes. *Appl. Environ. Microbiol.*, 72, 1719–1728.

Jastrow, J. D. and Miller, R.M. (1998). Soil aggregate stabilization and carbon sequestration: feedbacks through organomineral associations, in: R. Lal, J. M. Kimble, R. F. Follett, and B. A. Stewart (Eds.), *Soil processes and the carbon cycle* (pp. 207–223). Boca Raton, FL: CRC Press.

Joyce, S. and Smith, D. W. (2003). The effect of microorganisms on pulp mill effluent induced flocculation in receiving waters. *J. Environ. Eng. Sci.*, 2, 39-46.

Kaspar, H. F. and Tiedje, J. M. (1994). Anaerobic bacteria and processes. In R. D. Weaver (Ed.), Methods of soil analysis, part 2. Microbiological and biochemical properties. (Book Series No. 5,, pp. 223–243). Madison, WI: *Soil Science Society of America.*

Kato, S., Haruta, S., Cui, Z. J., Ishii, M., Yokota, A., and Igarashi, Y. (2004). *Clostridium straminisolvens* sp. nov., a moderately thermophilic, aerotolerant and cellulolytic bacterium isolated from a cellulose-degrading bacterial community. *Int. J. Syst. Evol. Microbiol.*, 54, 2043–2047.

Kato, S., Haruta, S., Cui, Z. J., Ishii, M., and Igarashi, Y. (2005). Stable coexistence of five bacterial strains as a cellulose-degrading community. *Appl. Environ. Microbiol.*, 71, 7099–7106.

Katznelson, H. (1965). Nature and importance of the rhizosphere. In K. F. Baker, and W.C. Synder (Eds.), Ecology of Soil Borne Plant Pathogens: *Prelude to Biological Control* (pp. 187–209). Berkeley, CA: University of California Press.

Kemper , B. T. and Rosenau, R. C. (1984). Soil cohesion as affected by time and water content. Soil Biol. Biochem., 7, 153-159.

Kim, S., Sessa, D. J., and Lawton, J. W. (2004). Characterization of zein modified with a mild cross linking agent. *Industrial Crops and Products,* 20, 291-300.

Kirk, J. L., Beaudette, L. A., Hart, M., Moutoglis, P., Klironomos, J. N., Lee, H., and Trevors, J. T. (2004). Methods of studying microbial diversity. *J. Microbiol. Meth.*, 58, 169–188.

Kreth, J., Hagerman, E., Tam, K., Merritt, J., Wong, D. T. W., Wu, B. M., Myung, N. V., Shi, W., and Qi, F. (2004). Quantitative analyses of *Streptococcus mutans* biofilms with quartz crystal microbalance, microjet impingement and confocal microscopy. *Biofilms*, 1, 277–284.

Kusel, K., Trinkwalter, T., Drake, H. L., and Devereux, R. (2006). Comparative evaluation of anaerobic bacterial communities associated with roots of submerged macrophytes growing in marine or brackish water sediment. *J. Exp. Mar. Biol. Ecol.,* 337, 49–58.

Lappin-Scott, H. M. and Costerton J. W. (1990). Starvation and penetration of bacteria in soils and rocks. *Experientia* 46: 807-812.

Lewenza, S., Conway, B., Greenberg, E. P., and Sokol, P. A. (1999). Quorum sensing in *Burkholderia cepacia*: identification of the LuxRI homologs CepRI. *J. Bacteriol.,* 181, 748–756.

Marilley, L. and Aragno, M. (1999). Phylogenetic diversity of bacterial communities differing in degree of proximity of *Lolium perene* and *Trifolium repens* roots. *Appl. Soil Ecol.* 13:127–136.

Matthies, C., Go¨ ßner, A., Acker, G., Schramm, A. and Drake, H. L. (2004). *Lactovum miscens* gen. nov. sp. nov., an aerotolerant, psychrotolerant, mixed fermentative anaerobe from acidic forest soil. *Res. Microbiol.,* 155, 847–854.

McClean, K. H., Winson, M. K., Fish, L., Taylor, A., Chhabra, S. R., Camara, M., Daykin, M., Lamb, J. H., Swift, S., Bycroft, B. W., Stewart, G. S., and Williams, P. (1997). Quorum sensing and *Chromobacterium violaceum*: exploitation of violacein production and inhibition for the detection of N-acylhomoserine lactones. *Microbiol.,* 143, 3703–3711.

Mueller, R. F. (1996). Bacterial transport and colonization in low nutrient environments. *Water Res*., 30, 2681–2690.

Mummey, D., Holben, W., Six, J., and Stahl, P. (2006). Spatial stratification of soil bacterial populations in aggregates of diverse soils. *Microbiol. Ecol*., 51, 404–411.

Muyzer, G. K. and Smalla, C. (1998). Application of denaturing gradient gel electrophoresis (DGGE) and temperature gradient gel electrophoresis (TGGE) in microbial ecology.Antonie Van Leeuwenhoek, 73, 127-141.

Nikolausz, M., Kappelmeyer, U., Székely, A. J., Márialigeti, K., and Kästner, M. (2006). Molecular assessment of the activity dynamics of rhizosphere microbial communities as a function of diurnal redox fluctuation. 11th International Symposium on Microbial Ecology (Abstract). Vienna, Austria

Nunan, N., Wu, K. J., Young, I. M., Crawford, J. W., and Ritz, K. (2003). Spatial distribution of bacterial communities and their relationships with the microarchitecture of soil. *FEMS Microbiol. Ecol*., 44, 203–215.

O'Donnell, A. G., Seasman, M., Macrae, A., Waite, I., and Davies, J. T. (2001). Plants and fertilisers as drivers of change in microbial community structure and function in soils. *Plant Soil*, 232, 135– 145.

Ofek, I., Sharon, N., and Abraham, S. N. (2007). Bacterial Adhesion. In M. Dworkin, S. Falkow, E. Rosenberg, K.-H. Schleifer, and E. Stackebrandt (Eds.), The Prokaryotes, Volume 2: *Ecophysiology and Biochemistry* (pp. 16-31). Springer-Verlag.

Palleroni, N. J. (1997). Prokaryotic diversity and the importance of culturing. Antonie Leeuwenhoek, 72, 3–19.

Pierson, L. S., Wood, D. W., and Pierson, E. A. (1998). Homoserine lactone-mediated gene regulation in plant-associated bacteria. *Ann. Rev. Phytopathol.*, 36, 207-225.

Prosser, J. I., Bohannan, B. J. M., Curtis, T. P., Ellis, R. J. Firestone, M. K.,

Freckleton, R. P., Green, J. L., Green, L. E., Killham, K., Lennon, J. J., Osborn, A. M., Solan, M. van der Gast, C. J., and Young, J. P. W. (2007). The role of ecological theory in microbial ecology. *Nature Rev. Microbiol.,* 5, 384-392.

Ramos, A. C. H. and McBride, M. B. (1996). Goethite dispersibility in solutions of variable ionic strength and soluble organic matter content. Clays and Clay Minerals, 44, 286-296.

Redfield, R. J. (2002). Is quorum sensing a side effect of diffusion sensing? *Trends Microbiol.,* 10, 365–370.

Rickard, A. H., McBain, A. J., Stead, A. T., and Gilbert, P. (2004). Shear rate moderates community diversity in freshwater biofilms. *Appl. Environ. Microbiol.,* 70, 7426-7435.

Sainju, U. M., Jabro, J. D., and Stevens, W. B. (2007). Soil CO2 emission and carbon content as affected by irrigation, 2 tillage, cropping system, and nitrogen fertilization. *J. Environ. Qual.*, (in press).

Schaefer, A. L., Taylor, T. A., Beatty, J. T., and Greenberg, E. P. (2002).Long-chain acyl-homoserine lactone quorum-sensing regulation of *Rhodobacter capsulatus* gene transfer agent production. *J. Bacteriol.,* 184, 6515–6521.

Schaudinn, C., Stoodley, P., Kainoviæ, A., O'Keeffe, T., Costerton, B., Robinson, D., Baum, M., Ehrlich, G., and Webster, P. (2007). Bacterial biofilms, other structures seen as mainstream concepts. *Microbe,* 2, 231-237.

Six , J, Frey, S. D., Thiet, R. K., and Batten, K. M. (2006). Bacterial and fungal contributions to carbon sequestration in agroecosystems. *Soil Sci. Soc. Am. J.,* 70, 555–569.

Six, J., Feller, C., Denef, K., Ogle, S. M., de Moraes SA, J. C., and Albrecht, A. (2002). Soil organic matter, biota and aggregation in temperate and tropical soils – effects of no-tillage. *Agronomie*, 22, 755-775.

Six, J., Elliott, T. and Paustian, K. (1999). Aggregate and soil organic matter dynamics under conventional and no-tillage systems. Soil Sci. Soc. Am. J., 63, 1350-1358.

Sudakevitz, D., Imberty, A., and Gilboa-Garber, N. (2002). Production, properties and specificity of a new bacterial L-fucose and D-arabinose-binding lectin of the plant aggressive pathogen *Ralstonia solanacearum* and its comparison to related plant and microbial lectins, *J. Biochem.* (Tokyo), 132, 353–358.

Thiet, R. K., Frey, S. D., and Six, J. (2006). Do growth yield efficiencies differ between soil microbial communities differing in fungal: bacterial ratios? Reality check and methodological issues. *Soil Biol. Biochem.,* 38, 837-844.

Tiedje, J. M., Sexstone, A. J., Parkin, T. B., Revsbech, N. P., and Shelton, D. R. (1984). In J. Tinsley, and J. F. Darbyshire (Eds.), Biological Processes and Soil Fertility, (vol II, pp. 197–212). The Hague: Martinus Nijhoff.

Tisdall, J. M. and Oades, J. M. (1982). Organic matter and waterstable aggregates in soils. *J. Soil Sci.*, 33, 141-163.

Tisdall, J. M. (1994). Possible role of soil microorganisms in aggregation in soils. *Plant Soil*, 159, 115–121.

Väisänen, R. K., Roberts, M. S., Garland, J. L., Frey, S. D., and Dawson, L. A. (2005). Physiological and molecular characterisation of microbial communities associated with different water-stable aggregate size classes. *Soil Biol. Biochem.,* 37, 2007-2016.

White, S. (1995). The physiology and biochemistry of prokaryotes. New-York, NY: Oxford University Press Inc.

Wintzingerode, F. V., Goëbel, U. B., and Stackebrandt, E. (1997). Determination of microbial diversity in environmental samples: pitfalls of PCR-based rRNA analysis. *FEMS Microbiol. Rev.,* 21, 213-229.

Wood, D. W. and Pierson III, L. S. (1996). The *phzI* gene of *Pseudomonas aureofaciens* 30–84 is responsible for the production of a diffusible signal required for phenazine antibiotic production. *Gene,* 168, 49–53.

Wood, D. W., Gong, F., Daykin, M. M., Williams, P., and Pierson III, L. S. (1997). *N*-acyl-homoserine lactone-mediated regulation of phenazine gene expression by *Pseudomonas aureofaciens* 30–84 in the wheat rhizosphere. *J. Bacteriol.,* 179, 7663–7670.

Wright, S. F. and Upadhyaya, A. (1996). Extraction of an abundant and unusual protein from soil and comparison with hyphal protein of arbuscular mycorrhizal fungi. *Soil Science,* 161, 575-586.

Xavier, J. B., Picioreanu, C., Rani, S. A., van Loosdrecht, M. C. M., and Stewart, P. S. (2005). Biofilm-control strategies based on enzymic disruption of the extracellular polymeric substance matrix – a modelling study. *Microbiology SGM* (UK), 151, 3817-3832.

Zhang, L., Murphy, P. J., Kerr, A., and Tate, M. E. (1993). *Agrobacterium* conjugation and gene regulation by *N*-acyl-L-homoserine lactones. *Nature,* 362, 446–448.

Zinder, S. H., and Salyers, A. A. (2001). Microbial ecology—new directions, new importance. In D. R. Boone and R. W. Castenholz (Eds.), Bergey's manual of systematic bacteriology, vol. 1: the *Archaea* and the deeply branching and phototrophic *Bacteria* (pp. 101–109). New York, N.Y.: Springer-Verlag.

In: Soil Ecology Research Developments
Editors: Tian-Xiao Liu, pp. 139-157

ISBN: 978-1-60021-971-9

Chapter 5

ARE RHIZOBIUM AND SOIL ENZYME ACTIVITIES GOOD INDICATORS OF HEAVY METAL SOIL CONTAMINATION?

S.I.A. Pereira[1], I.V. Castro[2] and E. Figueira[1]
[1]Centre for Cell Biology, Department of Biology, Campus Santiago, 3810-193 AVEIRO, Portugal
[2] Departamento de Ecologia, Recursos Naturais e Ambiente, EFN-INRB, Av. da República 2780-159 OEIRAS, Portugal

ABSTRACT

The widespread pollution of soils is an increasing urgent problem because of its contribution to environmental deterioration on a global basis. Several toxic compounds, such as heavy metals, often contaminate soils. The main sources of heavy metal pollution are mining, industries and application of metal-containing pesticides, fertilizers and sewage sludge. In recent decades there has been increasing concern with heavy metal, not only because of their toxicity to animals, plants and microorganisms, but also because they are highly toxic, mutagenic and/or carcinogenic to humans.

Due to their small size, which provides a large contact area that can interact with the surrounding environment, microorganisms are the first biota showing the impact of toxic compounds. Microorganisms being in intimate contact with the soil environment are considered to be the best indicators of soil pollution. In general, they are very sensitive to low concentrations of contaminants and provide a rapid response to soil perturbation.

Rhizobium spp. are ubiquitous gram-negative soil bacteria that have a profound scientific and agronomic significance due to their ability to establish nitrogen-fixing symbiosis with legumes, which is of major importance to the maintenance of soil fertility.

There is increasing evidence of the adverse effects of heavy metals on soil microbial processes, including on soil enzymatic activities. Soil enzymes are the driving force behind all the biochemical transformations occurring in the soil. Enzymes catalyse all

---

[1] Corresponding author address: Pereira, Sofia Isabel Almeida ,Centro de Biologia Celular, Departamento de Biologia, Campus de Santiago, Universidade de Aveiro, 3810-196 Aveiro, Portugal,E-mail: siapereira@portugalmail.com, Phone: 00351 234 370 782, Fax: 00351 234 865008.

biochemical reactions and are an integral part of nutrient cycling and soil fertility. Therefore, this chapter evaluated the impact of heavy metals on *Rhizobium* populations isolated from a lead mine which activity ceased 50 years ago. In order to reach this goal some physicochemical parameters that influence metal bioavailability was determined, as well as metal concentrations in soils. Soil enzyme activities are highly affected by soil conditions and their evaluation may provide useful information on soil microbial activity and survival. For this reason soil enzyme activities have been proposed as biological indicators of pollution, specially organic, but information about the influence of heavy metal on soil enzyme activities is scarcer. Thus, it was determined the activity of enzymes such as dehydrogenases,, hydrolases, phosphatases, catalase and lipase in heavy metal contaminated soils. Metal tolerance of *Rhizobium* isolates was also screened in artificial media supplemented with different metals (Pb, As, Cd, Cu, Co and Cr) and their tolerance related to soil contamination and enzyme activities.

This chapter can widen the knowledge about the pressure that soil microflora experience under the direct effect of different metals. *Rhizobium* and soil enzyme activities may be useful for the evaluation of agricultural soils pollution, which may be used on the improvement of soil productivity or on the reclaim of contaminated soils.

1. Introduction

Soil is an important natural resource that needs to be preserved and, if possible, its quality and productive ability improved. Doran and Parkin (1994) defined soil quality as the capacity to function within an ecosystem and sustain biological productivity, maintain environmental quality and promote plant, animal and human health. Soil biological and biochemical properties are responsive to small changes that occur in soil properties, thereby providing immediate and accurate information on changes in soil quality. For this reason, soil microbial activity has a direct influence in the ecosystem stability and in soil fertility (Smith and Papendick 1993). Microorganisms play a fundamental role on the biogeochemical cycles and are involved in forming the structure of a soil (Harris and Birch, 1989). The physical, chemical, and biological properties of the soil are all important for its behaviour (Arshad and Coen, 1992; Parr et al., 1992). Characterization of this behaviour should focus on the properties that are most sensitive to environmental stress (Dick and Gupta 1994; Elliott, 1997; Pankhurst et al., 1995; Vanhala and Ahtiainen 1994).

Soil equilibrium can easily be disturbed, especially by human intervention. Heavy metal contamination in soil is of major environmental concern on a world scale. Beside their natural occurrence, heavy metals may enter the ecological environment through anthropogenic activities, such as mining, smelting, sewage sludge disposal, application of pesticides and inorganic fertilizers and atmospheric deposition (Alloway, 1995a; Carrasco et al., 2005; Giller et al., 1989; Malik, 2004; McGrath et al., 1995; Robinson et al. 2001; Shen et al. 2005).

Since microorganisms are the most abundant and most genetically diverse living organisms adapted to almost all environments that exist on the Earth, they have been object of study in numerous works on environmental stress. Several authors have shown that metals adversely influence microorganisms (Shi et al., 2002), affecting their growth, morphology and activities (Bååth et al., 1998; Lakzian et al., 2002; Pereira et al., 2006;), including symbiotic N_2 fixation (Castro et al., 1997; Chaudhary et al., 2004). It is well known that metals are direct and/or indirectly involved in all aspects of microbial growth, metabolism and differentiation. Some metals such as Zn, Cu, Ni and Cr are essential or beneficial micronutrients for plants, animals and microorganisms, whereas others such as Cd, Hg and Pb have no known biological and/or physiological functions being toxic even at very low concentrations (Gadd, 1992).

The increasing occurrence of heavy metal-contaminated areas and the consciousness of their highly toxic effects to humans, as well as to animals, plants and microorganisms explain the growing concern about heavy metal pollution. Hence, the pressure to decontaminate heavy metal polluted soils has increased recently, however concern with the cost of soil remediation lead to explore not only cost effective technologies but also alternative monitoring tools. The immediate concern of rehabilitation practitioners is the availability and capacity or degradative potential of the autochthonous microbial communities. The adverse effects of heavy metals can lead to a reduction in biodiversity and resultant functions in the soil, for this reason it seems deem necessary to use more relevant ecological test species.

Rhizobium are ubiquitous bacteria that have a profound agronomic importance, since the symbiotic association between legumes and rhizobia is by far the most important contributor to the world's supply of biologically fixed N_2 (Somasegaran and Hoben, 1994). For this reason and taking into consideration the importance of legumes in animal and human consumption, some attention has been given to the effects that these elements exert on *Rhizobium* (Ibekwe et al., 1995). Several authors reported that leguminous species grown on contaminated soils, exhibited reduced yields and N content (Chaudhary et al., 2004), since rhizobia isolated from host plants root nodules were completely ineffective to fix N_2 (Giller et al., 1989; Hernandez et al., 2002). In others studies (Pereira et al., 2006) *Rhizobium* has shown to be a genus very sensitive to environmental stresses, evidencing to be a good indicator of soil contamination.

Contradictions on heavy metal ecotoxicity can be attributed to overgeneralization of the outcomes from short-term laboratory studies that focus on a single soil type under controlled conditions. Field data on the effects of heavy metal are limited, and most of the information has become from experiments using sludge containing several metals. However, heavy metal toxicity requires assays with sensitive, reliable and ecologically relevant biological tools. Parameters involving species abundance and diversity along with functional parameters can give a better understanding of heavy metal toxicity.

The presence of heavy metals in the soil may influence the biochemical process by affecting both microbial proliferation and enzyme activities (Gianfreda et al., 2005). Visser and Parkinson (1992) have suggested that the biological and biochemical properties that are most useful for detecting the deterioration of soil quality are those that are most closely related to nutrient cycles, including the activities of soil enzymes. Soil enzyme activities are the driving force behind all the biochemical transformations occurring in the soil since they catalyse all biochemical reactions and are an integral part of nutrient cycling and soil fertility (Bandick et al., 1999). Soil enzymes are believed to be primarily of microbial origin but also originate from plants and animals (Tabatabai, 1994). They are usually associated with viable proliferating cells, but enzymes can be either excreted from a living cell or released into soil solution from dead cells (Tabatabai, 1994). Soil heavy metal contamination exerts an influence on the microbiota, which manifests itself in changes in the enzyme activity (Baran et al., 2004; Gianfreda et al., 2005). There is growing evidence that soil biological parameters may have a potential as early and sensitive indicators of soil ecological stress and restoration (Dick and Tabatabai, 1992).

For this reason, one of the objectives of this chapter was to evaluate the impact of heavy metals on *Rhizobium* populations isolated from a real world pollution site in an abandoned mine deactivated 50 years ago. In order to reach this goal some physicochemical parameters that influence metal bioavailability were determined, as well as total heavy metal concentrations in soils. Metal tolerance of *Rhizobium* isolates was screened in artificial media supplemented with different metals (Pb, As, Cd, Ni, Cu and Co) and their tolerance related to soil contamination. Since soil enzymes are very sensitive to soil contaminants, such as heavy metals, soil enzyme activities have been proposed as biological indicators of pollution (Dick and Tabatabai, 1992; Nannipieri, 1994). Therefore, this chapter also aimed to determine the activity of soil enzymes such as dehydrogenases, hydrolases, phosphatases, catalase and lipase in order to evaluate the degree of soil contamination.

2. Heavy Metal Tolerance of *Rhizobium* Strains Isolated from Braçal Mine

The soil is a dynamic, multi-component system, whose properties are continually modified by microbial, chemical, geological and anthropogenic processes (Saeki et al., 2002). In addition, soil undergoes short-term fluctuations, such as variations in moisture, pH and redox conditions. These changes in soil properties may affect the form and bioavailability of metals. So, the amounts of metals accumulated in soils are dependent on the emission levels, the transport of the metal from the source to the accumulation site and the retention of the metal once it has reached the soil (Alloway, 1995b).

Three composite soils were obtained from a lead mine (Braçal) in Portugal (Aveiro region), deactivated 50 years ago. BA and BD soils were collected in the most contaminated area of the mine, while BC soil was collected in the periphery. This soil was used as control, since its contamination is lower (Figure 1).

Table 1 summarizes the physical and chemical properties of investigated soils. The pH values determined varied significantly ($P < 0.05$) between locations, being in most soils below the optimal pH (6.8) for rhizobia growth. BC and BA soils showed a quite acid pH (4.5 and 4.15 respectively), while the BD soil has a pH of 5.2. According to Ibekwe et al. (1997,1998), soil pH has a significant effect on rhizobia survival. These authors reported that in the more acid soils clover nodules were ineffective for N_2 fixation irrespective of soil metal content. Under lower pH conditions, rhizobia numbers and nodule number *per* plant were both significantly reduced. Broos et al. (2005) also reported that *Rhizobium leguminosarum* bv. *trifolii* survival was reduced at pH <5.5 and can be completely inhibited at pH < 4.5.

Total heavy metal concentrations (Pb, As, Cd, Zn, Ni, Cu, Cr and Co), measured after extraction with *aqua regia*, varied significantly between locations ($P < 0.05$) as shown in Figure 1. Generally, contaminated soils (BA and BD) presented higher concentrations of metals than the control soil (BC). However, Cr concentrations were higher in BC soil than in the other soils, but were below the higher EC limits permitted for heavy metals (CEC, 1986) The concentrations of non-essential metals (Pb and As) were higher in BA soil, while BD soil showed higher levels of essential metals (Ni, Co and Cu). Cadmium concentrations were similar in both soils (Figure 1). Maximum permitted levels imposed by the CEC (1986) were exceed for almost metals in BA and BD soils. (Figure 1).

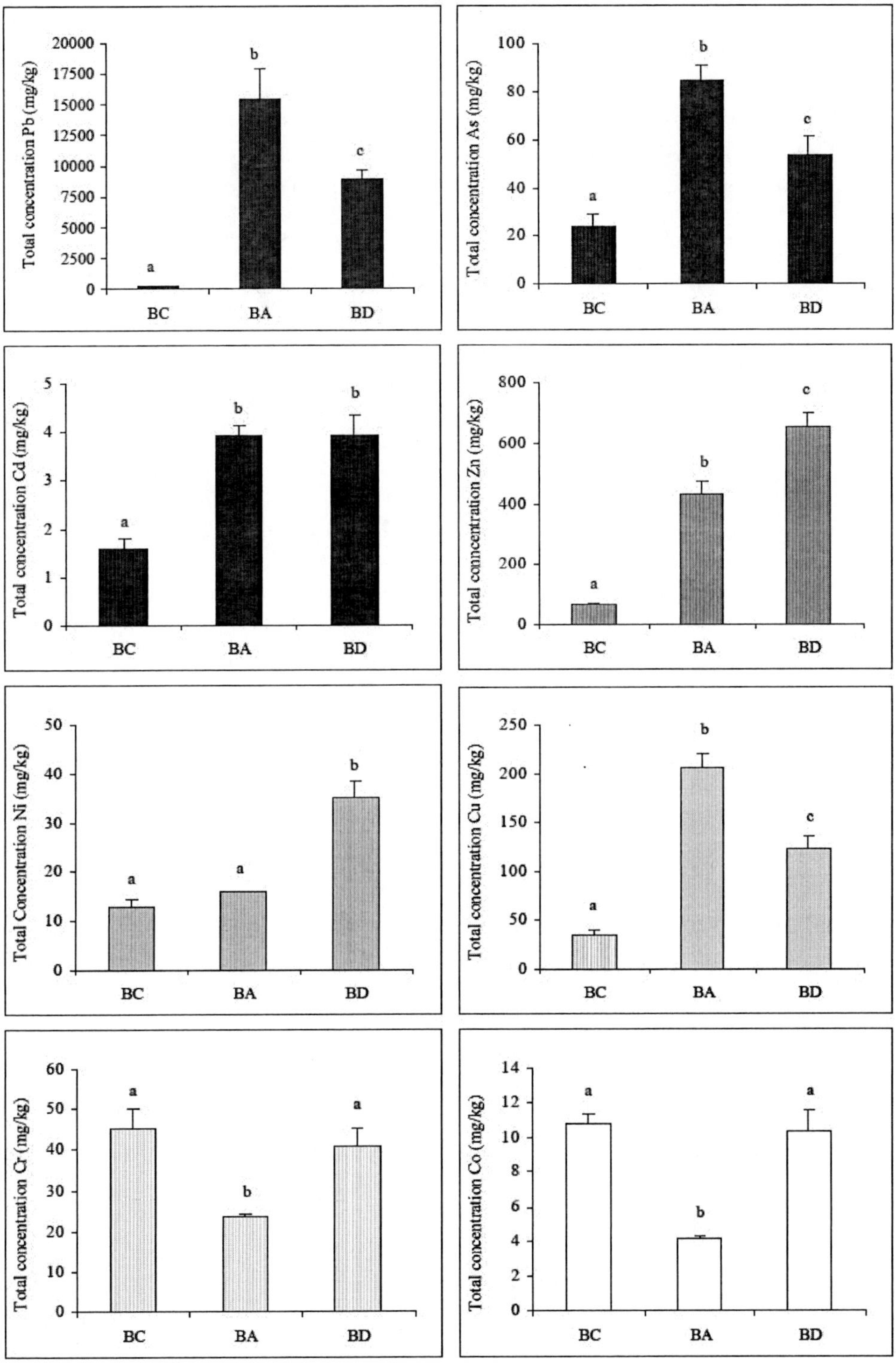

Figure 1. Total heavy metal concentration (Pb, As, Cd, Zn, Ni, Cu, Cr and Co) in the different soils. Data are mean ± SE of three replicate measurements. Different letters indicate means that are significantly different ($P < 0.05$) from each other.

As shown in Figure 1 BA and BD soils are extremely contaminated with Pb and As, imposing a enormous degree of stress in these soils.

Table 1. Physicochemical analysis of the soils from different locations. Data are the mean ± SE from three replicate measurements. Different superscripts letters (a-c) represent significant differences (P < 0.05) between values at same column

Soil	Physicochemical analysis			
	Water content (%)	pH	Redox potential (Eh)	Organic matter (%)
BC	16.4 ± 0.26 [a]	4.50 ± 0.08 [a]	127.7 ± 40.38 [a]	9.35 ± 0.35 [a]
BA	7.8 ± 0.18 [b]	4.15 ± 0.13 [b]	123.3 ± 4.04 [a]	4.72 ± 0.53 [b]
BD	24.1 ± 1.20 [c]	5.21 ± 0.005 [c]	121.7 ± 2.89 [a]	10.81 ± 1.06 [a]

Several authors reported that long-term metal deposition into soils results in high metal concentrations, which affects negatively soil microflora (Smith and Giller, 1992; Matsuda et al., 2002). In this work, we were able to isolate rhizobia from all soils, indicating that these bacteria were capable to survive under the metal concentrations determined. According to Ibekwe et al. (1997) survival can be related to the physical protection of clay minerals and organic matter or to the existence of microsites where metal contamination may be minimal. These "niches" may harbour rhizobia that are not resistant to heavy metals and may explain the presence of sensitive genotypes in contaminated soils.

Gross measurements of microbial diversity have been used to assess environmental stress (Atlas, 1984), but such studies are hampered by problems of sampling, extraction and culturing leading to bias towards certain groups within microbial communities. Pollution may lead to a decrease in microbial diversity due to the extinction of species which lack sufficient tolerance to the stress imposed, and enhanced population of other species which thrive under stress (Atlas, 1984). Microorganisms within species of the same genus or within strains of the same species can differ in their sensitivity to metals. Giller et al. (1993) demonstrated that *Sinorhizobium meliloti* was less sensitive, in terms of growth, to Cd than *R. leguminosarum* and *R. loti.* Therefore, the use of microbial populations which are easily culturable, ubiquitous, sensitive to several contaminants, and whose intraspecific diversity is enough to point out tolerance differences to contaminated soils, can be a way to overcome the problems enumerated above.

Ubiquitous species, such as rhizobia, allow a wider comparison of heavy metal effects between different environments. *Rhizobium* can probably be used as an effective tool in ecotoxicity assays since these bacteria are known to be sensitive to a wide variety of pollutants and are very important to soil fertility in a wide range of environments (Figueira, 2000; Figueira et al., 2005; Pereira et al., 2006; Vig et al., 2003).

In order to evaluate the influence of heavy metal contamination by a lead abandoned mine in the soil microflora, forty-one isolates of *Rhizobium leguminosarum* bv. *trifolii* were obtained from nodules of white clover (*Trifolium repens*, L.) plants grown in BA and BD contaminated soils and also in control soil (BC).

Based on the percentage of growth inhibition, the 41 isolates obtained were classified as: sensitive (growth inhibition – 80-100%), moderately tolerant (growth inhibition - 60-80%) and tolerant (growth inhibition - <60%).

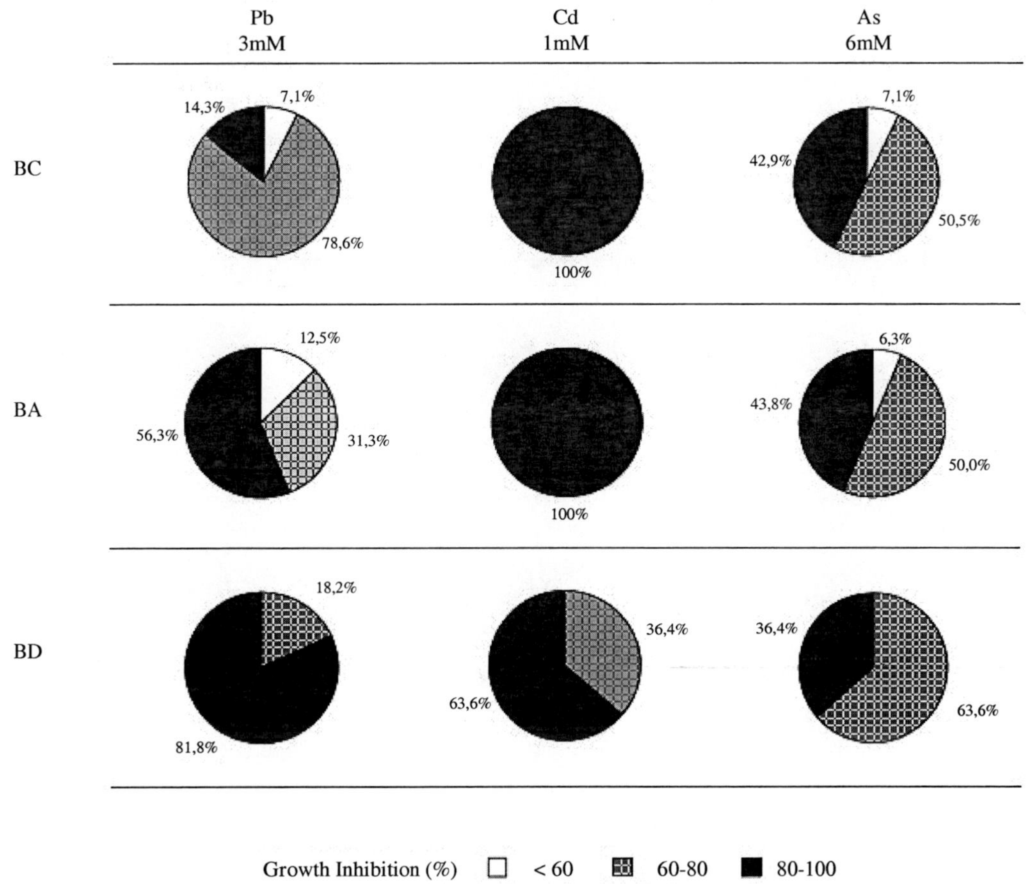

Figure 2. Percentage of isolates exhibiting a growth inhibition <60% - tolerant, 60-80% - moderately tolerant and 80-100% - sensitive in the different locations, for non-essential metals: Pb, Cd and As.

Figure 2 shows the percentage of sensitive, moderately tolerant and tolerant isolates in each location, for the concentration where all isolates still showed growth, to non-essential metals (Pb – 3 mM, As – 6 mM and Cd – 1 mM). Generally, isolates showed high sensitivity to Cd, since growth was severely inhibited at 1 mM for all isolates of BC and BA soils and to 2/3 of BD soil, and moderately tolerance to As, as in the three soils the percentage of sensitive isolates to 6 mM As was similar (between 36.4% and 43,8%). Pb was the metal where tolerance differences between soils were more marked. BC isolates showed the highest tolerance, since only 14.3% of them were sensitive to Pb. *Rhizobium* BD population showed the lowest tolerance, being more than 80% of the isolates sensitive to 3 mM Pb.

Figure 3 shows the percentage of sensitive, moderately tolerant and tolerant isolates in each location for the essential metals (Ni – 0.5 mM, Cu – 0.5 mM and Co – 0.5 mM). Once again, BC isolates showed a moderately tolerance to all essential metals tested. On the other hand, BA isolates showed high sensitivity to all metals.

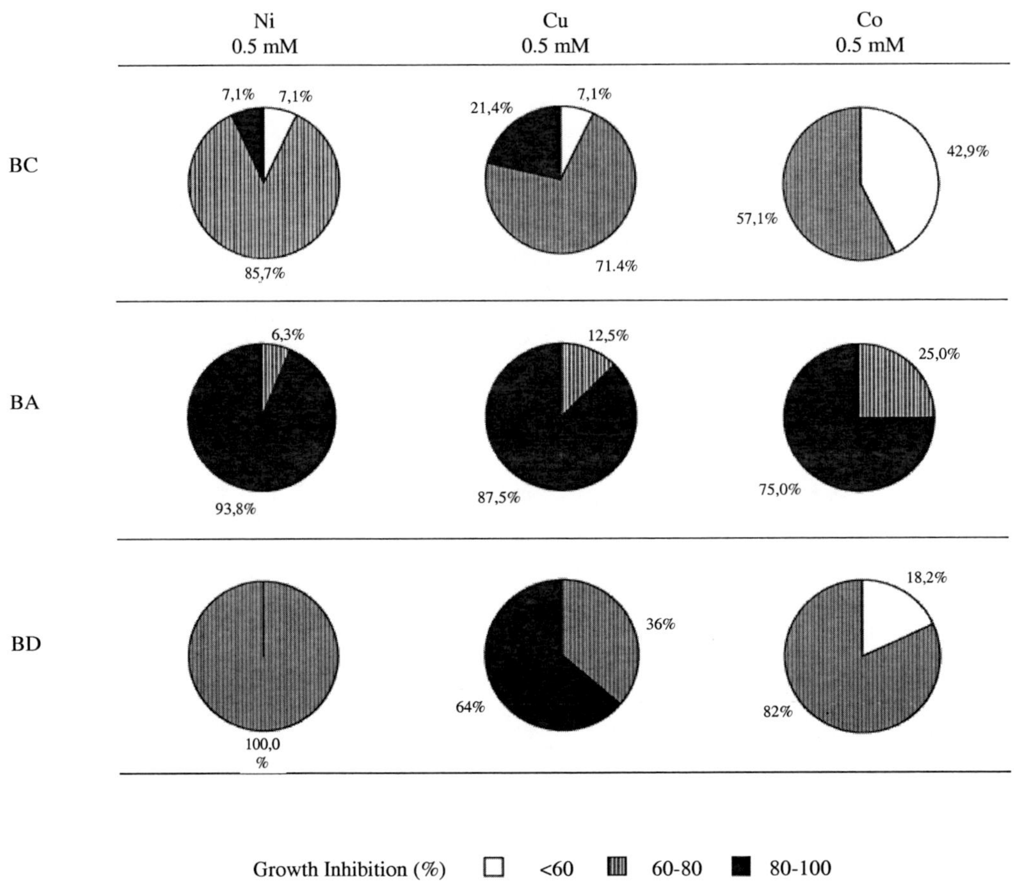

Figure 3 .Percentage of isolates exhibiting a growth inhibition <60% - tolerant, 60-80% - moderately tolerant and 80-100% - sensitive in the different locations, for essential metals: Ni, Cu and Co.

Pb concentrations were extremely high in BA and BD soils, while in control soil the level of this metal was 50 times lower. The activity of the mine ceased 50 years ago, however the inputs of metals resulting from the old smelter and from the tunnels that access to the mine continue to happen. Due to their localization towards the tunnels accessing to the mine and the old smelter, BA and BD soils are easily waterlogged with water from the smelter and the tunnels, which easily became inundated when there is heavy rainfall or floods from a nearby river. These events take place frequently during winter, inputting heavy metals, even now, 50 years after the mine deactivation. BC soil was collected from the periphery of the mine, where the inputs of metals at the present are low.

Nevertheless, *Rhizobium* population from BC soil shows higher tolerance to Pb and As than the populations from BA and BD soils. Probably when the mine was laborating, BC isolates suffered the influence of higher heavy metal concentrations, which originate a selective pressure leading, at that time, to the appearance of genotypes with high lead tolerance, fat already described by Bååth (1992) and Bååth *et al.* (1998) who considered the possibility of heavy metals to exert a selective pressure in bacteria of contaminated soils, resulting in the presence of more tolerant organisms. These tolerant genotypes persisted until now in the population, evidencing an overlap of time between actual heavy metal concentrations and *Rhizobium* tolerance, probably due to the time needed for the population to adapt to the new conditions.

Interestingly, results show that *Rhizobium* population with higher resistance to Pb (BC isolates) also have tolerance to other metals such as As, Cu, Ni ad Co. This fact suggests that the selective pressure induced by Pb allowed the development of resistance mechanisms that bestow tolerance to different metal stresses, which is corroborated by Díaz-Raviña et al. (1994) and Nies (1992), who reported the existence of heavy metal multiple tolerance patterns in soil microorganisms.

3. Soil Enzyme Activities

Soil enzymes are important to several crucial functions. They are intimately involved in the cycling of nutrients, they affect the efficiency of fertilization, reflect the microbiological activity in soil and act as indicators of soil change (Gianfreda et al., 2005; Giller et al., 1998). Soil enzymes are the catalysts of important metabolic process functions including the decomposition of organic inputs and the detoxification of xenobiotics (Gianfreda et al., 2005).

Soil enzymes activity is used as a sensitive indicator of the effect of pollutants, including metals in soils (Bayer et al., 1982; Dick, 1997; Giller et al., 1998; Top et al., 1999). As previously pointed out (Brookes, 1995), applications of European Community standards for heavy metal concentrations result in significant negative impacts on microbial biomass and activity, indicating the greater sensitivity of the soil to these impacts in comparison with plants or animals. Important questions are what constitutes a significant environmental impact on soils and when is reclamation complete. Indicators are needed, not only as surrogates for reflecting the functionality of soils, but also to guide reclamation. The soil microbial component and soil enzyme activities are attractive as indicators for monitoring disturbance or pollution of soils because of their central and crucial role in the functions of the soil ecosystem.

Enzymes may rapidly respond to the changes caused by both natural and anthropogenic factors (Gianfreda et al., 2005). The strong inhibition of the activities of a variety of enzymes has been reported in metal polluted soils over the past years (Doelman and Haanstra, 1986; Marzadori et al., 1996; Mathur et al., 1980; Tyler, 1974; Tabatabai, 1977) and these effects vary considerably.

Heavy metals may inhibit enzyme activities by masking catalytically active groups, having denaturing effects on the conformation of proteins, competing with the natural ions involved in the formation of enzyme-substract complexes (Gianfreda et al., 2005; Hinojosa et al., 2004; Nannipieri, 1994), or by affecting the synthesis of the enzymes within the microbial cells . For these reasons soil enzymes activities have been suggested as suitable indicators of soil quality since they have been considered sensitive indicators to measure the degree of soil degradation in both natural and agro-ecosystems, being thus well suited to measure the impact of pollution on the quality of soil (Dick, 1997; Giller et al., 1998; Trasar-Cepeda et al., 2000).

Dehydrogenase (DHA) activity typically occurs in all viable microbial cells (Gianfreda et al., 2005). Thus, its measurement is usually related to the presence of viable microorganisms and their oxidative ability has been often used as a functional indicator of soil health. DHA is related to a group of intracellular enzymes that are present in active soil microorganisms. This enzyme is found in all living organisms and takes part in many metabolic reactions involved in oxidative energy transfer in microbes. As dehydrogenases are not active as extracellular enzymes in the soil, the management of DHA has been used as a good overall indicator of microbial activity and of the capacity of microbes to oxidize soil organic matter (Bolton et al., 1985; Dick, 1997). Several studies have demonstrated that dehydrogenase enzyme activity of microorganisms is among the most sensitive parameters for the evaluation of toxicity (Shen et al., 2005).

In this chapter, dehydrogenase activity was determined using a method developed by Casida et al. (1964) using as subtract triphenyltetrazolium chloride (TTC) and glucose. After 24 hours of incubation triphenylformazan formed (TPF) from TTC was extracted with methanol. According to Obbard et al. (2001) substrate-induced DHA was found to be a sensitive assay for determining metal effects on the physiologically active soil microbial biomass. The precision of the assay allows evaluating, separately, effects due to pH, contaminant concentration and other soil properties.

DHA activity varied significantly ($P< 0.05$) between control soil and BA and BD soils (Figure 4), being the DHA activity in BC soil 2 times higher. This finding is in agreement with (Trasar-Cepeda et al. 2000) who reported less dehydrogenase activity in metal-contaminated soil than in a similar uncontaminated soil, and with Marzadori et al. (1996) who stated that dehydrogenase activity was inhibited by the toxic effects of heavy metals, particularly Pb. Dar (1996) also reported a decrease in dehydrogenase activity at 50 mg Cd $\cdot$ kg^{-1} in a laboratory study.

DHA activity appears to depend on the type of pollutant; for example, it is high in soils polluted with pulp and paper mill effluents and low in soils polluted with fly ash (Trasar-Cepeda et al., 2000). Nevertheless, several studies have demonstrated that dehydrogenase enzymatic activity of microorganisms is among the most sensitive parameters for evaluation of toxicity (12), and that may be the reason why dehydrogenases activity are a frequently used test for determining the influence of various pollutants (heavy metals, pesticide, crude oil, etc.) on the microbiological quality of soil (Margesin et al., 2002).

The fluorescein diacetate hydrolysis (FDAH) assay measures the hydrolytic enzyme activities of microbial populations and can provide an estimate of overall microbial activity in an environmental sample (Adam and Duncan, 2001; Bandick et al., 1999; Burns, 1982). The assay is considered non-specific because it is sensitive to the activity of several enzyme classes including urease, lipases, phosphatases, glucosidase, esterases and proteases (Bandick et al, 1999; Hayano and Tubakai, 1985). The enzymes responsible for Fluorescein diacetate (FDA) hydrolysis are plentiful in the soil environment, and are involved in the decomposition of many types of tissues (Adam and Duncan, 2001). The ability to hydrolyse FDA thus seems widespread, especially among the major decomposers, bacteria and fungi. Generally more tan 90% of the energy flow in a soil system passes through microbial decomposers, therefore an assay which measures microbial decomposing activity will provide a good estimate of total microbial activity (Adam and Duncan, 2001).

In our study FDA assay was performed according to Adam and Duncan (2001). However, no relationship where detected between soil contamination and FDA assay, since all soils presented similar activities of these enzymes (Figure 4).

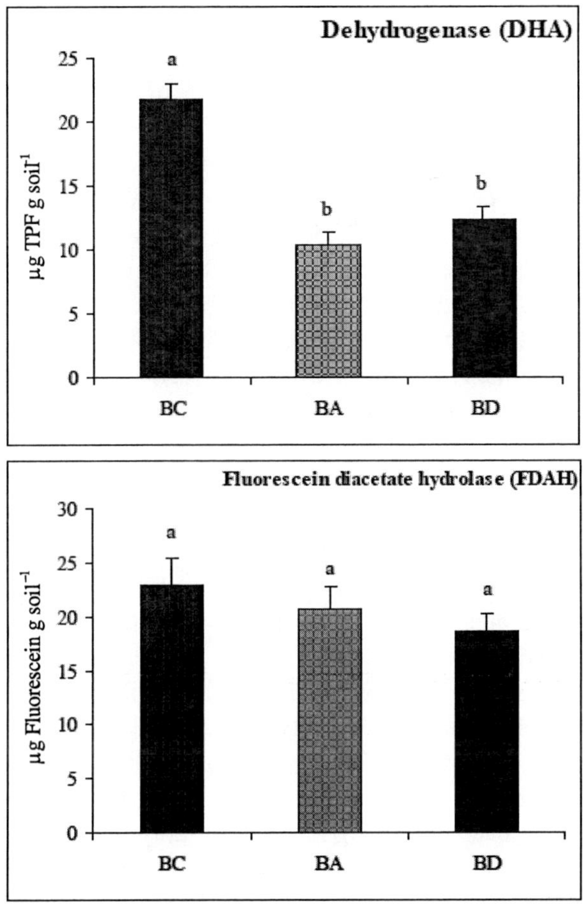

Figure 4 .Dehydrogenase activity (µg TPF g soil^{-1}) determined over a period of 24 hours; and FDA activity (µg Fluorescein g soil^{-1}) determined over a period of 20 minutes, in BC, BA and BD soils. Data are mean ± SE of 30-40 replicate measurements. Different letters indicate means that are significantly different ($P < 0.05$) from each other.

Catalase is an intracellular enzyme found in all aerobic bacteria and in most facultative anaerobes. Catalase activity in soils is considered an indicator of aerobic microbial activity and has been related to both the number of aerobic microorganisms and to soil fertility (Garcia and Hernández, 1997).

In this work catalase activity varied significantly ($P<0.05$) between control soil (BC) and BA and BD soils. The soil with higher organic matter content (BD) was the same where catalase activity was lower (Figure 5). This is in disagreement with Kizilkaya et al. (2004) who related increases of catalase activity to the higher soil organic matter content. Heavy metal contamination negatively influences catalase activity, since catalase is a metaloenzyme, and the presence of heavy metals such as Ni and Cu inhibits its activity (White and White, 1997). Kizilkaya et al. (2004) reported that catalase activity is reduced by heavy metal contamination in agricultural soils. On the other hand, under heavy metal stress catalase activity of microorganisms, such as *Rhizobium* (Corticeiro et al., 2006), increase markedly. Thus, catalyse activity appears to be influenced in different ways by heavy metals, explaining why the results between BA and BD soils are difficult to interpret.

Phosphatases are broad enzymes that catalyze the hydrolysis of both esters and anhydrides of H_3PO_4 (Eivazi and Tabatabai, 1977). These enzymes are responsible for soil organic phosphorous mineralization and the release of inorganic phosphorous needed by microorganism and plants. The enzymes are classified as acid and alkaline phosphatases due to their optimal activities at (< 6.5) and (11) pH ranges, respectively. Phosphatases activities have been described as a good indicator for heavy metal contamination (Wyszkowska, 2002). However, Al-Khafaji and Tabatabai (1979), Bardgett et al. (1994) and Yeates et al. (1994) described that acid phosphatase activity was less affected by heavy metals than arylsulfatase. Acid and alkaline phosphatase activities were determined according to Eivazi and Tabatabai (1977). Results showed in Figure 5 indicate that acid phosphatase activity varied significantly ($P<0.05$) between locations. However, it was not detected activity of alkaline phosphatase in all of the soils tested. According to Gianfreda et al. (2005) soil enzyme activities are usually significantly correlated to soil pH. The investigated soils showed low pH values (Table 1), which could affect the activity of alkaline phosphatase. Arguably, the field soil pH would not directly affect the enzyme assay as the assays are run using buffers at their optimal pH. However, the long-term effect of low pH would probably cause shifts in microbial community composition and size and this, in turn, would affect enzyme dynamics at the time of sampling (Hinojosa et al., 2004). Furthermore, according to Eivazi and Tabatabai (1977) acid phosphates is predominant in acid soils and alkaline phosphatase is predominant in alkaline soils. Acid phosphate activity was higher in the most contaminated soils. However, according to Wyszkowska (2002) the activity of acid and alkaline phosphatases decreased in soils contaminated with Cr (IV), and in fact our work shows that acid phosphatase activity is negatively correlated with Cr levels. On the other hand, phosphatases can be inhibited not only by heavy metals (Tyler, 1974), but also by inorganic phosphate, which produces a feedback inhibition of this enzyme (Nannipieri et al., 1979). Actually, in this study acid phosphatase activity is higher in the BA soil with the lower level of phosphorus (7 mg kg^{-1}) and lower in BC, which has a higher level of phosphorus (131 mg kg^{-1}).

Lipase activity was determined according to Margesin et al. (2002). Lipase activity is often associated to oil degradation; however information about its behaviour in the presence of metals is limited. In this work, lipase activity was similar in all soils (Figure 5), showing that its activity is not differentially influenced by heavy metals concentrations present in the three soils.

4. Conclusion

Due to their small size, which provides a large contact area that can interact with the surrounding environment, microorganisms are the first biota showing the impact of toxic compounds. For this reason they are considered to be the best indicators of soil pollution. Metals exert a selective pressure on organisms, resulting in microbial populations with higher tolerance, but with lower diversity, when compared to unpolluted neighbouring areas (Bååth,

1992; Bååth et al., 1998). The soil microbial population presents variability, which, in a non-stressed environment, would not bring any advantage.

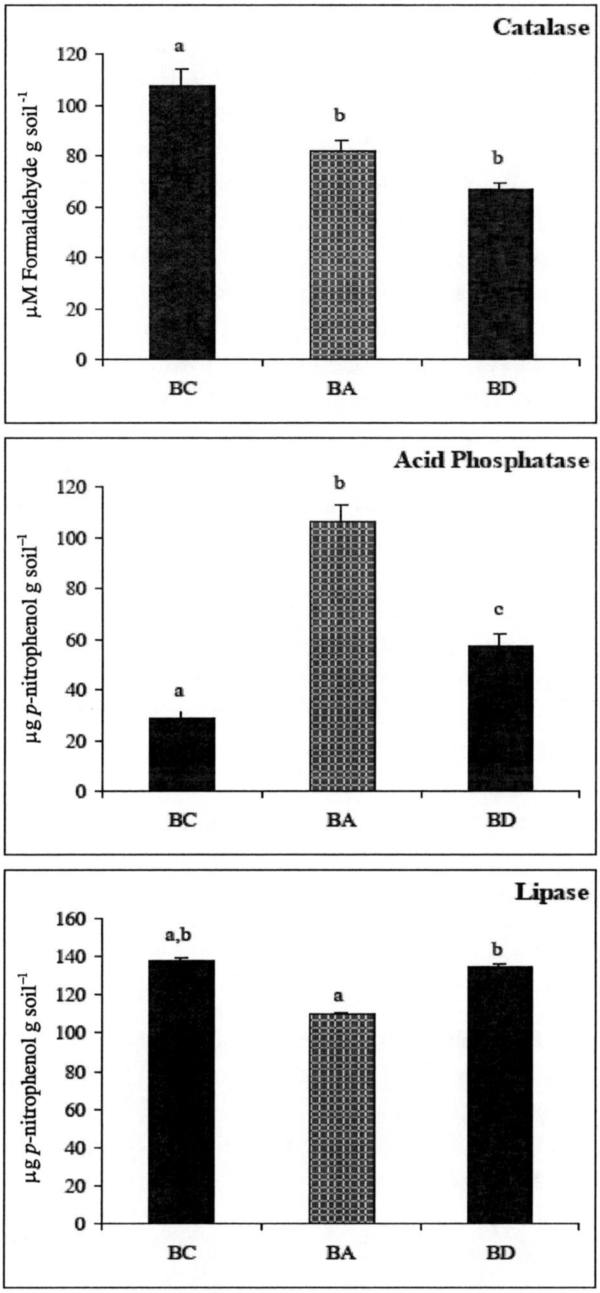

Figure 5. Catalase activity (μM Formaldehyde g soil^{-1}) determined over a period of 1 hour; Acid phosphate activity (μg *p*-nitrophenol g soil^{-1}) determined over a period of 40 minutes; and Lipase activity (μg *p*-nitrophenol g soil^{-1}) determined over a period of 10 minutes, in BC, BA and BD soils. Data are mean ± SE of 30-40 replicate measurements. Different letters indicate means that are significantly different ($P < 0.05$) from each other.

However, when the conditions are changed and this population is submitted to a selective pressure, the genotypes presenting tolerance ability can survive, hence yielding a tolerant population, with lower genetic variability. Hinojosa et al. (2004) referred that indicators of soil quality and health, to be useful and practical, must meet certain criteria.

These include: sensitivity to perturbation or contamination; a relationship to soil function; reproducibility; low temporal and spatial variability; and have simple sampling and analytical methods. Both enzymes activity and *Rhizobium* tolerance showed to meet some of these criteria. Our results evidenced that dehydrogenases and catalase, which are intracellular oxirreductases, showed a good correlation with the heavy metal contamination of soils. Nevertheless, there are several works (Gianfreda et al., 2005; Kizilkaya et al., 2004) alerting to several factors (pH, organic matter, clay etc.) that affect soil enzymatic activity, and they should be taken into consideration, if we want to extrapolate or compare values from environments with distinct edaphic, climatic or contamination features. *Rhizobium* isolates showed a wide range of heavy metals tolerance, confirming the responses showed to lower levels of heavy metal contamination (Pereira et al, 2006) and other environmental stresses (Figueira, 2000), which indicates that, in addition to their ubiquity, culturability, and involvement in the N cycle, *Rhizobium* populations are a good indicator of soil quality and heath. However, it should be taken in consideration that rhizobia populations take some time to adapt to the habitat conditions, which can limit their ability as immediate indicators of the soil's health. Even so, this feature presents the advantage of reflecting the circumstances that conditioned the population's evolution in the past.

A study, like this chapter, that simultaneously determine the soil enzyme activity and the tolerance *Rhizobium* populations towards the heavy metal contamination, will provide an indication of not only the perturbations that a specific pollutant is causing in the present, but also the effects induced in the past, which is of extreme importance in long-term contaminations or with pollutants that are recalcitrant like heavy metals, and may be useful for the evaluation of natural and agricultural soils pollution, and for the improvement of soil productivity or the reclaim of contaminated soils.

5. References

Adam, G; Duncan, H. Development of sensitive and rapid method for the measurement of total microbial activity using fluorescein diacetate (FDA) in a range of soils. *Soil Biology and Biochemistry*, 2001 33, 943-951.

Al -Khafaji, AA; Tabatabai, MA. Effects of trace elements on arylsulfatase activity in soils. *Soil* Science, 1979 127, 129–133.

Alloway, BJ. Introduction. In: Alloway BJ editor. *Heavy Metals in Soils*. New York: Blackie Academic and Professional; 1995a. 3-9.

Alloway, BJ. Soil processes and the behaviour of metals. In: *Heavy Metals in Soils*. New York: Blackie Academic and Professional; 1995b. 11-37.

Arshad, MA; Coen, GM. Characterization of soil quality: physical and chemical criteria. *American Journal of Alternative Agriculture*, 1992 7, 25–33.

Atlas, RM. Use of microbial diversity measurements to assess environmental stress. In: Klug MJ, Reddy CA editors. *Current Perspectives in Microbial Ecology*. Washington: American Society for Microbiology; 1984; 540–545.

Bååth, E. Measurement of heavy metal tolerance of soil bacteria using thymidine incorporation into bacteria extracted after homogenization-centrifugation. *Soil Biology and Biochemistry*, 1992 24, 1167-1172.

Bååth, E; Díaz-Ravina, M; Frostegård, Å; Campbell, CD. Effect of metal-rich sludge amendments on the soil microbial community. *Applied and Environmental Microbiology*, 1998 64, 238-245.

Bandick, AK; Dick, RP. Field management effects on soil enzyme activities. *Soil Biology and Biochemistry*, 1999 31, 1471-1479.

Baran, S; Bielinska, JE; Oleszczuk, P. Enzymatic activity in an air.eld soil polluted with polycyclic aromatic hydrocarbons. *Geoderma*, 2004 118, 221–232.

Bardgett, RD; Speir, TW; Ross, DJ; Yeates, GW; Kettles, HA. Impact of pasture contamination by copper, chromium, and arsenic timber preservative on soil microbial properties and nematodes. *Biology and Fertility of Soils*, 1994 18, 71–79.

Bayer, H; Mitterer, M; Schinner, F. Der Ein.uss von Insektiziden auf mikrobiogene Prozesse in Ah-Materialien eines landwirtschaftlich genutzten Bodens. *Pedobiologia*, 1982 23, 311–319.

Bolton Jr, H; Elliot, LF; Papendick, RI; Berdicek, DF. Soil microbial biomass and selected soil enzyme activities: effects of fertilization and cropping practices. *Soil Biology and Biochemistry*, 1985 14, 423-427.

Brookes, PC. The use of microbial parameters in monitoring soil pollution by heavy metals. *Biology and Fertility of* Soils, 1995 19, 269-275.

Broos, K; Bryens, H; Smolders, E. Survival of rhizobia in soil is sensitive to elevated zinc in the absence of the host plant. *Soil Biology and Biochemistry*, 2005 37, 573-579.

Burns, RG. Enzyme activity in soil: location and a possible role in microbial ecology. *Soil Biology and Biochemistry*, 1982 34, 121-125.

Carrasco, JA; Armario, P; Pajuelo, E; Burgos, A; Caviedes, MA; López, R; Chamber, MA; Palomares AJ. Isolation and characterisation of symbiotically effective *Rhizobium* resistant to arsenic and heavy metals after the toxic spill at the Aznalcóllar pyrete mine. *Soil Biology and Biochemistry*, 2005 37,1131-1140.

Casida, LE; Klein, DA; Santaro, T. Soil desidrogenase activity. *Soil Science*, 1964 96, 371-376.

Castro IV; Ferreira EM; McGrath, SP. Effectiveness and genetic diversity of *Rhizobium leguminosarum* bv. *trifolii* isolates in Portuguese soils polluted by industrial effluents. *Soil Biology and Biochemistry*, 1997 29, 1209-1213.

Chaudhary, P; Dudeja, SS; Kappor, KK. Effectivity of host *Rhizobium leguminosarum* symbiosis in soils receiving sewage water containing heavy metals. *Microbiology Research*, 2004 159, 121-127.

CEC - Commission of the European Communities. Council Directive on the protection of the environment, and in particular of the soil, when sewage sludge is used in agriculture. *Official Journal of European Community*, 1986 L181. Annex 1A, 10.

Corticeiro, S; Lima, A; Figueira, E. The importance of glutathione in oxidative status of *Rhizobium leguminosarum* bv. *viciae* under Cd exposure. *Enzyme and Microbial Technology*, 2006 40, 132-137.

Dar, GH. Effects of cadmium and sewage-sludge on soil microbial biomass and enzyme activities. *Bioresource Technology*, 1996 56, 141-145.

Díaz-Raviña, M; Bååth, M; Frostegård, A. Multiple heavy metal tolerance of soil bacterial communities and its measurement by thymidine incorporation technique. *Applied and Environmental Microbiology*, 1994 60, 2238-2247.

Dick, WA; Tabatabai, MA. Potential uses of soil enzymes. In: Metting B editor. *Soil Microbial Ecology*. New York: Marcel Dekker; 1992; 95–127.

Dick, RP; Gupta, VVSR. A conceptual model for the role of abiotic soil enzymes in a microbial ecology: A potential analogue for soil quality. In: Pankhurst CE, Doube BM, Gupta VVSR, Grace PR editors. *Soil Biota: Management in Sustainable Farming Systems*. Adelaide: Commonwealth Scientific and Industrial Research Organization Information. 1994. 167–168.

Dick, RP. Soil enzyme activities as integrative indicators of soil health. In: Pankhurst CE, Doube BM, Gupta VVSR editors. *Biological indicators of soil health*. Wallingford, USA: CAB International; 1997; 121-156.

Doelman, P; Haanstra, L. Short- and long –term effects of heavy metals on urease activity in soils. *Biology and Fertlity of Soils*, 1986 2, 213-218.

Doran, JW; Parkin, TB. Defining and assessing soil quality. In: Doran JW editor. *Defining Soil Quality for Sustainable Environment*. Madison: SSSA Special Publication; 1994; 3-23.

Eivazi, F; Tabatabai, MA. Phosphatases in soils. *Soil Biology and Biochemistry*, 1977 9, 167-172.

Elliott, ET. 1997. Rationale for developing bioindicators of soil health. In: Pankhurst CE, Doube BM, Gupta, VVSR editors. *Biological Indicators of Soil Health*. Wellington: CAB International; 1997;. 49–78.

Figueira, E. Aspectos da tolerância salina em *Pisum sativum* L.: influência da nutrição azotada. PhD Thesis. Aveiro University. 2000.

Figueira, EMAP; Lima, AIG; Pereira, SIA. Cadmium tolerance plasticity in *Rhizobium leguminosarum* bv. *viciae*: glutathione as a detoxifying agent. *Canadian Journal of Microbiology*, 2005 51: 7–14

Gadd, GM. Metals and microorganisms: A problem of definition. *FEMS Microbiology Letters*, 1992 100, 197-204.

García, C; Hernández, T. Biological and biochemical indicators in derelict soils subjected to erosion. *Soil Biology and Biochemistry*, 1997 29, 171-177.

Gianfreda, L; Rao, MA; Piotrowska, A; Palumbo, G; Colombo, C. Soil enzyme activities as affected by anthropogenic alterations: intensive agricultural practices and organic pollution. *Science of Total Environment*. 2005 341, 265-279.

Giller, KE; McGrath, SP; Hirsch, PR. Absence of nitrogen fixation in clover grown on soil subject to long-term contamination with heavy metals is due to survival of only ineffective *Rhizobium*. *Soil Biology and Biochemistry*, 1989 21, 841-848.

Giller, KE; Nussbaum, R; Chaudri, AM; McGrath, SP. *Rhizobium meliloti* is less sensitive to heavy-metal contamination in soil than *R. leguminosarum* bv. *trifolii* or *R. loti*. *Soil Biology and Biochemistry*, 1993 25, 273-278.

Giller, KE; Witter, E; McGrath, SP. Toxicity of heavy metals to microorganisms and microbial process in agricultural soils: a review. *Soil Biology and Biochemistry*, 1998 300, 1389-1414.

Harris, JA; Birch, P. Soil microbial activity in opencast coal mine restoration. *Soil Use and Management*, 1989 5, 155-160.

Hayano, K; Tubaki, K. Origin and properties of b-glucosidase activity of tomato field soil. *Soil Biology and Biochemistry*, 1985 17, 53-557.

Hernandez, T; Moral, R; Perez-Espinosa, A; Moreno-Caselles, J; Perez-Murcia, MD; Garcia, C. Nitrogen mineralization potential in calcerous soils amended with sewage sludge. *Bioresource Technology*, 2002 83, 213-219.

Hinojosa, MB; Carreira, JA; García-Ruíz, R; Dick, RP. Soil moisture pre-treatment effects on enzyme activities as indicators of heavy metal-contaminated and reclaim soils. *Soil Biology and Biochemistry*, 2004 1559-1568.

Ibekwe, AM; Angle, JS; Chaney, RL; van Berkum, P. Sewage sludge and heavy metal effects on nodulation and nitrogen fixation of legumes. *Journal of Environmental Quality*, 1995 24, 1199-1204.

Ibekwe, AM; Angle, JS; Chaney, RL; van Berkum, P. Differentiation of clover *Rhizobium* isolated from biosolids-amended soils with varying pH. *Soil Science Society American Journal*, 1997 61, 1679-1685.

Ibekwe, AM; Angle, JS; Chaney, RL; van Berkum, P. Zinc and cadmium effects on Rhizobia and white clover using chelator-buffered nutrient solution. *Soil Science Society American Journal*, 1998 62, 204-211.

Kizilkaya, R; Askin, T; Bayrakli, B; Saglam, M. Microbiological characteristics of soils contaminated with heavy metals, *European Journal of Soil Science*, 2004 40, 95-102.

Lakzian, A; Murphy, P; Turner, A; Beynon, JL; Giller, KE. *Rhizobium leguminosarum* bv. *viciae* populations in soils with increasing heavy metal contamination: abundance, plasmid profiles, diversity and metal tolerance. *Soil Biology and Biochemistry*, 2002 34, 519-529.

Malik, A. Metal bioremediation through growing cells. *Environmental International*, 2004 30, 261-278.

Margesin, R; Zimmerbauer, A; Schinner, F. Monitoring of bioremediation by soil biological activities. *Chemosphere*, 2002 40, 339-346.

Mathur, SP; MacDougall, JI; McGrath, M. Levels of activities of some carbohydrates, protease, lipase and phosphates in organic soils of differing copper content. *Soil Science*, 1980 129, 376-385.

Marzadori, C; Ciavatta, C; Montecchio, D; Gessa, C. Effetcs of lead pollution on different soil enzyme activities. *Biology and Fertility of Soils*, 1996 22, 53-58.

Matsuda A; Moreira, FMS; Siqueira, JO. Tolerância de rizóbios de diferentes procedências ao zinco, cobre e cádmio. *Pesquisa Agropecuária Brasileira*, 2002 37, 343-355.

McGrath, SP; Chaudri, AM; Giller, KE. Long-term effects of metals in sewage sludge on soils, microorganisms and plants. *Journal of Industrial Microbiology* 1995 14, 94-104.

Nannipieri, P; Pechozzini, F; Arcada, PG; Pioranelli, C. Changes in amino acids, enzyme activities and biomass during soil microbial growth. *Soil Science*, 1979 127, 26–34.

Nannipieri, P. The potential use of soil enzymes as indicators of productivity, sustainability and pollution. In: Pankhurst CE, Doube BM, Gupta VVSR, Grace PR editors. *Soil Biota, Management in Sustainable Farming Systems*. Australia: CSIRO Publication; 1994; 238–244.

Obbard, JP. Ecotoxicological assessment of heavy metals in sewage sludge amended soils. *Applied Geochemistry*, 2001 16, 1405-1411.

Pankhurst, CE; Hawke, BG; McDonal, HJ; Kirkby, CA; Buckerfield, JC; Michelsen, P ; O'Brien, KA; Gupta, VVSR; Doube, BM. Evaluation of soil biological properties as potential bioindicators of soil health. *Australian Journal of Experimental Agriculture*, 1995 35, 1015–1028.

Parr, JF; Papendick, RI; Hornick, SB; Meyer, RE. Soil quality: attributes and relationship to alternative and sustainable agriculture. *American Journal of Alternative Agriculture*, 1992 7, 5–11.

Pereira, SIA; Lima, AIG; Figueira, EMAP. Screening possible mechanisms mediating cadmium resistance in *Rhizobium leguminosarum* bv. *viciae* isolated from contaminated portuguese soils. *Microbial Ecology*, 2006 52, 176-186.

Robinson, B; Russell, C; Hedley, M; Clothier, B. Cadmium adsorption by rhizobacteria: implications for New Zealand pastureland. *Agriculture Ecomical Environmental*, 2001 87, 315-321.

Saeki, K; Kunito, T; Oyaizu, H; Matsumoto, S. Relationships between bacterial tolerance levels and forms of copper and zinc in soils. *Journal of Environmental Quality*, 2002 31, 1570-1575.

Shen, G; Lu, Y; Zhou, Q; Hong, J. Interaction of polycyclic aromatic hydrocarbons and heavy metals on soil enzyme. *Chemosphere*, 2005 61. 1175-1182.

Shi, W; Bischoff, M; Turco, R; Konopka, A. Long-term effects of chromium and lead upon the activity of soil microbial communities. *Applied and Soil Ecology*, 2002 21, 169-177.

Smith, SR; Giller, KE. Effective *Rhizobium leguminosarum* biovar *trifolii* present in five soils contaminated with heavy metals from long-term applications of sewage sludge or metal mine spoil. *Soil Biology and Biochemistry*, 1992 24, 781-788.

Smith, LJ; Papendick, RI. Soil organic matter dynamics and crop residue management. In: Metting B editor. *Soil Microbial Ecology*. New York: Marcel Dekker; 1993

Somasegaran, P; Hoben, HJ. Handbook for Rhizobia. Berlin: Springer-Verlag; 1994.

Tabatabai, MA. Effects of trace elements on urease activity. *Soil Biology and Biochemistry*, 1977 9, 9-13.

Tabatabai, MA. Soil enzymes. In: Weaver RW, Angle JS, Bottomly PS editors. *Methods of Soil Analyses, Part 2, Microbiological and Biochemical Properties*. USA: Soil Science Society of America; 1994; 775–833.

Top, EM; Maila, MP; Clerinx, M; Goris, J; De Vos, P; Verstraete, W. Methane oxidation as method to evaluate the removal of 2,4-D from soil by plasmid-mediated bioaugumentation. *FEMS Microbiology Ecology*, 1999 28, 203–213.

Trasar-Cepeda, C; Leirós, MC; Seoane, S; Gil-Sotres, F. Limitations of soil enzymes as indicators of soil pollution. *Soil Biology and Biochemistry*, 2000 32, 1867-1875.

Tyler, G. Heavy metal pollution and soil enzyme activity. *Plant and Soil*, 1974 41, 303-311.

Vanhala, PT; Ahtiainen, JH. Soil respiration, ATP content and photobacterium toxicity test as indicators of metal pollution in soil. *Environmental Toxicology and Water Quality: An International Journal*, 1994 9, 115–121.

Vig, K; Megharaj, M; Sethunathan, N; Naidu, R. Bioavailability and toxicity of cadmium to microorganisms and their activities in soil: a review. *Advances in Environmental Research*, 2003 121–135.

Visser, S; Parkinson, D. Soil biological criteria as indicators of soil quality: soil microorganisms. *American Journal of Alternative Agriculture*, 1992 7, 33-37.

White, JS; White, DC. Source book of enzymes. Boca Raton- Florida: CRC Press LC; 1997

Wyszkowska, J. Soil contamination by chromium and its enzymatic activity and yielding. *Polish Journal of Environmental Studies*, 2002 11, 79-84.

Yeates, GW; Orchard, VA; Speir, TW; Hunt, JL; Hermans, MCC. Reduction in soil biological activity following pasture contamination by copper, chromium, arsenic timber preservative. *Biology and Fertility of Soils*, 1994 18, 200–208.

In: Soil Ecology Research Developments
Editors: Tian-Xiao Liu, pp. 159-178
ISBN: 978-1-60021-971-9

Chapter 6

SUPPRESSIVE MECHANISMS OF USED PUMICE TO BACTERIAL WILT OF TOMATO AND THEIR APPLICATION INTO BIOLOGICAL CONTROL IN HYDROPONIC PUMICE CULTURE

Koki Toyota
Graduate School of Bio-Applications and Systems Engineering,
Tokyo University of Agriculture and Technology, 2-24-16,
Naka, Koganei, Tokyo 184-8588, Japan

ABSTRACT

Suppressiveness of pumice, which had been used for 13 years for continuous cropping, was confirmed against bacterial wilt of tomato caused by *Ralstonia solanacearum*, in comparison with unused pumice. Since there were significant differences in some of the chemical and biological properties, suppressive mechanisms of the used pumice were investigated. Contribution of pH, EC and higher amounts of salts, such as Ca, to the suppression mechanisms appeared to be very low. In contrast, microbial biomass and respiration were significantly higher in the used pumice than in the unused pumice and the suppressiveness of the used pumice disappeared after sterilization by autoclaving and gamma irradiation. These results suggested that biological factors rather than chemical factors may be involved in the suppression mechanisms of bacterial wilt in the used pumice. The result of substrate-induced respiration inhibition method indicated that bacteria, rather than fungi, may be the predominant microbial community in the used pumice. Only one isolate, designated as *Burkholderia* sp. W3, showed a suppressive effect on bacterial wilt among 50 dominant bacterial colonies obtained from tomato roots grown in used pumice. When the strain W3 was inoculated into autoclaved used pumice, suppressive effect completely recovered. The bacterial communities of tomato roots evaluated by PCR-DGGE were different between the unused and the used pumice, and roots grown in the used pumice showed more diverse bacterial community. The band corresponding to W3 was not observed in the unused pumice, but there was the band in the used pumice, suggesting that W3 was an initially minor bacterium in tomato roots, but became a major colonizer after repeated cropping. These results may suggest that W3 is involved in one of the major mechanisms in the suppression of bacterial wilt in the used pumice. Unused pumice inoculated with W3 showed higher resistance to bacterial wilt, compared with that without inoculation, and this resistance was further enhanced by the addition of xlyose and glucose. It was further confirmed in small pot and greenhouse experiments that the application of biocontrol agent with substrates, such as lysine, available for the antagonist and not for the pathogen enabled more stable disease suppression in unused pumice. Possible methods to make unused pumice suppressive to bacterial wilt are presented.

INTRODUCTION

In modern greenhouses, soilless culture methods are employed. In Japan, while total cultivation area decreases year by year, cultivation area for hydroponics is increasing at a rate of 5 to 10% per year and has now exceeded 1000 ha (Japan Greenhouse Horticulture Association). Advantages of hydroponics are 1) higher productivity, 2) reduced intensity/cost of cultivation management, 3) elimination of soil-borne plant pathogens (Zinnen, 1988). But, in these systems there is sometimes a high risk for the establishment of plant pathogens, especially *Oomycetes*, which are well adapted to aquatic life (Folman *et al.* 2004). Once introduced in the nutrient solution of hydroponics, which is weak in buffering capacity to microbial attacks, pathogenic organisms may spread rapidly through the system and infect the roots more vigorously than in the soil culture. Therefore, different control methods for diseases, especially for the pathogens attacking from the root, are required for safe production, such as disinfection of pathogens in nutrient solutions using heat, filtration, radiation, chemicals or biological control (Komada 1994; Zhang and Tu, 2000; Ehret *et al.* 2001; Karras *et al.* 2007).

Tomato (*Lycopersicon esculentum* Mill.) is the major crop occupying ca. 50% of total hydroponics (Japan Greenhouse Horticulture Association). In tomato cultivation, the following diseases are considered to be serious: bacterial wilt caused by *Ralstonia solanacearum*, Fusarium wilt by *F. oxysporum* f. sp. *lycopercisi*, Fusarium crown and root rot by *F. oxysporum* f. sp. *radicis-lycopersici* and Pyhtophthora blight caused by *P. capsici* (Takeuchi 1995). To control these diseases, a variety of methods are already developed and applied, but techniques with low cost and that are environmentally friendly are still required. As we found that used pumice was suppressive to bacterial wilt, Fusarium wilt and damping-off of tomato, we report the phenomenon, suppression mechanisms and possible application into hydroponics under near-commercial conditions.

MATERIALS AND METHODS

Pumice and Pathogens Used

Two types of pumice were used: 1) unused (new) pumice (Ecoporous^{R}, Ebara Corp., Tokyo, Japan; particle size distribution: > 2mm 78% , 1 to 2 mm 21%, < 1mm 1%), 2) used pumice in which Boston lettuce was cultivated for 13 years (particle size distribution: > 2mm 40% , 1 to 2 mm 51%, < 1mm 9%). Pumice is a medium used in hydroponics, instead of rockwool, sand or palm husk, etc. (Savvas et al. 2003; Van Der Gaag and Wever 2005).

Ralstonia solanacearum YU1Rif43 (Toyota and Kimura 1996), Fusarium oxysporum f. sp. lycopersici race 2 880621 and Pythium aphanidermatum OPU431 were used. The bacterial strain was overnight cultured at 30C in 10^{-1} strength nutrient broth medium (Eiken Chemical Co., Ltd., Tokyo, Japan). F. oxysporum strain was cultured in Czapek medium (Dhingra and Sinclair 1995) for three days at 30C and microconidia was obtained by passing the culture through adsorbent cotton. Pythium strain was cultured for 2 days at 30C on 10^{-1} strength PDA agar medium (1/10 PDA) (Eiken Chemical Co., Ltd., Tokyo, Japan) and the plate with fungal mycelia was directly used as an inoculum.

Cultivation of Tomato in Small Pot Experiments

Pumice (100g dry basis) was put into a vinyl pot 9 cm in diameter and inoculated with bacterial wilt pathogen at a density of 5 x 10^4 cfu g^{-1} pumice. In case of Fusarium disease, pumice was added at a density of 1 x 10^5 spores g^{-1} pumice and a whole 1/10 PDA plate was mixed with 100 g pumice for Pythium disease. Unused pumice was added with $CaCO_3$ at a rate of 15 g kg^{-1}. Then, six of two-day- old tomato seeds (Lycopersicon esculentum Mill. cv. Momotaro) were transplanted into a pot. Pots were prepared in triplicates per treatment and grown in a Biotron (LPH200, Nippon Medical and Chemical Instruments Co., Ltd., Osaka, Japan; day:night=12h:12h, 285 μmol m^{-2} s^{-1} (photon flux), at 30C. Watering was done two times a day using Otsuka liquid fertilizer A formula [Otsuka Chemical Co., Ltd., Osaka, Japan; pH 3.8, N 260, P 120, K 405, Ca 230, Mg 60, Mn 1.5 B 1.5, Fe 2.7, Cu 0.03, Zn 0.09, Mo 0.03 (mg L^{-1})] to adjust its moisture content to pF 2.0 (nearly field capacity). The number of wilted plants was recorded at 2-day intervals until 30 days after seeding on the following basis: 0; no wilting symptoms, 1; 0 to 25 % of plant showing wilting, 2; 26 to 50%, 3: 51 to 75%, 4: 76 to 100%.

Measurement of R. Solanacearum YU1Rif43 in Pumice

Since the pathogen was resistant to rifampicin, the number of YU1Rif43 was counted by the dilution plate method using 10^{-1} strength nutrient agar supplemented with cycloheximide, rifampicin and polymyxin B at a concentration of each 100 mg L^{-1}.

Substrate Induced Respiration Method to Estimate Dominant Microbes in Pumice

Bacterial and fungal contribution to total microbial biomass was separated using antibiotics (Anderson and Domsh 1973). Preincubated used pumice for 7 days at 30C at pF 2.0 was dispensed into a 30 ml glass vial in triplicates and then amended with either each 500 μg g^{-1} soil of rifampicin and kanamycin or 500 μg g^{-1} soil of cycloheximide for the inhibition of bacterial and fungal activities, respectively, together with 2 mg g^{-1} soil of glucose and 0.5 mg g^{-1} soil of asparagines. CO_2 emission was measured at 0, 2, 4, 6 and 8 h after addition using a TCD-GC (GC-8A, Shimadzu, Kyoto, Japan), equipped with a Porapaq Q column (3 mm x 2m, 80/100 mesh, Shimadzu Good Laboratory Component Center, Tokyo, Japan). Average CO_2 emission rates were calculated between 0 and 8 h.

Microbial Community Structure Analysis of Tomato Roots by PCR-DGGE

Tomato roots were gently collected from the pots after 30 days of cultivation. DNA was extracted from a portion (0.1 g) of them using a bead beater (Bead Smash-12, Wakenyaku Co. Ltd., Kyoto, Japan) with 1 ml of extraction buffer (0.1 M Tris-HCl (pH 8.0), 40 mM EDTA (pH 8.0), 0.2 M NaCl, 20 g L^{-1} SDS) and the supernatant was collected and 400 μl of 7.5 M NH_4OAc was added to it and then kept on ice for 5 minutes and centrifuged for 3 min at 13,200 x g and to the supernatant 70 % of volume of ethanol is added and kept in -80°C for 1 h and centrifuged for 13,200 x g for 10 min and washed with 70 % ethanol and finally centrifuged and suspended in TE buffer. PCR amplification for DGGE was performed in a 50 μL volume containing 1 μl template DNA, Gene Taq universal buffer (Nippon Gene Co., Ltd., Tokyo, Japan), 6.25 nM each dNTP, 30 pmol each primer (F968GC, 1401L2: Muyzer et al. 1993) and 1.25 U of Gene Taq FP (Takara). The temperature program was as follows: a denaturing step at 94°C for 3 min, followed by 30 cycles of 94°C for 1 min, 55°C for 1 min, 72°C for 1.5 minn, and final extension step of 72°C for 7 min. DGGE was performed using a Bio-Rad Dcode^{TM} mutation analysis system (Bio-Rad Laboratories, Inc., Tokyo, Japan). Electorophoresis was done using a 6% (w/v) polyacrylamide gel in 1 x TAE buffer (40 mM Tris, 20 mM acetic acid and 1 mM EDTA, pH 8.0) under 100 volts at 60°C for 14 h. The polyacrylamid gels were made with parallel denaturing gradients 40-70% (100% denaturant contains 7 M urea and 40% formamide).

Chemical and Other Biological Analysis of Pumice

Electric conductivity (EC, pumice:water=1:5), pH (pumice:water=1:2.5), total C and N, exchangeable cations were measured using the conventional methods (Method of Soil Analysis). Exchangeable cations were extracted with 1 M ammonium acetate (pH 7.0) and measured using an atomic absorption spectrophotometer (Z-5010, Hitachi Ltd., Tokyo, Japan).

Microbial numbers were measured by the direct count method using the ethidium bromide staining (Roser 1980) and microbial biomass by the chloroform fumigation and extraction method (Vance et al. 1987).

Biological Control of Bacterial Wilt in Pumice Culture Using A Combination of Biocontrol Agents and Carbon Substrates

Different biocontrol agents isolated in our laboratory were tested for their disease suppressing property in combination with different kinds of substrates. Pseudomonas fluorescens MelRC2Rif (Toyota and Ikeda 1997), P. fluorescens FN2, Rhizobium monglense CF5b, Aquaspirillum arcticum F3a, Pseudomonas citronellolis F3b, Ralstonia pickettii

K20□Bacillus megaterium N2S6 (Nishijima et al. 2005) were added into unused pumice at a density of 5 x 10^7 cfu g^{-1} pumice, to which R. solanacearum YU1Rif43 were inoculated at a density of 5 x 10^4 cfu g^{-1} pumice and added with different carbon sources at a rate of 1 mg g^{-1}. One hundred gram of the unused pumice containing biocontrol agent-pathogen-substrate mixture was put into a vinyl pot with 9 cm diameter and six of two days old tomato seeds (Lycopersicon esculentum Mill. cv. Momotaro) were transplanted into a pot. Pots were prepared in triplicates per treatment and grown in a Biotron (LPH200, Nippon Medical and Chemical Instruments Co., Ltd., Osaka, Japan; day:night=12h:12h, 285 μmol m^{-2} s^{-1} (photon flux), at 30C). Substrates were also added after 2 weeks of seeding.

Cultivation of Tomato in A Greenhouse

Twelve two days old tomato seeds (Lycopersicon esculentum Mill. cv. House Momotaro) were transplant into a 0.05 m^2 pot containing 3.5 kg pumice (equivalent to 7L volume) inoculated with biocontrol agents and the pathogen at densities of 5 x 10^7 and 5 x 10^4 cfu g^{-1} pumice, respectively. Seeding was done in September 14, 2005 and September 24, 2006. The pots planted with tomato were put on a heater mat (Takii Co., Ltd., Kyoto, Japan) and grown in a greenhouse. Twelve seedlings were thinned to 3 after one month, then to one after two months and cultivated until the end of February each year. Substrates were added into each pot at a rate of 1 mg g^{-1} of pumice in 2005 and 0.25 mg g^{-1} of pumice in 2006 every two weeks from just after seeding. Watering was done using Otsuka liquid fertilizer A formula one to two times a day to adjust its moisture content to pF 2.0. Fifty % of concentration of the liquid fertilizer was used until the first flowering and then 75% of that until initial enlargement of fruits.

Statistics

Statistical analysis (mainly ANOVA) was done using the software Excel statistics 2002 (Social Survey Research Information, Tokyo, Japan).

Results and Discussion

Disease Suppressiveness of Used Pumice

When unused (new) and used pumice in which crops were continuously grown for 13 years were inoculated with the same amount of R. solanacearum and then tomato was grown, almost all tomato plants died in the unused pumice, but most of them survived in the used pumice (Figure 1). This tendency was observed consistently in six times repeated experiments. The used pumice was also suppressive to damping-off caused by Pythium aphanidermatum and Fusairum wilt caused by Fusarium oxysporum f. sp. lycopersici (Figure 1). It was considered that virgin pumice is susceptible to different belowground plant pathogens and then it became naturally suppressive to the diseases following repeated cultivations.

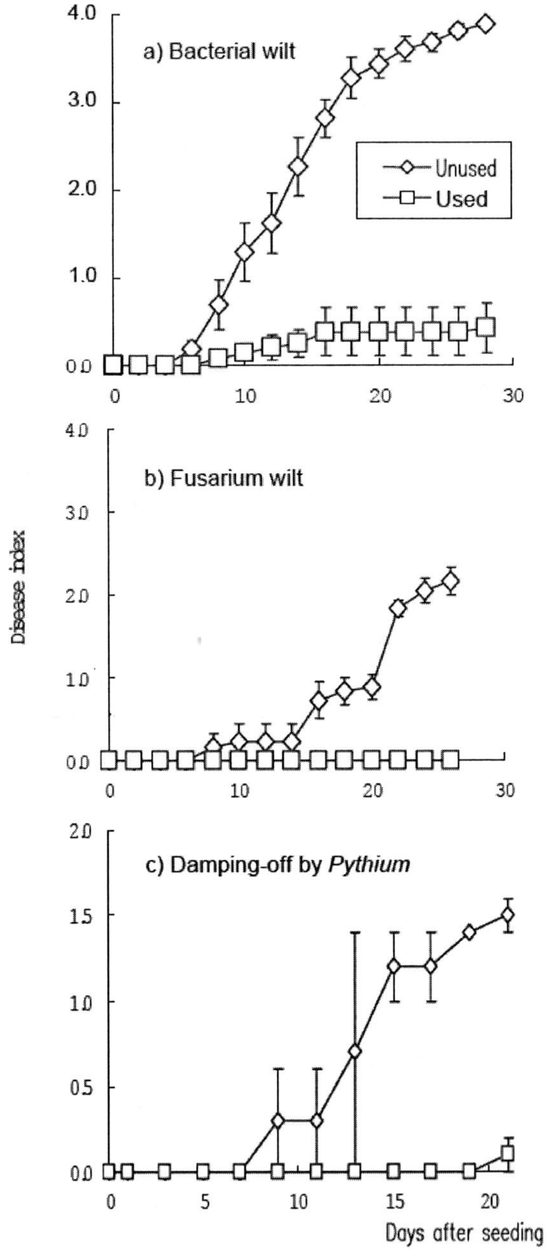

Figure 1. Disease index of bacterial wilt of tomato (a), Fusarium wilt (b) and damping-off caused by *Pythium aphanidermatum* (c) in unused and used pumice artificially infested with the pathogen. Date are means ±SE of six independent experiments (a) or means ±SE of single three replicates experiment (b,c).

Then, the physicochemical and biological properties of the unused and used pumices were compared. After repeated cultivation, total C and N, cations, especially Ca and Mg, accumulated, resulting in increased microbial number and biomass. Respiration was also markedly higher in the used pumice (138 ng CO_2-C g^{-1} day^{-1}) than in the unused pumice (0.7 ng CO_2-C g^{-1} day^{-1}). Microbial biomass and activity were markedly enhanced in the repeated cultivation, compared to microbial number. A similar finding was reported by Postma et al. (2000), in which it was demonstrated that used rockwool was suppressive to disease caused by Pythium aphanidermatum. They considered that the suppression mechanisms were due to certain members, such as antagonists, of microflora enriched through repeated cultivation and emphasized that although "clean" and "sterile" environment is addressed in hydroponics, microflora plays an important role in minimize the risk of Pythium disease. These results may indicate that microbial buffering capacity functions to make any media, such as rockwool and pumice, resistant to the attack by different belowground pathogens.

Suppression Mechanisms of Used Pumice –Contribution of Chemical Factors

When unused pumice was mixed with different ratios of used pumice, the disease index of bacterial wilt decreased with increasing ratios of the used pumice. The number of the pathogen R. solanacearum YU1Rif43 also decreased with increasing ratios of the used pumice (Figure 2). To estimate the suppression mechanisms of the used pumice to bacterial wilt, tomato plants were grown in non-treated and gamma-irradiated used pumice (Figure 3). Compared with the disease index (DI) 1.0 of the used pumice, it was 3.0 in the gamma-irradiated used pumice, suggesting that the major suppression mechanism was of biological origin. Since DI of the unused pumice showed always 4 (all the plants died), suggesting that even gamma-irradiated used pumice possessed a low degree of disease suppression. To estimate chemical factors involved in the disease suppression, unused pumice was added with different rates of Ca and a combination of Ca, Mg and K as their chloride forms (Figure 4). While the addition of Ca did not affect DI of bacterial wilt significantly at a rate of 200 mg kg^{-1} pumice, Ca significantly decreased DI at rates of 500 and 800 mg kg^{-1} pumice.

A combination with Mg and K further decreased DI at both rates, although the different was not significant due to high variations. When Ca was added with unused pumice at more than 1000 mg kg^{-1} as $CaCl_2$, the pathogen died quickly and no disease occurred, but both of tomato germination and their subsequent growth were inhibited. Ca toxicity to growth and chemotaxis of bacterial cells is reported (Sakai et al. 1995, 2003). In contrast, suppression of bacterial wilt by Ca is also reported (Yamazaki and Hoshina 1995; Yamazaki et al. 1999). In their study, Ca was added into a nutrient solution at rates of 4.4 to 20.4 mmol Ca^{2+} L^{-1} as a form of calcium nitrate. Our study suggested that suppressive effects of Ca to bacterial wilt might be further enhanced with a combination of Mg and K. When 10% of used pumice was mixed with 90% unused pumice, Ca content of the mixture pumice was 125 mg kg^{-1} and DI decreased less than 2.0 (Figure 2). However, the addition of Ca at a rate of 200 mg kg^{-1} did not decrease DI at all (Figure 4), suggesting that the suppression mechanisms of the used pumice may be due not to Ca effects, although Ca itself possesses disease suppression property.

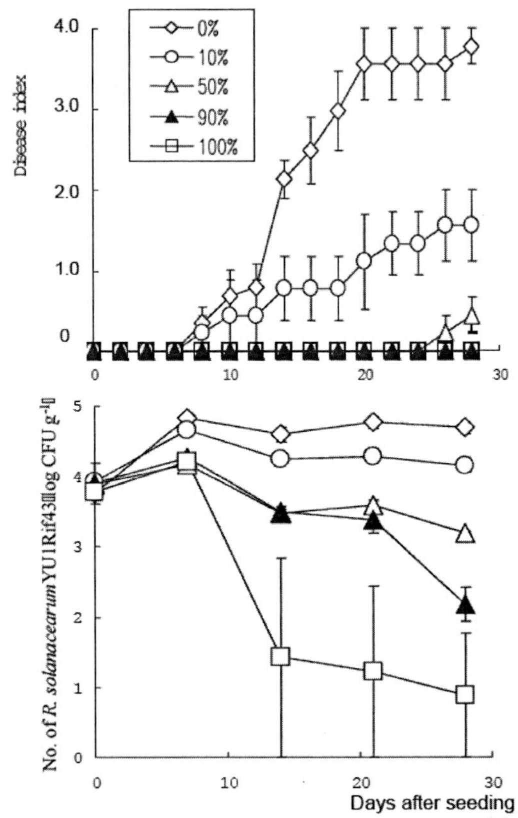

Figure 2. Disease index (upper) and the number of the pathogen in unused pumice mixed with different ratios of used pumice. 0% : only unused pumice, 100%: only used pumice. Date are means ±SE of triplicates.

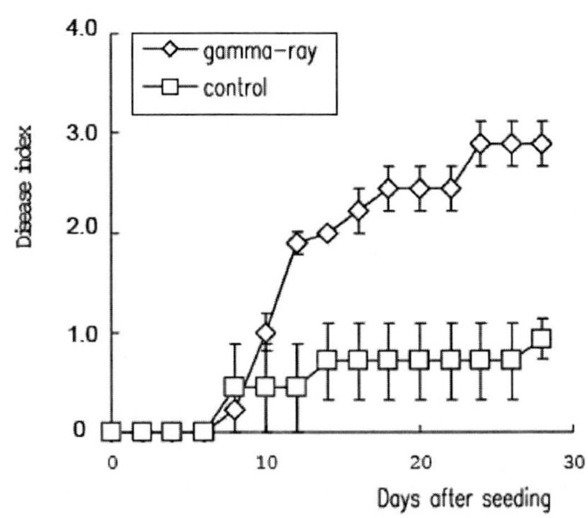

Figure 3. Effect of gamma-irradiation of used pumice on disease index of bacterial wilt of tomato. Date are means ±SE of triplicates .

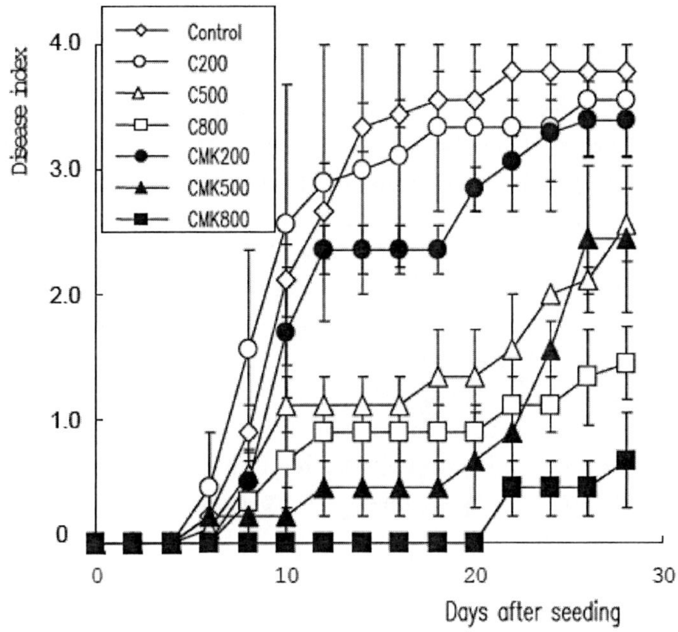

Figure 4. Effect of the additions of Ca, Mg and K to unused pumice on disease index of bacterial wilt of tomato C200□Ca 200mg kg-1as CaCl2. C500□Ca 500mg kg-1, C800□Ca 800mg kg-1. CMK200□Ca 200mg kg-1, Mg and K each 16mg kg-1 as MgCl2 and KCl. CMK500□Ca 500mg kg-1, Mg and K each 46mg kg-1. CMK800□Ca 800mg kg-1, Mg and K each 76mg kg-1. Data are means ±SE of triplicates.

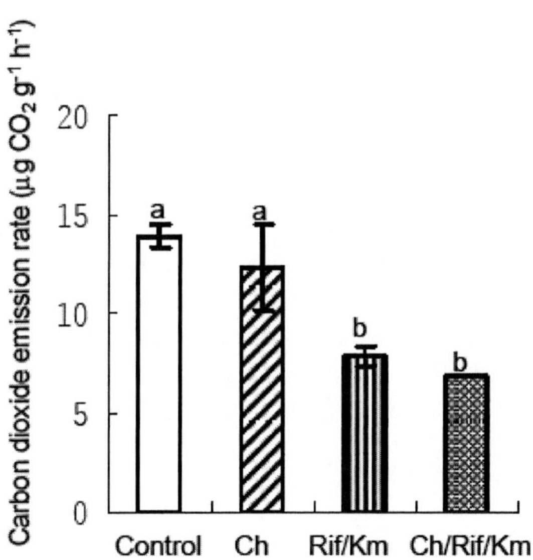

Figure 5. Estimate of dominant microbes in used pumice by the substrate induced respiration method. Ch:cycloheximide, Rif/Km:rifampicin/kanamycin. Data are means ±SE of triplicates.

Suppression Mechanisms of Used Pumice –Contribution of Biological Factors

Dominant microbial groups were estimated by the SIR method (Figure 5). The addition of cycloheximide into the used pumice did not decrease CO_2 emission rate significantly, suggesting that fungal contribution to total microbial biomass may be small. In contrast, the addition of rifampicin and kanamycin significantly decreased CO_2 emission rate roughly by 50%. It was concluded that pumice is originally colonized by little microbes and then microbes, especially bacteria, were enriched with repeated cultivation, which provide substrates to microbes in the pumice through root exudations and plant residues.

Then, dominant bacteria colonizing tomato roots grown in unused and used pumice were isolated and screened for their disease suppressing potential against bacterial wilt. Phylogenic analysis suggested that bacterial community of tomato roots was diverse both in unused and used pumice, and that there seemed to be no specific group detected only in the used pumice. When 50 strains isolated from the used pumice were tested for growth suppression of *R. solanacearum* YU1Rif43 in a root exudates medium, five strains suppressed the growth of the pathogen markedly. Then, five strains were inoculated into unused pumice infested with the pathogen, only one strain W3, nearest to *Burkholderia* sp., suppressed DI significantly (Figure 6). Interestingly, the other strains did not decreased DI, suggesting that most rhizobacteria did not interfere with the pathogen. W3 completely recovered the disease suppressing property of used pumice by inoculation into sterilized used pumice (Figure 7). Microbial community structures were compared of tomato roots grown in different treatments of pumice media using PCR-DGGE (Figure 8). The band corresponding to the strain W3 was not detected in the tomato roots grown in unused pumice, but was present in those in used pumice. When tomato was grown in unused pumice inoculated with W3, DI decreased significantly (and the band for W3 was detected. These results may indicate that W3 is involved in disease suppression of bacterial wilt in used pumice and unused pumice inoculated with the strain. However, single inoculation of W3 was not so effective at reducing DI compared to used pumice (Figure 9). Considering physical and chemical conditions of unused and used pumices, it was considered that W3 seemed not to perform well in the unused pumice because of low C content. Thus, W3 were inoculated into unused pumice together with carbon sources, such as glucose and xylose. It was observed that disease suppressive properties of W3 were further enhanced by the addition of glucose or xylose.

Glucose is utilizable to *R. solanacearum* YU1Rif43, but xylose is not. Therefore, we imagined that xylose be more effective at reducing bacterial wilt, but the result showed that glucose and xylose were similarly effective, suggesting that utility of carbon substrate by the antagonist be more important in successful biological control.

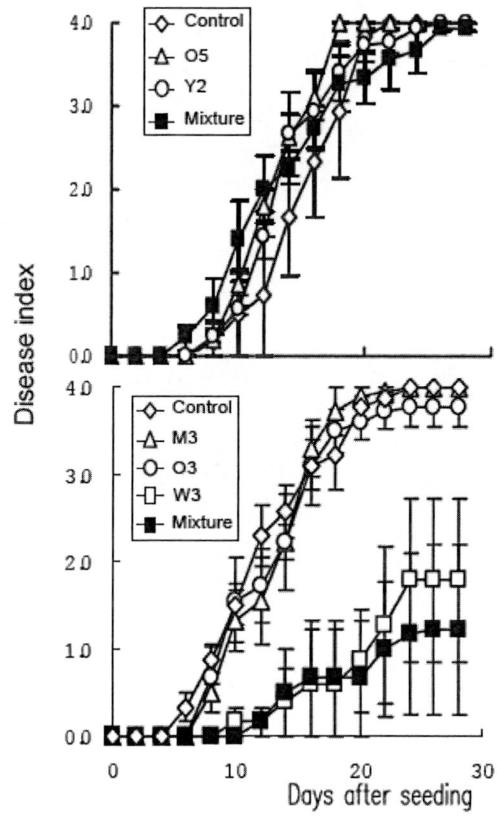

Figure 6. Effect of the inoculation of different antagonist candidates to unused pumice on bacterial wilt of tomato. Data are means ±SE of triplicates.

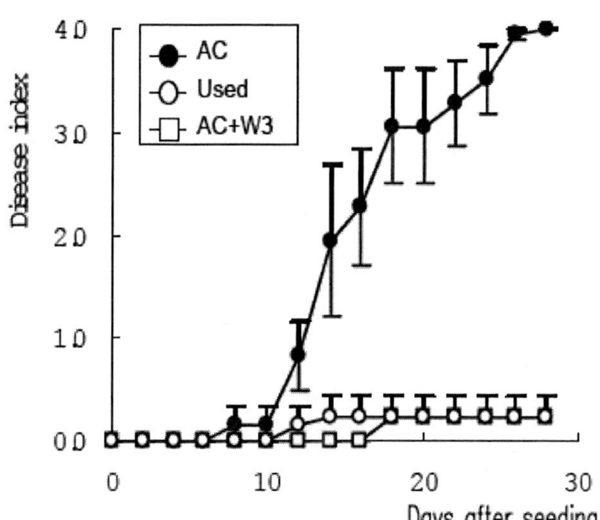

Figure 7. Effect of inoculation of W3 strain to autoclaved used pumice on bacterial wilt of tomato. Data are means ±SE of triplicates.

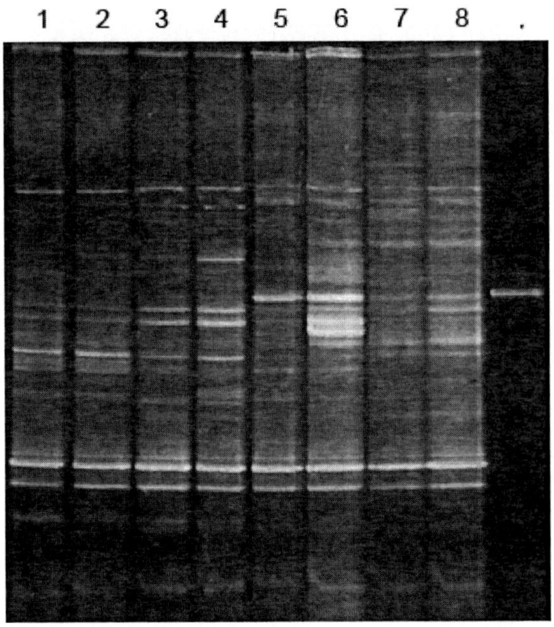

Figure 8. DGGE analysis of bacterial community structures of tomato roots grown in unused pumice (lanes 1,2), unused pumice inoculated with W3 (lanes 3,4), autoclaved used pumice inoculated with W3 (lanes 5,6) and used pumice (lanes 7,8). W: pure culture of W3 strain.

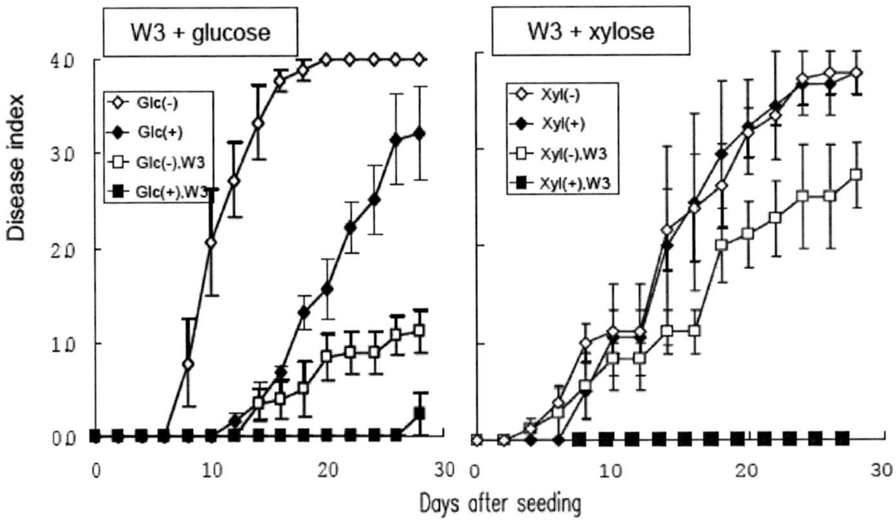

Figure 9. Effect of the inoculation of W3 strain and the addition of glucose (left) and xylose (right) to unused pumice on bacterial wilt of tomato. Data are means ±SE of triplicates.

Biological Control of Bacterial Wilt in Pumice Culture Using a Combination of Biocontrol Agents and Carbon Substrates

Using different bacterial agents, their suppressive effects on bacterial wilt of tomato were examined in the presence and absence of carbon substrates added at a rate of 1 mg g^{-1} pumice. Compared to single inoculation, combination of bacterial inoculation and substrate addition decreased DI more remarkably, especially MelRC2Rif, FN2, N2S6 strains (Figure 10). Among the substrates added with bacteria, lysine and glutamine were most suppressive to bacterial wilt. Similar results were previously reported (Posas et al. 2007), in which bacterial wilt of tomato was markedly suppressed by the addition of glucose, lysine, serine, glutamine and arginine into soil and concomitantly the pathogen declined more rapidly. In this study, glucose and xylose were not suppressive, when the substrates were added into pumice infested with the pathogen, but both amino acids, lysine and glutamine, suppressed bacterial wilt remarkably (Figure 11). It is interesting to note that the addition of glucose into soil made the soil suppressive to bacterial wilt, but that into pumice did not. This fact may suggest that indigenous microbes that utilize glucose and perform as antagonists are present in soil, but such types of microbes are not present in pumice since microbial abundance is too low in unused pumice. Lysine was added to pumice along with different bacterial agents (Figure 12). Combinations of lysine with MelRC2Rif and FN2 were consistently effective at decreasing bacterial wilt, while those with F3 and CF5b were not so effective, although DIs significantly decreased. These results suggested that a combination of lysine and biocontrol agents, rather than a single application of lysine, is the best management in pumice culture to minimize the risk to bacterial wilt of tomato minimum.

Trials in A Greenhouse

Tomato was grown in a greenhouse using 0.05 m^2 pot containing 3.5 kg pumice, near commercial size, to which the pathogen was inoculated and antagonists and lysine were added at 5 x 10^7 cfu g^{-1} pumice and 1 mg g^{-1} pumice, respectively. Lysine was added at seeding and after that every two week at a rate of 1 mg g^{-1}. In small pot experiments described above, no inhibitory effect was found to the growth of tomato, but in the larger size of pots, tomato growth was significantly reduced in the lysine added pots (Figure 13). Symptoms of bacterial wilt was observed only in control(+), MelRC2Rif and FN2 plots at a disease incidence of 8%, 31% and 6%, respectively, but not in control(-), MelRC2Rif+lysine and FN2+glucose plots, suggesting that a combination of the antagonists and the substrates be effective at reducing bacterial wilt incidence. This result was supported by the number of MelRC2Rif and the pathogen: the former was higher and the latter lower in the MelRC2Rif+lysine plot than in the MelRC2Rif plot. It is common knowledge that certain amino acids can inhibit plants in millimolar amounts and lysine is categorized into amino acids with relatively higher inhibitory effects (Vurro et al. 2006). Therefore, we tested the concentrations of lysine that are effective at reducing bacterial wilt, but not inhibitory to tomato growth and found that the addition of lysine at a rate of 1 mg g^{-1} pumice was never inhibitory to tomato for one month cultivation period in the small pot experiments. Lysine addition, however, was inhibitory to tomato growth in a large pot experiment in 2005. Thus, the amount of lysine was further reduced and its effect on disease suppression properties was evaluated (Figure 14).

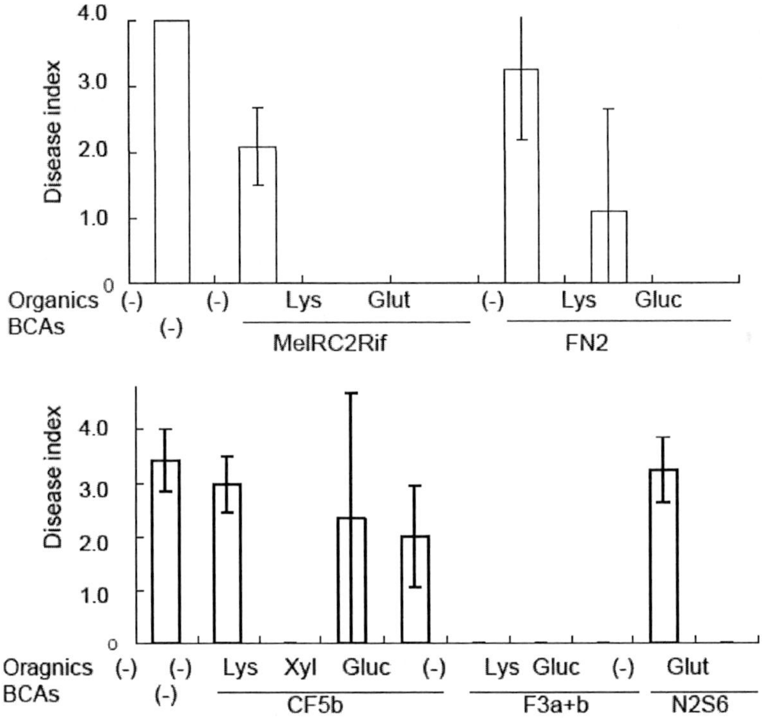

Figure 10. Effect of inoculation of different biocontrol agents (BCAs) in combination of organic compounds on bacterial wilt of tomato. Lys: Lysine, Glut: Glutamine, Gluc: Glucose, Xyl: xylose. Data are means ±SE of triplicates.

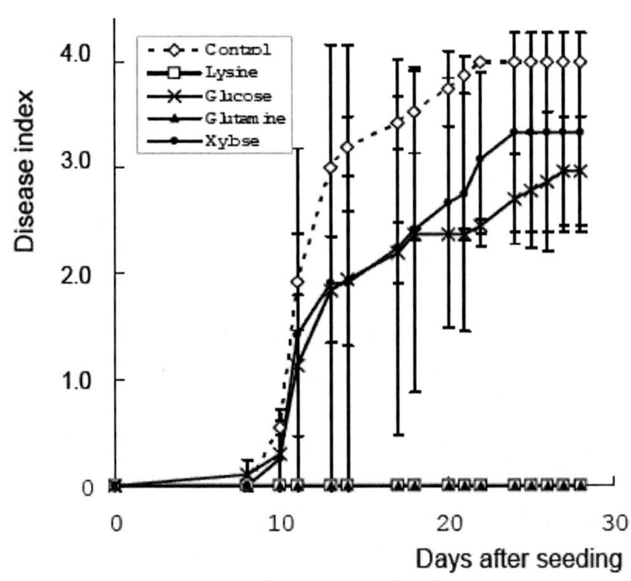

Figure 11. Effect of addition of organic compounds at a rate of 1mg g-1 to unused pumice on bacterial wilt of tomato. Data are means ±SE of triplicates.

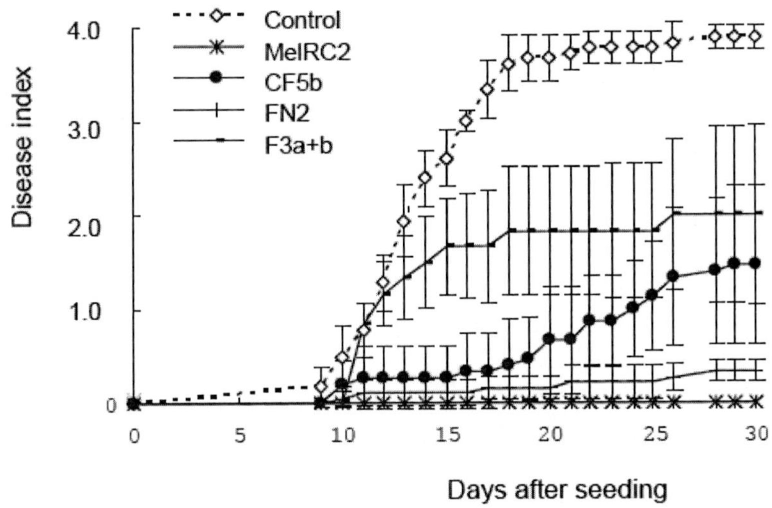

Figure 12. Effect of addition of lysine at a rate of 1mg g-1 together with different BCAs on bacterial wilt of tomato. Data are means ±SE of triplicates.

Even at a rate of 0.25 mg g^{-1} pumice, lysine addition significantly suppressed bacterial wilt if it was added at the same timing as the inoculation of the pathogen. When the pathogen was firstly inoculated and lysine was added one to two weeks later, no disease suppression was observed by lysine. In 2006, tomato was grown using the same 0.05 m^2 pot, the same conditions as 2005 experiment, to which lysine was added at a rate of 0.25 mg g^{-1} at the seeding and two weeks intervals after that, together with a biocontrol agent K20. A single inoculation of K20 significantly decreased bacterial wilt incidence and a combination of K20 and lysine and the addition of only lysine further decreased bacterial wilt incidence. The tendency was supported by tomato growth and yield. Tomato yield was about 2500 g $plant^{-1}$ in control(-), K20, K20+lysine and lysine plots and 1400 g $plant^{-1}$ in control(+) and the difference was highly significant. This absolute yield was not so high in terms of commercial level, but this is considered due to our not-sophisticated cultivation technique. An important result was considered that the yield of lysine, K20 and K20+lysine plots was not different from that of control(-), suggesting little inhibitory effect of these treatments on tomato yield. This was further supported by the quality of tomato evaluated by sugar, lycopene and vitamin C contents: there was no significant difference in the parameters between control(-) and lysine, K20 and K20+lysine plots. The cost of lysine was calculated to be 100 yen (0.8 $) per plant when lysine is added to each plant at a rate of 0.25 mg g^{-1} every two weeks during 24 weeks. But this cost is based on the price of 4,300 yen (35.8$) per 500 g of the reagent lysine (special grade, Wako Pure Chemical Industries, Osaka, Japan). In contrast, alternative lysine for an additive of pig fodder costs only 500 yen (4.2$) per 1 kg (personal communication), and its effect on bacterial wilt is now under investigation in comparison with pure lysine. These results suggested that lysine addition to pumice every two weeks markedly suppress bacterial wilt and does not inhibit tomato growth and yield. Therefore, it was concluded that lysine may be an effective and economical means by which bacterial wilt be controlled in hydroponics.

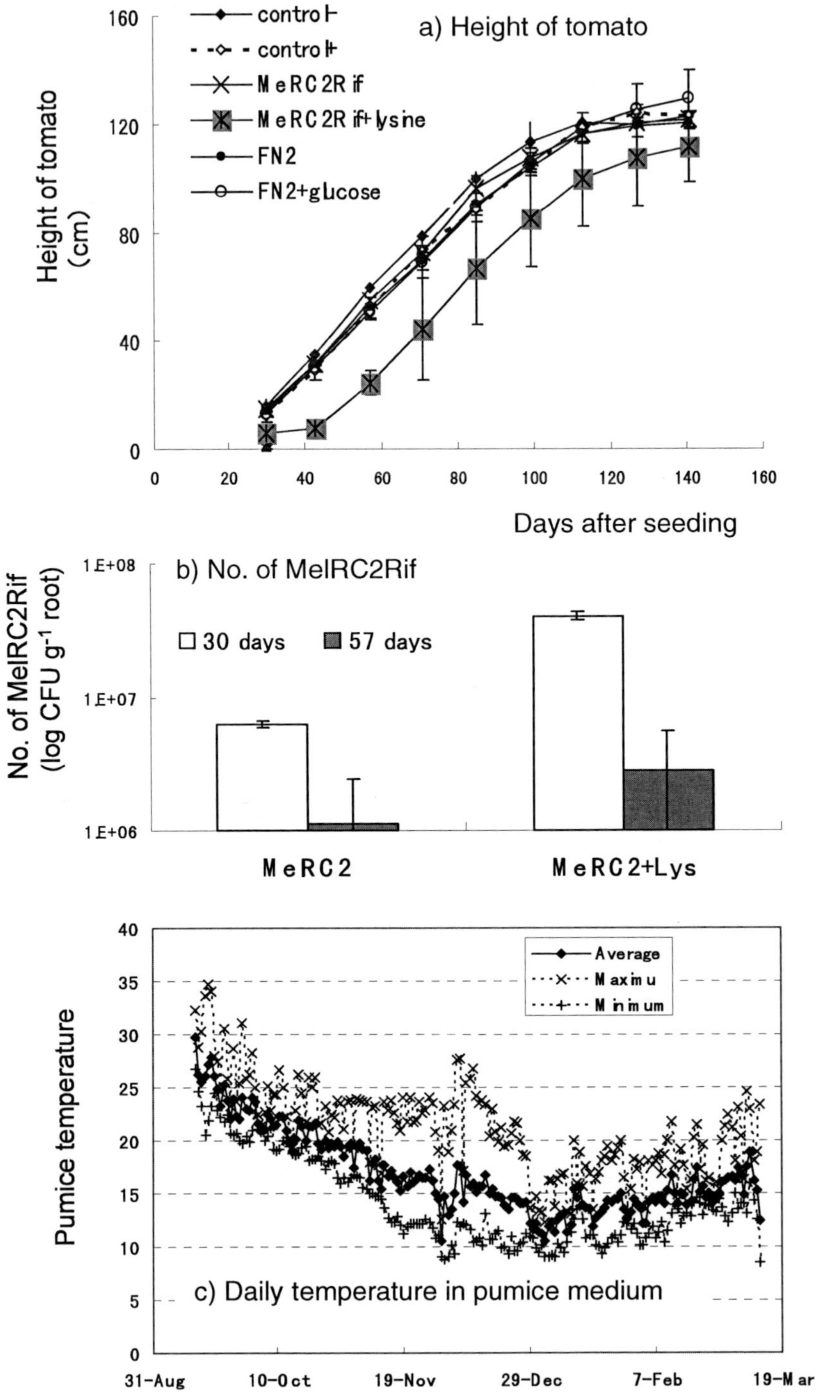

Figure 13. Effect of the addition of organic compounds at a rate of 1 mg g-1 with BCAs on tomato growth and colonization of BCA in a greenhouse in 2005-2006. Data are means ±SD (five replicates).

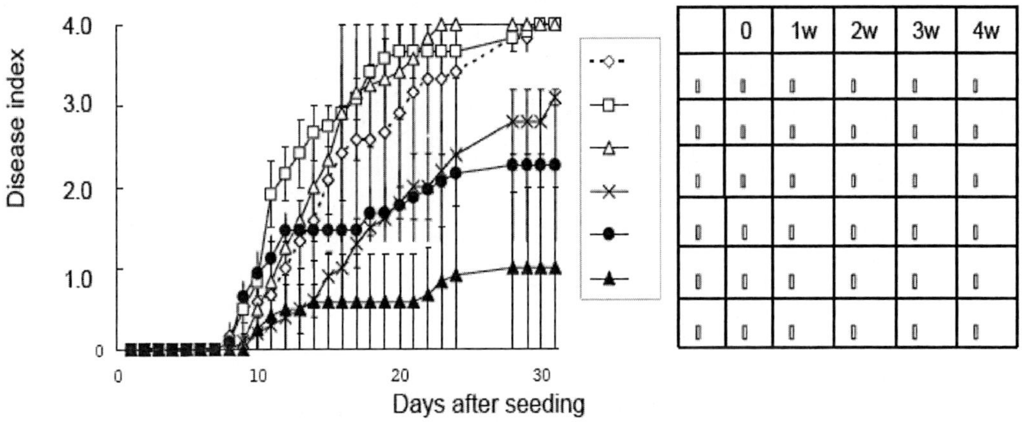

Figure 14. Effect of the addition of lysine at a rate of 0.25 mg g-1 in different timings to unused pumice on bacterial wilt of tomato. Data are means ± SE of triplicates.

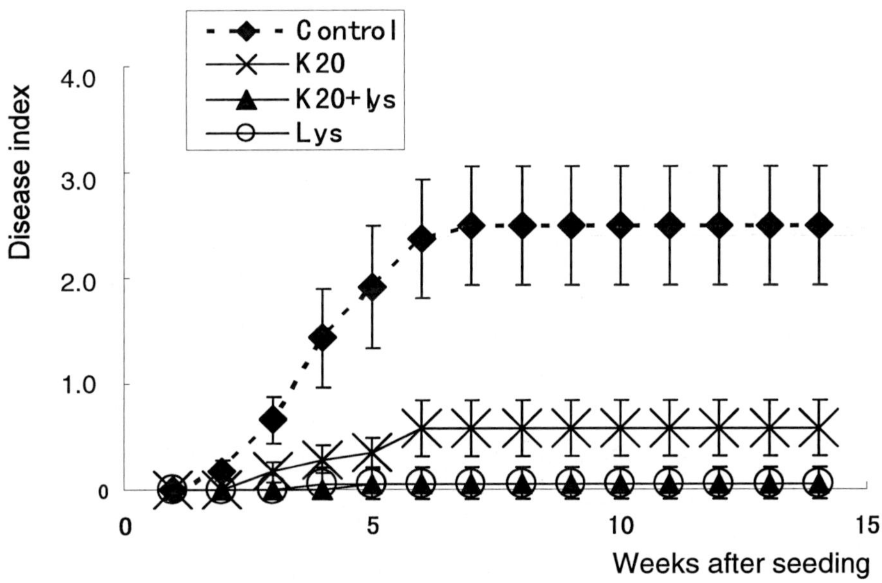

Figure 15. Effect of the addition of lysine at a rate of 2.5 mg g-1 and K20 on bacterial wilt of tomato grown in a greenhouse in 2006-2007. Data are means ±SE of six replicates.

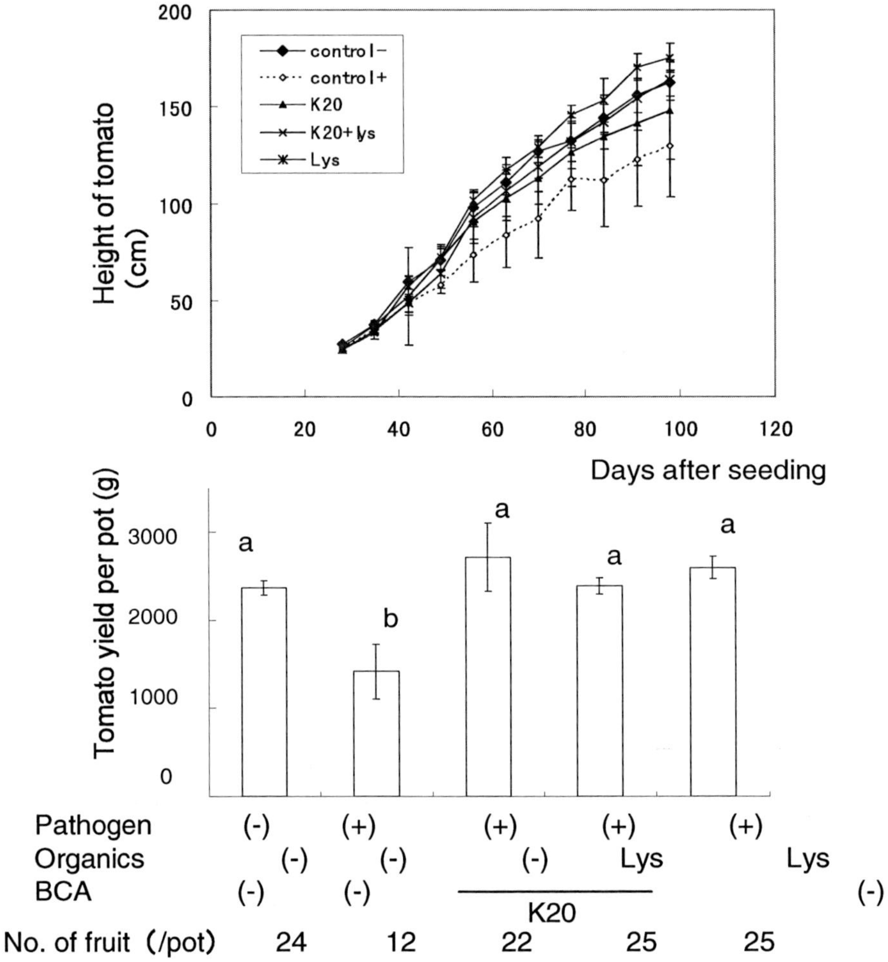

Figure 16. Growth of tomato and their yields in different treatments in 2006-2007 greenhouse experiment. Date are means ±SE of six replicates.

Conclusion

In hydroponics, the cleaner environment has been considered important to avoid risks by diseases. But the present study demonstrated that a pumice medium may become suppressive against different media-borne diseases by repeated cultivation in the same medium possibly due to accumulation of organic matters and consequent enrichment of certain types of antagonistic microorganisms. In addition, it was clarified that addition of lysine into the pumice medium to enhance microbial activity may be effective at reducing bacterial wilt with low cost.

ACKNOWLEDGEMENT

Pumice used in this study was provided by Ebara Corp. (1-6-27, Kohnan, Minato, Tokyo 108-8480) and the "Ecovegetable®" system, which gave us a chance to study biological control in hydroponics, is commercially available. The author thanks Dr. Kazunari Masuda and Mr. Tetsuo Kuroda, as representatives of Ebara Corp., and Mr Tomoo Matsuoka, Ms Mami Ide and Mr. Takeshi Igawa for their excellent contribution to the study. *Fusarium oxysporum* f. sp. *lycopersici* and *Pythium aphanidermatum* were kindly gifted by Dr. Tsutomu Arie and Dr. Motoaki Tojo. This work was supported by a Research project (No. 1765) for utilizing advanced technologies in agriculture, forestry and fisheries.

REFERENCES

Anderson JPE, Domsch KH (1973) Quantification of bacterial and fungal contributions to soil respiration. *Arch. Microbiol.*, 93, 113-127

Dhingra OD, Sinclair JB (1995) Basic Plant Pathology Methods Second Edition. CRC Press. Inc., Florida

Ehret DL, Alsanius B, Wohanka W, Menzies JG, Utkhede R (2001) Disinfection of recirculating nutrient solutions in greenhouse horticulture. *Agronomie*, 21, 323-339

Igawa T, Ide M, ToyotaK, Masuda K, Kuroda T: Effects of additions of lysine and biocontrol agents into hydroponics using pumice medium on bacterial wilt of tomato. Submitted to *Soil Microorganisms.* (in Japanese with English summary)

Karras G, Savvas D, Patakioutas G, Pomonis P, Albanis T (2007) Fate of cryomazine applied in nutrient solution to a gerbera (*Gerbera jamesonii*) crop grown in a closed hydroponic system. *Crop Prot.*, 26, 721-728

Komada H (1994) Strategies and prospects on biocontrol of soilborne diseases in soilless culture. *Shokubutuboeki*, 48, 9-13 (in Japanese)

Matsuoka T, Toyota K, Masuda K, Kurod T (2004) Potential of continuously cropped pumice for suppressing bacterial wilt of tomato and its mechanisms. *Soil Microorganisms*, 58, 25-32 (in Japanese with English summary)

Matsuoka T, Toyota K, Sato K, Masuda K, Kurod T (2005) Suppressive mechanisms of repeatedly used pumice to bacterial wilt of tomato. *Soil Microorganisms,* 59, 15-20 (in Japanese with English summary)

Muyzer G, De Waal EC, Uitterlinden AG (1993) Profiling of complex microbial populations by denauturing gradient gel electrophoresis analysis of polymerase chain reaction-amplified genes coding for 16SrRNA. *Appl. Environ. Microbiol.*, 59, 695-700

Nishijima T, Toyota K, Mochizuki M (2005) Predominant culturable *Bacillus* species in Japanese arable soils and their potential as biocontrol agents. *Microbes Environ.*, 20, 61-68

Posas MB, Toyota K, Islam TMD (2007) Inhibition of bacterial wilt of tomato caused by *Ralstonia solanacearum* by sugars and amino acids. *Microbes Environ.*, (in press)

Postma J, Willemsen-de Klein MJEIM, van Elsas JD (2000) Effect of the indigenous microflora on the development of root and crown rot caused by *Pythium aphanidermatum* in cucumber grown on rockwool. *Phytopathology*, 90, 125-133

Roser DJ (1980) A general purpose fluorescent strain for nucleic acid in bacteria and eucaryotes and its use in microbial ecology studies. *Soil Biol. Biochem.,* 12, 329-336

Sakai M, Futamata H, Urashima Y, Matsuguchi T (1995) Effect of cations on the growth of fluorescent pseudomonad isolates from spinach roots grown in soils with different salinity levels. *Soil Sci. Plant Nutr.*, 41, 605-611

Sakai M, Futamata H, Kanazawa S (2003) Effects of high concentrations of inorganic salts on swarming ability in fluorescent *Pseudomonas* strains. *Biosci. Biotech. Biochem.*, 67, 1479-1484

Savvas D, Karagianni V, Kotsiras A, Demopoulos V, Karkamisi I, Pakou P (2003) Interactions between ammonium and pH of the nutrient solution supplied to gerbera (*Gerbera jamesonii*) grown in pumice. *Plant Soil*, 254, 393-402

Takeuchi T (1995) Control of bacterial wilt of tomato in hydroponics. Nogyo oyobi Engei, 70, 1121-1126 (in Japanese)

Toyota K, Ikeda K (1997) Relative importance of motility and antibiosis in the rhizoplane competence of a biocontrol agent *Pseudomonas fluorescens* MelRC2Rif. *Biol. Fertil. Soils*, 25, 416-420

Toyota K, Kimura M (1996) Growth of the bacterial wilt pathogen *Pseudomonas solanacearum* introduced into soil colonized by individual soil bacteria. *Soil Biol. Biochem.*, 28, 1489-1494

Tu JC, Paradopoulos AP, Hao X, Zheng J (1999) The relationship of Pythium root rot and rhizosphere microorganisms in a closed circulating and an open system in rockwool culture of tomato. *Acta Hortc.*, 481, 577-583

Vance ED, Brooks PC, Jenkinson DS (1987) An extraction method for measuring soil microbial biomass C. *Soil Biol. Biochem.*, 19, 703-707

Van Der Gaag DC, Wever G (2005) Conduciveness of different soilless growing media to *Pythium* root and crown rot of cucumber under near-commercial conditions. *Eur. J. Plant Pathol.*, 112, 31-41

Vurro M, Boari A, Pilgeram AL, Sands DC (2006) Exogenous amino acids inhibit seed germination and tubercle formation by *Orobanche ramose* (Broomrape): Potential application for management of parasitic weeds. *Biol. Control.*, 36, 258-265

Yamazaki H, Hoshina T (1995) Calcium nutrition affects resistance of tomato seedlings to bacterial wilt. *HortScience*, 30, 91-93

Yamazaki H, Kikuchi S, Hoshina T, Kimura T (1999) Effect of calcium concentration in nutrient solution before and after inoculation with *Ralstonia solanacearum* on resistance of tomato seedlings to bacterial wilt. *Soil Sci. Plant Nutr.*, 45, 1009-1014

Zang W, Tu JC (2000) Effect of ultraviolet disinfection of hydroponic solutions on *Pythium* root rot and non-target bacteria. *Eur J. Plant Pathol.*, 106, 415-421

Zinnen TM (1988) Assessment of plant diseases in hydroponic culture. *Plant Disease*, 72, 96-99.

In: Soil Ecology Research Developments
Editors: Tian-Xiao Liu, pp. 179-196
ISBN: 978-1-60021-971-9

Chapter 7

LIMITS IMPOSED BY CD AND NI ON GROWTH AND METALS ACCUMULATION IN TWO HALOPHYTES SPECIES: *SESUVIUM PORTULACASTRUM* AND *MESEMBRYANTHEMUM CRYSTALLINUM*

Tahar Ghnaya[1], Inès Slama[1], Baligh Tray[2], Dorsaf Messedi[1], Claude Grignon[3], Mohamed Habib Ghorbel[2] and Chedly Abdelly[1]

[1] Laboratoire d'Adaptation des Plantes aux Stress Abiotiques, Institut National de Recherche Scientifique et Technique, B.P. 95, 2050 Hamam-Lif, Tunisia

[2] Nutrition azotée et protéines de stress,
Faculté des Sciences de Tunis, 1060 Tunis, Tunisia

[3] Biochimie et Physiologie Moléculaire des Plantes, Agro-M INRA, 34060 Montpellier CEDEX 1, France

ABSTRACT

Phytoextraction, which is the use of plant to extract heavy metals from polluted soils, could be limited by the high toxicity of these elements to plant development. Many data suggest that heavy metals-induced nutritional disturbance is one of the major causes of the plant growth restriction. The present work aims to check the validity of this hypothesis in two halophytes species: *Sesuvium portulacastrum* and *Mesembryanthemum crystallinum* cultivated in the presence of Cd and Ni. Seedlings were grown for 30 days in split-root conditions. Five treatments were applied: in the first two treatments, one half of the root system was immersed in a basal medium (B), while the other half was in the same solution supplemented with 100 μM Cd^{2+} (B/Cd plants) or 50 μM Ni^{2+} (B/Ni plants). In the other treatments, the two halves of the root system were immersed either in a metal-free medium (B/B control plants) or in basal medium containing 100 μM Cd^{2+} (Cd/Cd plants) or 50 μM Ni^{2+} (Ni/Ni plants). At the harvest, dry weight as well as the Cd^{2+}, Ni^{2+}, K^{+}, Ca^{2+} and Fe concentrations in tissues were determined. As compared to Cd/Cd and Ni/Ni treatments, culture on dual medium (B/Cd and B/Ni) alleviated significantly the effects of Cd^{2+} and Ni^{2+} on growth, attenuated the leaf toxicity symptoms and led to appropriate shoot K^{+} and Fe amounts in spite of relatively high Cd^{2+}

and Ni^{2+} concentrations. Ca^{2+} status was not modified by Cd^{2+} or Ni^{2+} in *Sesuvium*, but it was decreased by Cd/Cd and Ni/Ni treatments in *Mesembryanthemum*. However, for the latter species, B/Cd and B/Ni plants showed appropriate Ca^{2+} shoot amounts. Hence, our results indicate that nutritional disturbances induced by Cd^{2+} or Ni^{2+}, contributed largely to the growth restriction in both halophytes leading to limitation of heavy metals extracted in the shoots. So, we suggest the possibility to enhance the capacity of both species to extract these metals by increasing nutrient availability in soil.

Keywords: Cd and Ni accumulation, halophytes, nutritional disturbances, phytoextraction, split root system.

INTRODUCTION

Human activities, particularly industry, urbanism and agricultural practices, have resulted in increased mobilisation and deposition of potentially toxic heavy metals, posing a major threat to the environment and human health (Cunningham *et al.* 1995; Lu *et al.* 2004; Ait Ali *et al.* 2004). In the United States for example, Approximately 63% of the sites on the National Priority List (NPL) for the treatment of contaminated soils include contamination with toxic heavy metals (Hazardous Waste Consultant, 1996), indicating the extensiveness of this problem. Some heavy metals (Zn, Ni, Co) are essential micro nutrients for plant growth. However, excess of these metals and the presence of non essential heavy metals like Cd, Pb, Hg, even at low concentrations, may be toxic and always induce growth reduction and leaf chlorosis (Sanità di Toppi and Gabbrielli, 1999; Ghnaya *et al.* 2005; Wojas *et al.* 2007). In Tunisia, saline depressions, which are less populated, often constitute sites of accumulation of industrial and urban effluents contaminated by heavy metals. Preliminary studies achieved in various saline industrial sites in Tunisia showed that these zones are contaminated by Cd, Ni and Pb (Nouairi *et al.* 2002; Ghnaya *et al.* 2004).

These elements can not be biodegraded and must be extracted from contaminated sites. The clean up of heavy metals contaminated soils is one of the most difficult tasks for environmental engineering. For these reasons, several techniques have been investigated for the removal of these pollutants from polluted soils; but such efforts necessitated intensive labor and were costly (Chen *et al.* 2000). Interest in using plants for soil rehabilitation has increased due to natural capacity of particular species to accumulate various heavy metals. For phytoextraction high biomass yielding plants are to be chosen, which are able to transfer the heavy metals efficiently and rapidly from soil via the root system to the aerial part (Arduini *et al.* 2004). Thus, the two major factors that determine the total amount of metal extracted by plants are: (a) the metal concentrations in dry biomass and (b) the total biomass produced by the plant. (Ghosh and Singh, 2005). Thus, plants used in this process should be fast growing, easily propagated and able to accumulate the target metals.

Most plants used for metal accumulation are crop plants, including sunflower (*Helianthus annus*), corn (*Zea mays*), pea (*Pisum sativum*) and mustard (*Brassica juncea*). These plants are glycophytes (salt sensitive species) and cannot be used in phytoextraction of metals from salty zones characterized by a higher salinity levels. Halophytes, supporting high salinity, are promising candidates for soil desalination by producing higher biomass concomitant to elevated salt concentration in harvestable parts. For example, Glenn *et al.* (1999) showed that

Atriplex nummularia can achieve biomass yield of 20 to 30 t $ha^{-1}year^{-1}$ and have been shown to accumulate up to 40% NaCl in their dry matter. Recent studies suggest that halophytes may be useful for phytoremediation of heavy metal contaminated salty soils (Glenn *et al.* 1999; Williams *et al.* 1994; Jordan *et al.* 2002, López-Chuken and Young, 2005; Ghnaya *et al.* 2005).

Furthermore, heavy metals limit plants' growth and decrease its development rate by affecting various aspects of plant physiology. They restrict photosynthesis, decrease chlorophyll content, as well as induce oxidative damages (Olmos *et al.* 2003; Ederli *et al.* 2004; Chaoui and Ferjani, 2005). Alteration of the plant water status has been reported in some studies (Barcelo and Poschenrieder, 1990; Perfus-Barbeoch *et al.* 2002). Mineral disturbances induced by the interaction of heavy metals with essential macro and microelements are important causes of toxicity in plants (Larbi *et al.* 2002, Kim *et al.* 2003, Ederli *et al.* 2004, Ghnaya *et al.* 2005). Thus, the reduction of the growth observed in plants subjected to heavy metals often corresponds to the resultant of several parameters related to direct action (toxic effect linked to an accumulation of heavy metals in tissues) and/or to indirect effect (limitation of mineral and water nutrition's).

However, little is known about the implication of each of these two components in the growth limitation. The split root system, ensuring a load of the shoots with heavy metals and an appropriate nutrient supply, constitutes an adequate method to determine the implication of the nutritional disturbance in the growth-decrease under heavy metal treatment.

In a previous study, we showed that shoot-Cd^{2+} concentration varied between 350 and 700 µg/gDW in both species when cultivated in the presence of Cd^{2+} (Ghnaya *et al.* 2005). However, this accumulation was accompanied with a growth reduction and mineral nutrition (K, Ca and Fe) disturbances. The present study aimed at determining whether heavy metals (Cd and Ni) limit *S. portulacastrum* and *M. crystallinum* growth through impairment of some essentials nutrients acquisition (Ca^{2+}, K^{+} and Fe), or through a toxic effect linked to an excessive metal accumulation in shoots. Thus we conducted a split-root experiment to separate the two factors.

Materials and Methods

Plant Material and Growth Parameters

Sesuvium portulacastrum L. (Aizoaceae), a dicotyledonous halophyte commonly known as sea purslane, was propagated by cutting. Three cm long-stem segments with one node and two opposite leaves were taken from mother plants growing under natural condition, in a mixture of sandy soil and organic matter, and irrigated with tap water. They were disinfected for 5 minutes in saturated calcium hypochlorite solution, and rinsed abundantly with distilled water. They were then placed for 7 days in an aerated Hewitt (1966) solution diluted 10 times, supplemented with 4.5 µM Fe EDTA (Jacobson, 1951) and micronutrients (Arnon and Hoagland 1940). Rhizogenesis took place after one week. Seedlings of *Mesembryanthemum crystallinum* L. (dicotyledonous halophyte from the Aizoaceae family, commonly known as common ice plant) were obtained by germination. The seeds were collected from Thina (Sfax, 300 km south of Tunis), sterilised by dipping them into a 10 % H_2O_2 solution during 20 min.

They were then washed with distilled water and sown on perlite imbibed with distilled water. They remained in the dark, at room temperature for 7 days. The rooted cutting (*Sesuvium*) and the seedlings (*Mesembryanthemum*) were transferred for 15 days on 1/10 Hewitt solution supplemented with micronutrient and with 4.5 μM Fe in a glass house with mean temperatures (night-day) of 18-25 °C and relative humidity 80-70%. Thereafter, the medium was changed for non-diluted Hewitt solution, with micronutrients and 45 μM Fe plus 100 mM NaCl (Basal medium (B)). For split root experiments, each plant was maintained between two 750 ml plastic containers filled with aerated solutions, with one half of the root system plunging in each container.

Split-root experiment aimed at determining the implication of nutritional disruptions in growth inhibition under heavy metals (Cd^{2+} and Ni^{2+}) stress. Five treatments were carried out and schematized in Figure 1: In the first and second treatments, one half of roots were immersed in basal medium, and the other half in the same medium supplemented with 100 μM Cd^{2+} (B /Cd treatment) or 50 μM Ni^{2+} (B/Ni treatment). In the others treatments, the two halves of root system were immersed either in metal free basal medium (B/B) or in basal medium with 100 μM Cd^{2+} (Cd/Cd) or 50 μM Ni^{2+} (Ni/Ni). The culture solution was renewed weekly. Two harvests were made at the beginning of the treatment, and 30 days later.

At the harvests, shoots were successively rinsed three times in cold water and blotted between two layers of filter paper. Roots were dipped in a 0.01 M HCl cold solution to eliminate any external heavy metals adsorbed at the root surface, then rinsed three times with cold distilled water (Aldrich *et al.* 2003) and blotted with filter-paper. The fresh weight was immediately determined, and the dry one was measured after 48 hours of desiccation in an oven at 60 °C.

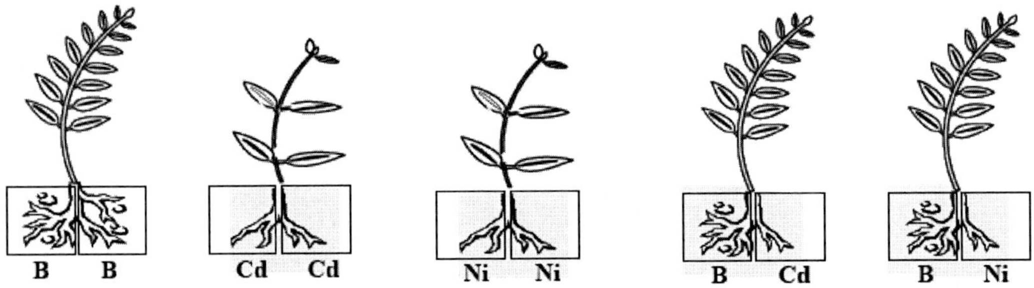

Figure 1. schematic representation of different treatments used in this experiment. B: Basal medium; Cd basal medium supplied with 100 μM Cd^{2+}, Ni basal medium supplied with 50 μM Ni^{2+}. For each treatment, 8 replicates corresponding to single plants are used for both species.

Cation Analysis

Desiccated samples were grounded to a fine powder using a porcelain mortar and pestle, then digested in 4/1 (v/v) $HNO_3/HClO_4$ mixture. Cd, Ni, Ca and Fe concentrations were determined by atomic absorption spectrometry (VARIAN, Spectra AA 220 FS). K^+ amounts were determined in the same homogenate by flame spectrometry (Corning photometer).

Statistical Analysis

Analyses of variance (ANOVA) with orthogonal contrasts and mean comparison procedures were used to detect differences between treatments. Mean separation procedures were carried out using the multiple range tests with Fisher's least significant difference (LSD) ($P< 0.05$).

RESULTS

Plant Morphology and Growth

After 15 days, chlorosis was visible on young leaves of *Sesuvium* plants subjected to Cd/Cd and Ni/Ni treatments. Two weeks later, chlorosis was accentuated and other toxicity symptoms, petiole necrosis and leaf fall, were evident in these plants. *Mesembryanthemum* plants submitted to Cd/Cd treatment showed a generalized chlorosis; while plants submitted totally to Ni^{2+} (Ni/Ni plants) showed a pronounced stunting. However, no visual toxicity symptoms appeared in shoots of both species subjected to B/Cd and B/Ni treatments. Globally, B/Cd and B/Ni treatments attenuated the toxicity symptoms found in Cd/Cd and Ni/Ni plants.

Cd/Cd and Ni/Ni treatments reduced the shoot and the root dry matter production in both species (Figure 2A, B) as compared to control (B/B). These effects were less marked (not significant) for *Sesuvium* (Figure 2 A) and particularly pronounced in *Mesembryanthemum* (Figure 2 B). In *Sesuvuim*, B/Cd and B/Ni plants produced significantly more biomass than B/B plants. In *Mesembryanthemum*, plants cultivated on dual medium (B/Ni and B/Cd), produced a similar biomass when compared to control (B/B) and two fold of those produced by Ni/Ni and Cd/Cd plants.

Cd^{2+} Accumulation

In shoots, Cd^{2+} accumulation differed between the two species. Accumulation in *Mesembryanthemum* shoots was twice of *Sesuvium* for both Cd/Cd and B/Cd treatments (Figure 3A). In plants subjected to Cd/Cd treatment, Cd^{2+} shoot-concentration reached 193 and 323 µg/g DW, respectively in *Sesuvium* and *Mesembryanthemum* and was higher than those measured in B/Cd plants which attained 85 and 185µg/g DW respectively in *Sesuvium* and *Mesembryanthemum.* In spite of the alleviation of the Cd-effects on growth observed in B/Cd plants as compared to Cd/Cd ones, the quantities of Cd^{2+} deposited in shoots (Cd^{2+}-shoot concentration * shoot dry weight) in the two types of treatment were similar for each species (Figure 3A). Thus the reduced Cd^{2+} concentrations showed in the shoot of B/Cd plants resulted from a dilution by growth.

In the root of both species, Cd^{2+} was more accumulated than in the shoots. The root part of B/Cd plants immersed in solution containing Cd^{2+} and those of Cd/Cd showed the highest Cd^{2+} concentrations (Figure 3B). Moreover, the root part of B/Cd developed in Cd^{2+}-free medium and having any direct contact with Cd^{2+} showed a relatively high Cd^{2+} concentration reaching 330 µg/gDW in both species.

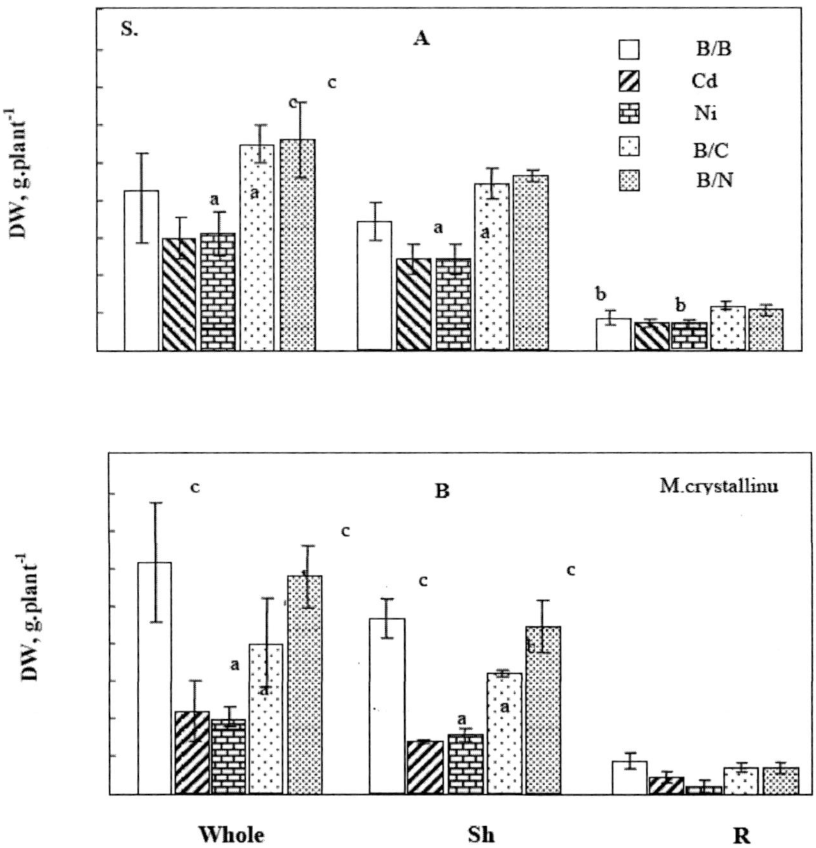

Figure 2. Changes in whole plant, shoots and roots dry weight (g.plant^{-1}) produced by *S. portulacastrum* (A) and *M. crystallinum* (B) grown 30 days in a split-root system. B: standard nutrient solution; Cd: standard nutrient solution supplemented with 100 µM Cd^{2+}; Ni: standard nutrient solution supplemented with 50 µM Ni^{2+}. Means of 8 replicates. Bars marked with same letter are not significantly different at p= 0.05.

Ni^{2+} Accumulation

Ni^{2+} was less accumulated than Cd^{2+} in the different plant tissues. Indeed, in *Sesuvium,* this metal was not detected in the shoot of plants subjected to B/Ni treatment and did not exceed 12 µg/g DW in Ni/Ni treated plants (Figure 4A). In *Mesembryanthemum,* shoot-Ni^{2+} concentrations reached 49µg/g DW in B/Ni plants, which represents 43% of those measured in Ni/Ni plants. Ni^{2+} was not detected in the root part of B/Ni plants of *Sesuvium* developed in basal medium. In the other half of roots immersed medium containing 50 µM Ni, Ni^{2+} concentration represent 32% as compared to those of Ni/Ni plants (Figure 4B).

As compared to *Sesivium, Mesembryanthemum* accumulated more Ni^{2+} in the roots. Indeed, this parameter reached 1000 µg/g DW in the root developed in medium containing 50 μM Ni in both Ni/Ni and B/Ni plants. For B/Ni plants, the root part developed in Ni-free medium contained up to 325 μg Ni/g DW.

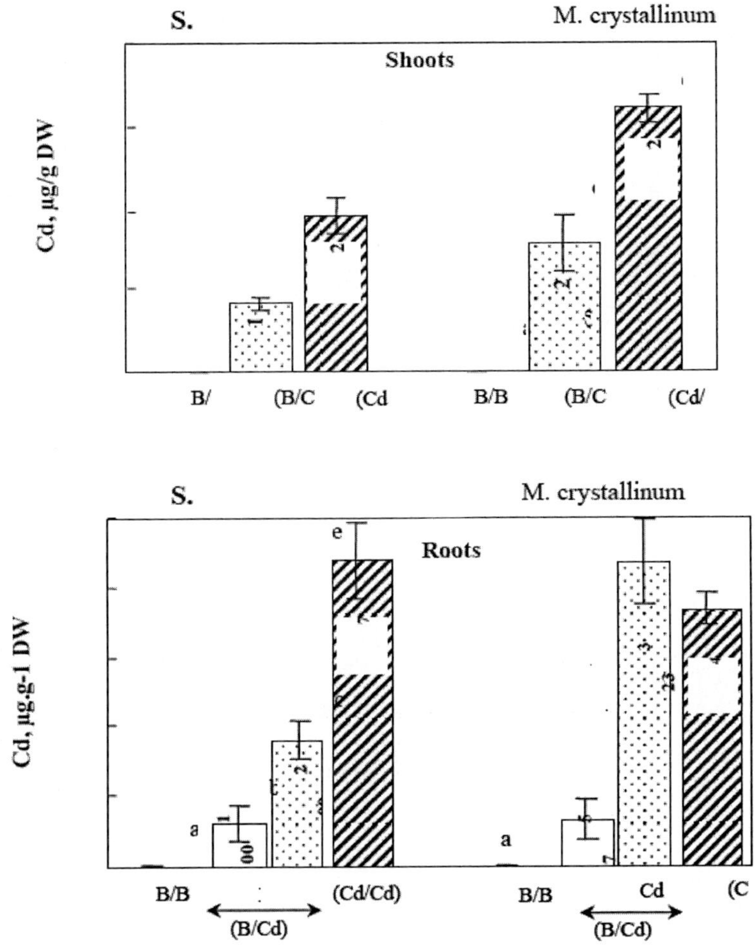

Figure 3. Changes in Cd concentration (μg.g^{-1} DW) in the shoots and the roots of *S. portulacastrum* and *M. crystallinum* plants subjected to B/B (two halves of the root system in basal medium), B/Cd (one half of the root system in basal medium and the other half in basal medium added with 100 μM Cd^{2+}) and Cd/Cd (two halves of the root system in basal medium added with 100 μM Cd^{2+}) treatments. The numbers inside the bars are the amounts of Cd accumulated in the shoots and the roots (μg.plant^{-1}). Means of 8 replicates. Bars marked with same letter are not significantly different at p= 0.05.

Potassium Status

Addition of 100 μM Cd^{2+}, (Cd/Cd treatment), or 50 μM Ni^{2+}, (Ni/Ni treatment), in the medium culture reduced significantly the shoot K^+ concentration in both species (Figure 5A).

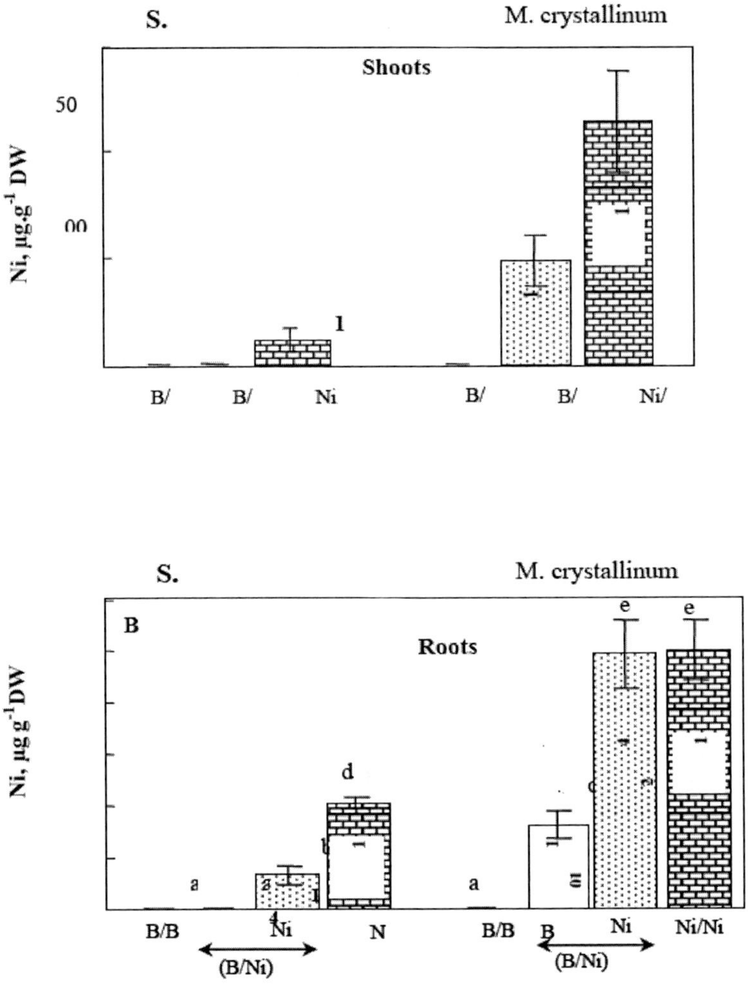

Figure 4. Changes in the Ni concentration (µg.g^{-1} DW) in the shoots and the roots of *S. portulacastrum* and *M. crystallinum* plants subjected to B/B (two halves of the root system in basal medium), B/Ni (one half of the root system in basal medium and the other half in basal medium added with 50 µM Ni^{2+}) and Ni/Ni (Two halves of the root system in basal medium added with 50 µM Ni^{2+}) treatments. The numbers inside the bars are the amounts of Ni accumulated in the shoots and the roots (µg.plant^{-1}). Means of 8 replicates. Bars marked with same letter are not significantly different at p= 0.05.

However these effects were more pronounced in *Mesembryanthemum*. Indeed, compared to B/B plants, this decrease reached respectively 26 and 30 % in Cd/Cd and Ni/Ni plants in *Sesuvium* and it was 55 and 60 % in *Mesembryanthemum*.

As compared to Cd/Cd or Ni/Ni plants, culture on dual medium led to significant increase of shoot K^+ status. In both species, K^+ shoot-concentration was similar in B/B and B/Cd or B/Ni plants. Changes in the mean quantities of K^+ (mmol.plant^{-1}) accumulated in shoots are represented in Table 2. K^+ amounts measured in B/Cd or B/Ni plants were 2 to 5 fold higher than those of Cd/Cd or Ni/Ni plants, respectively in *Sesuvium* and *Mesembryanthemum*.

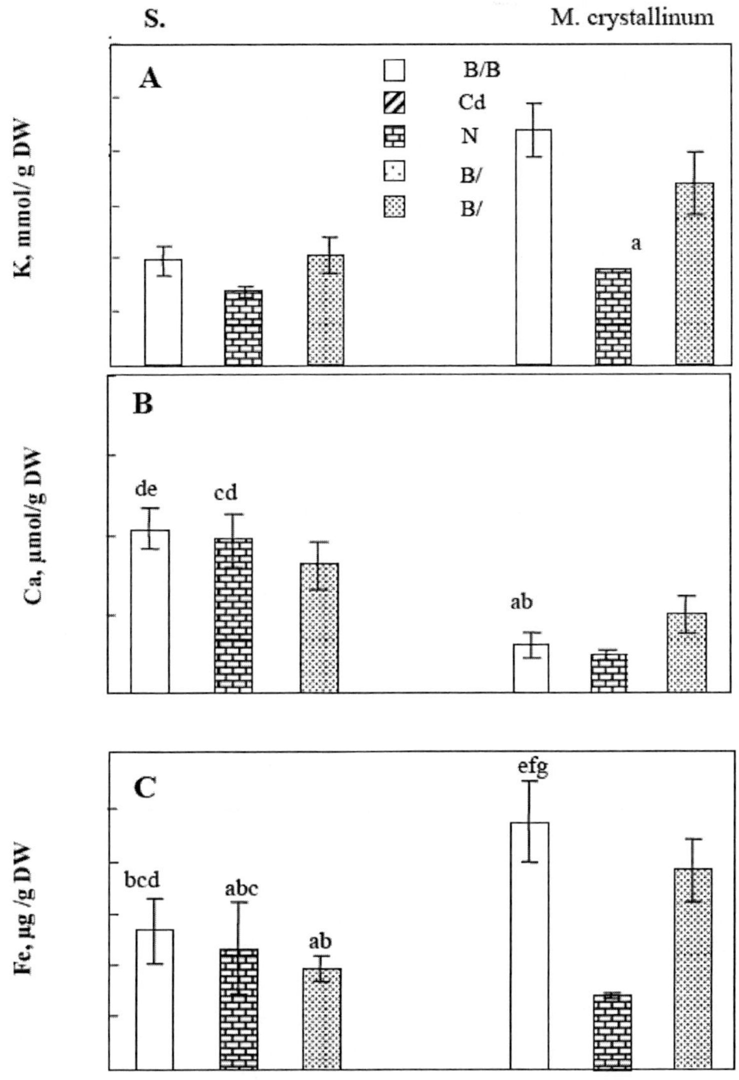

Figure 5. Changes in (A) potassium (mmol.g^{-1}DW), (B) calcium (µmol.g^{-1}DW) and (C) iron (µg.g^{-1} DW) concentrations in the shoots of *S. portulacastrum* and *M. crystallinum* with different treatments. Means of 8 replicates. Bars marked with same letter are not significantly different at p= 0.05.

Plants subjected to the Split root system-treatments showed an adequate K^+ shoot provision in *Sesuvium.* However, in *Mesembryanthemum*, B/Cd and B/Ni plants accumulated in their shoots respectively 60 and 75% of K^+ amounts measured in B/B plants (Table 2).

Calcium Nutrition

The presence of Cd^{2+} or Ni^{2+} in the culture medium (Cd/Cd and Ni/Ni treatments) did not modify significantly shoot Ca^{2+} concentration in both species (Figure 5B). Analysis of Ca^{2+} amounts allocated to shoots showed specific differences (Table 2). In *Sesuvium*, no significant difference was found in shoot calcium amounts between control and Cd/Cd or Ni/Ni plants. However, in *Mesembryanthemum* this parameter was strongly reduced in Cd/Cd and Ni/Ni plants as compared to B/B ones. Indeed, Ca^{2+} amounts represent only 26 to 43 % of the control value, respectively in Ni/Ni and Cd/Cd plants. In the last species, culture on dual medium (B/Cd and B/Ni) led to an appropriate shoot Ca^{2+} provision (Table 2).

Iron Status

Shoot Fe concentration decreased significantly in *Mesembryanthemum* plants subjected to Ni/Ni and Cd/Cd treatments (Figure 5C). In *Sesuvium*, the reduction was significant only under Cd/Cd treatment. In the first species, split root system increased significantly both shoot Fe concentration (Figure 5C) and amount (Table 2) as compared to Cd/Cd and Ni/Ni treatments. In *Sesuvium,* culture on divided roots (B/Cd and B/Ni) led to an increase in the shoot-iron amounts (Table 2), although B/Ni treatment did not modify Fe shoot concentration (Figure 5C). In general, the part of root system developing in free metal (Cd and Ni) medium in B/Cd and B/Ni ensures an adequate iron nutrition in both species, although the presence of Cd or Ni through the other part .

Discussion

Plant biomass measured at the final harvest depends on the initial size of the plant (before the beginning of the treatments) and on its growth activity during the treatment period. Relative growth rate (RGR), i. e. the rate of increase in total dry weight per unit of plant dry weight, is recommended to evaluate the effect of environmental constraints on the growth activity. Analysis of the RGR values based on whole plant biomass (Table 1) showed that the addition of Cd^{2+} or Ni^{2+} through the whole root system (Cd/Cd and Ni/Ni treatments) decreased significantly the rate of biomass production. In *Mesembryanthemum*, RGR shifts from high ($0.097d^{-1}$) to low ($0.06\ d^{-1}$) value in plant subjected to Cd/Cd and Ni/Ni treatments. *Sesuvium* growth activity was less sensitive to Cd^{2+} and Ni^{2+} as compared to *Mesembryanthemum.* Indeed, in this species (*Sesuvium*), the decrease of RGR in Cd/Cd and Ni/Ni plants does not exceed 16 % as compared to control plants B/B. These data confirm the higher sensitivity of *Mesembryanthemum* reported in a previous study (Ghnaya *et al.* 2005). However, in both halophytes, plants cultivated in dual medium (B/Cd) and (B/Ni) maintained normal growth activity although one half of the root system was developed in medium containing Cd^{2+} and Ni^{2+}. In addition, the maintenance of growth activity showed in plants cultivated on split-root system was accompanied with an alleviation of visual toxicity symptoms appeared in the shoots of Cd/Cd and Ni/Ni plants, despite the relatively elevated shoot-metal concentrations (Figure 4 and 5). Hence, both species can cope with high heavy metal levels (especially Cd^{2+}) in the photosynthetic organs, when a part of root system is maintained in free-metal solution. These data suggest that the restriction of growth showed in plants having the whole root systems in medium added with Cd^{2+} or Ni^{2+} could results from the effects of metals (Cd^{2+} and Ni^{2+}) on the absorption and translocation of water and essentials elements from roots toward the shoots.

Table 1. Changes in the RGR (Relative Growth Rate) of *S. portulacastrum* and *M. crystallinum* exposed during 30 days to different treatments. RGR measures the quantity of biomass deposited by 1 g of biomass per unit of time. It was estimated as Δln(DW)/Δt, where DW is the dry weight, ln stands for natural logarithm and Δ represents the difference between final and initial value (Hunt, 1990). Means of 8 replicates. Values marked with same letter are not significantly different at p= 0.05

	RGR (d^{-1})	
M. crystallinum Treatments		*S. portulacastrum*
B/B	0.097 ± 0.01(b)	0.06 ± 0.01(b)
B/Cd	0.082 ± 0.012(b)	0.07 ± 0.003(c)
B/Ni	0.096 ± 0.006(b)	0.07 ± 0.01(c)
Cd/Cd	0.06 ± 0.015(a)	0.05 ± 0.01(a)
Ni/Ni	0.06 ± 0.005(a)	0.05 ± 0.01(a)

Table 2. Changes in K^+ (mmol $plant^{-1}$), Ca^{2+} (mmol $plant^{-1}$) and Fe (µmol $plant^{-1}$) amounts accumulated in the shoots of *S. portulacastrum* and *M. crystallinum* submitted to different heavy metals treatments. Means of 8 replicates. Values marked with same letter are not significantly different at p= 0.05.

	S. portulacastrum				
	B/B	B/Cd	B/Ni	Cd/Cd	Ni/Ni
Element					
K (mmol)	1.66 ± 0.5(b)	2.06 ± 0.2 (cb)	2.26 ± 0.2 (c)	0.86± 0.15(a)	0.87 ± 0.16(a)
Ca (mmol)	0.67 ± 0.2 (a)	1.13 ± 0.2 (b)	0.72 ± 0.22(a)	0.56 ± 0.1(a)	0.48 ± 0.09(a)
Fe (µmol)	8.14 ± 3(dc)	13.21 ± 3.2(ef)	7.5 ± 0.9(bc)	2.75 ± 0.5(a)	5.35 ± 1(ab)
	M. crystallinum				
	B/B	B/Cd	B/Ni	Cd/Cd	Ni/Ni
K (mmol	5.9 ± 1.2(c)	3.56 ± 1.1 (b)	4.44 ± 1.1 (cb)	0.86 ± 0.3 (a)	0.813 ± 0.1(a)
Ca (mmol	0.33 ± 0.09(c)	0.27± 0.06(bc)	0.52 ± 0.04 (d)	0.14 ± 0.03(ab)	0.08 ± 0.01 (a)
Fe (µmol)	22.35 ± 4.5(g)	10.7± 3.2(de)	17.6 ± 2.5(fg)	3.32 ± 0.5 (ab)	2.32 ± 0.3 (a)

Indeed, one of the known toxic effects of heavy metals on plant physiology is the alteration of the water balance, which is probably due to damages caused to membrane integrity, turgor potential and cell wall elasticity (Barcelo and Poschenrieder, 1990, Perfus-Barbeoch *et al.* 2002). Some of our data support the depression action of Cd^{2+} and Ni^{2+} on cellular turgor, when plants were cultivated completely in the presence of Cd^{2+} (Cd/Cd) or Ni^{2+} (Ni/Ni). Plants showed shoot dehydration in both halophytes (not shown). However, split-root system treatments restored adequate shoot water provisioning, demonstrating that the half of root system grown in free metal medium is able to provide sufficient water to sustain the whole plant growth and thus mitigated the effects of absorbed metal ensured by the other half.

On the other hand, excepting few hyperaccumulators species which are able to accumulate toxic heavy metals in the shoots at relatively high concentrations without growth reduction, other species can not survive even at very low toxic metals levels accumulation in photosynthetic organs. Our data showed that reduced growth activity estimated by the RGR values in Cd/Cd and Ni/Ni plants were accompanied by the highest Cd^{2+} (Figure 6A) and Ni^{2+} (Figure 6B) shoot concentrations. However, plants grown on dual medium showed high growth potentialities concomitant with a relatively high Ni^{2+} and particularly Cd^{2+} concentrations in leaf tissues. Thus, growth inhibition showed in plants subjected totally to Cd^{2+} and Ni^{2+} resulted not only from the direct alteration of photosynthesis process by Cd^{2+} and Ni^{2+} transported to the shoots, but this effect could imply additionally the action of metals on essential nutrients acquisition. In fact, the relationship between RGR and the nutrients (K^+, Ca^{2+} and Fe) amounts in the shoots (Figure 7) demonstrated that the high growth potentialities observed in B/B (control), B/Cd and B/Ni plants were accompanied by the highest K^+, Ca^{2+} and Fe uptake. Conversely, Cd/Cd and Ni/Ni plants expressed the lowest growth activities and the reduced K^+, Ca^{2+} and Fe amounts in the shoots. On the basis of these data, we suggest that the reduction of biomass production found in Cd/Cd and Ni/Ni plants may results, at least partially, from the restriction of K^+, Ca^{2+} and Fe uptake by both metals Cd and Ni.

Nutritional disturbances have been reported in several species submitted to heavy metal stress. For example, a drastic decrease in shoot potassium concentrations was observed in white lupin (Zornoza *et al.* 2002), pea (Sandalio *et al.* 2001) and *Pinus sylvestris* (Kim *et al.* 2003) plants under cadmium stress. Heavy metals could act indirectly or directly on nutrient absorption, several explanations were given. Cd^{2+} may limit K^+ absorption by complexing ATP and reducing energy availability for the absorption of this nutrient. In the same context, Kim *et al.* (2002) concluded that heavy metals compete with Ca^{2+} for transport into root cells of rice plants. They showed that the level of heavy metals bound to cell wall is much lower than that of the total cellular level, so, this competition was not for the binding sites on cell wall, but for the transport across the plasma membrane. Additionally, it has been demonstrated that Cd^{2+} could permeates through Ca^{2+}-channels in animal (Olivi and Bressler, 2000) and plant (White 2000, Krämer *et al.* 2007) cells. Furthermore, Clemens et al. (1998) demonstrated that the plant transporter LCT1 mediates the uptake of both Ca^{2+} and Cd^{2+} in yeast. In higher plants, Cd^{2+} was assumed to enter root cells via either the higher affinity uptake system for Fe2+ or low affinity system for Ca^{2+} uptake (Lombi *et al.* 2002; Zhao *et al.* 2002; Roosens *et al.* 2003). More recently, Suzuki (2005) found that elevated Ca^{2+} concentration in the culture medium can greatly alleviated the toxicity of Cd^{2+} in *Arabidopsis* even at high (200 μM) concentration in the medium, which is consistent with the competition theory between the two cations during influx.

In addition, we observed that when root system was completely immersed in solution containing metals, especially Cd^{2+}, young leaves of both species suffered from chlorosis. This symptom, which is a typical response of all plant to the lack of iron in the medium, was attributed by (Wallace *et al.* 1992, Larbi *et al.* 2002) to heavy metals-induced Fe deficiency. In this experiment some data were consistent of these explications. In fact, figure 5 shows that when the entire root system was developed in the presence of Cd or Ni, shoot Fe concentrations were significantly reduced. Some researchers explicated the inhibition of Fe absorption in the presence of heavy metals in the medium by the competition between iron and metallic cations in root plant cells transporters. Therefore, it was shown in *Arabidopsis* that uptake of both Cd^{2+} and Fe may occur via members of a plant metals transporters family with homology to *Nramp* genes that are inducible by Fe starvation (Thomine *et al.* 2000).

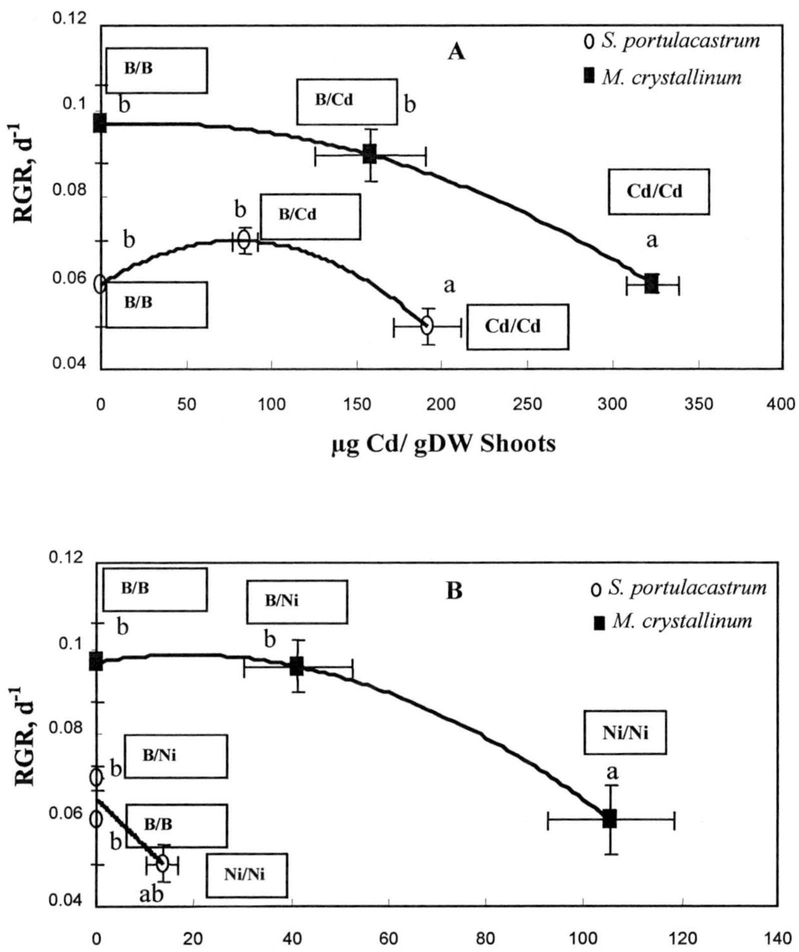

Figure 6. Relationship between RGR (d-1) and Cd (A) or Ni (B) shoots concentrations in S. portulacastrum and M. crystallinum. Means of 8 replicates. Statistical analysis was linked to the RGR, Values marked with same letter are not significantly different at p= 0.05.

This hypothesis was confirmed by the experiments of Cohen *et al.* (1998) and those of Awad and Römheld (2000) who founded that Fe-deficiency would stimulate Cd^{2+} uptake respectively by the roots of pea and wheat plants. In this context, other researches showed that in plants, Cd^{2+} was assumed to enter root cells via either the higher affinity uptake system for Fe^{2+} or low affinity system for Ca^{2+} (Roosens *et al,* 2003, Lombi *et al,* 2002, Zhao *et al.* 2002).

Our results also showed that the split-root system, ensuring an adequate Ca^{2+}, K^{+} and Fe supply by one half of the root system, and a load of leaves with Cd^{2+} and at a less degree with Ni^{2+} by the other root half, improved completely growth. Thus, relatively high leaf heavy metals (Ni^{2+} and Cd^{2+}) contents could be compatible with an optimal growth activity when shoots were sufficiently alimented with essential nutrients. Thus, we suggest that high Ca, K and Fe availabilities in the medium could alleviate the toxicity induced by heavy metals and enhance the capacity of plants to extract Cd^{2+} and Ni^{2+} by improving growth concomitant to a relatively elevated Cd^{2+} and Ni^{2+}-shoots concentration. With respect to metals (Cd^{2+} and Ni^{2+}) shoot accumulation, and since extracted metals amount depend in the same time on shoot metals concentration and shoot biomass, plants cultivated on dual medium accumulated approximately the same amounts of metals in the shoots as compared to plants submitted completely to metals.

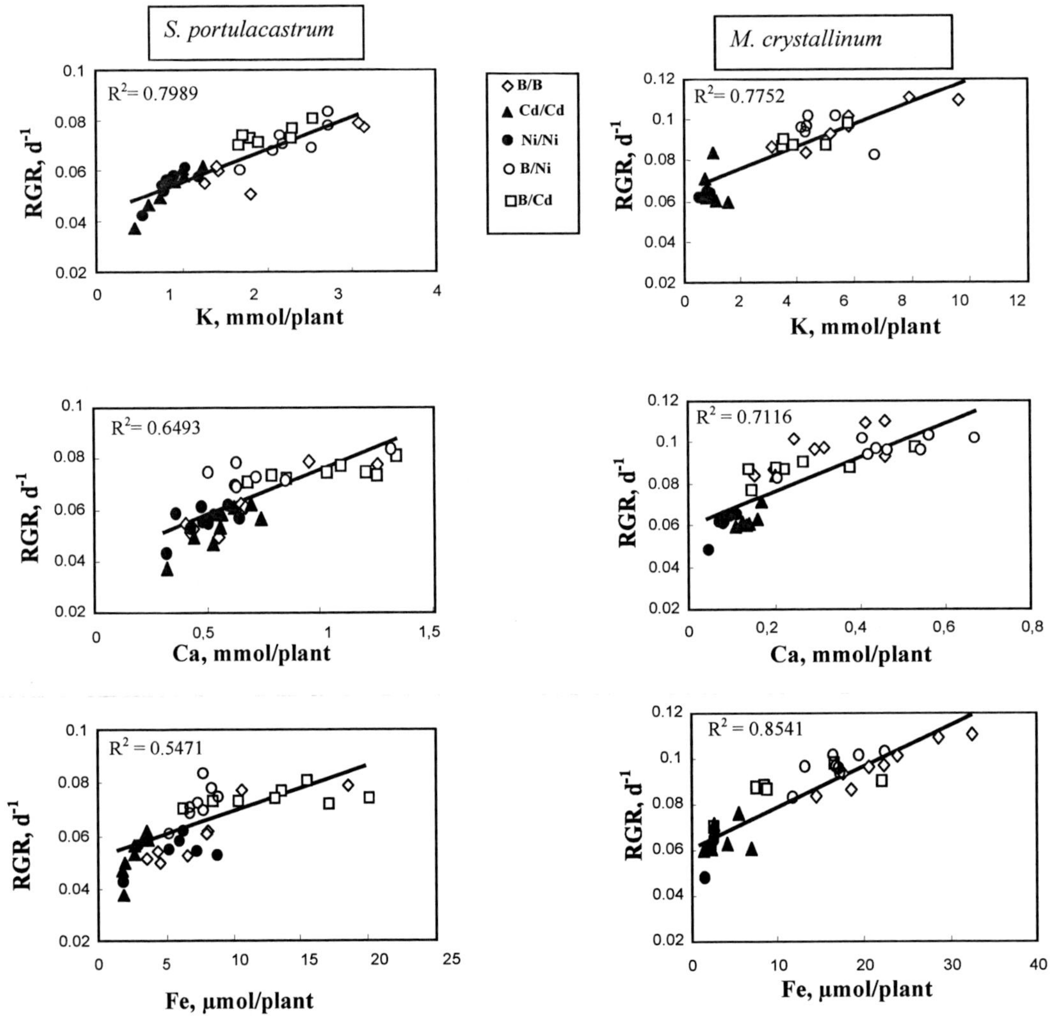

Figure 7. Relationship between RGR (d-1) and K+ (mmol/ plant), Ca2+ (mmol/ plant) and Fe (µmol/ plant) shoot amounts. Means of 8 replicates.

Hence, the reduced Cd^{2+} and Ni^{2+} concentrations in the shoots of B/Cd and B/Ni plants resulted principally from the dilution by growth. Nevertheless, since these data were related to a short duration-experiment, we suppose that under a long term experiment, the quantity of metal deposit in the shoots of these plants will be more important, because plants grown on dual medium maintained their vigor without any symptoms of toxicity and are capable to continue their life cycles. Thus, nutritional disturbances induced by heavy metals could limit plant extraction capacities through its effects on plant growth. Some available data support this assumption. For instance, it has been reported in some studies that high availability of K^+, Ca^{2+}, Fe^{2+} and Mg^{2+} improved growth of plants subjected to Pb^{2+} and Cd^{2+} although the presence of these cations in shoot-tissues (Cohen *et al.* 1998; Kim *et al.* 2002; Suzuki, 2005).

In the others hand, our data showed also that the root part of B/Cd plants which have any direct contact with metal presented an important Cd^{2+} concentration in both species (Figure, 3B). This result suggests a possible recirculation of this metal through the phloem from the shoots towards roots. Ni^{2+} recirculation was also probable in *Mesembryanthemum* but not in *Sesuvium*. This result cannot correspond to a contamination of the roots developing in free metal solution by an exogenous source of Cd^{2+} or Ni^{2+}. Indeed, in *Sesuvium*, B/Ni plants contained 160 µg Ni /g DW in the half of roots growing on basal solution supplied with 50 µM Ni^{2+} and no trace of this metal in the other half of roots maintained in the Ni^{2+} free solution.

The load of Cd^{2+} in phloemic flux was shown in some studies. For example, a linear relationship was found between the Cd^{2+} concentration of the soil and those of wheat grain, suggesting that this metal passed through the xylem towards phloemic circulation (Herren and Feller, 1997). In the same context, Zn^{2+} introduced into the cut stem of wheat plants was shown to be removed from the transpiration stream into the peduncle, loaded into the phloem, and transported to the maturing wheat grains (Herren and Feller, 1996). To our knowledge, no data are available concerning the recirculation of heavy metals from shoots toward the roots and its implication in the plant responses to heavy metal stress.

In conclusion, this study showed that the culture on dual medium improve generally plant supply with K^{+}, Ca^{2+} and Fe and led to an optimal growth activities, in spite of a relatively high Cd^{2+} and at less degree Ni^{2+} concentrations in the shoots. These data showed clearly that nutritional disturbances induced by the presence of Cd^{2+} or Ni^{2+} in the culture medium, contributed largely to the growth restriction in both halophytes and limit the phytoextraction capacities of both species. Additionally, our results indicate that Cd^{2+} and Ni^{2+} accumulation in the shoots is compatible with a good growth activity and suggest the possibility to enhance plant capacity to extract heavy metals by increasing nutrient availability in soil.

Acknowledgments

This work was achieved in the Laboratory of Plant Adaptation to Abiotic Stresses. It was supported by the PRC project *"Utilisation des halophytes pour la réhabilitation et la valorisation des sols salins"* of the Tunisian Ministry of Scientific Research, Technology and Competence Development.

References

Aldrich, M.V., Gardea-Torresdey J.L., Peralta-Videa, J.R., Parsons J.G.: Uptake and reduction of Cr(VI) to Cr(III) by mesquite (*Prospis spp.*): chromate-plant interaction in hydroponics and solid media studied using XAS. *Envron. Sci. Technol.* 37: 1859-1864, 2003.

Arduini, I., Masoni, A., Mariotti, M, Ercoli L.: Low cadmium application increase miscanthus growth and cadmium translocation. *Environ. and Exp Bot.* 52: 89-100, 2004.

Arnon, D.I., Hoagland, D.R.: Crop production in artificial solutions and in soils with special reference to factors affecting yields and absorption of inorganic nutrients. *Soil Sci.* 50: 463-484, 1940.

Awad, F., Römheld, V.: Mobilization of heavy metals from contaminated calcareous soils by plant born, microbial and synthetic chelators and their uptake by wheat plants. *J. Plant Nutr.* 23: 1847-1855, 2000.

Barceló, J., Poschenrieder, C.: Plant water relations as affected by heavy metal stress: a review. *J. Plant Nutr.* 13: 1-37, 1990.

Chaoui, A., Ferjani, E.: Effects of cadmium and copper on an antioxidant capacities, lignification and auxin degradation in leaves of pea (*Pisum sativum* L.) seedlings. *C. R. Biol.* 328: 21-31, 2005.

Chen, H.M., Zheng, C.R., Tu, C., Shen, ZG.: Chemical methods and phytoremediation of soil contaminated with heavy metals. *Chemosphere* 41: 229-234, 2000.

Cohen, C.K., Fox, T.C., Garvin, D.F., Kochian, L.V.: The role of iron-defeciency stress responses in stimulating heavy-metal transport in plants. *Plant Physiol.* 116: 1063-1072, 1998.

Cunningham, S.D., Berti, W.R., Huang, J.W.: Phytoremediation of contaminated soils. *Trends Biotec.* 13: 393-397, 1995.

Ederli, L., Reale, L., Ferranti, F., Pasqualini, S.: Responses induced by high concentration of cadmium in *Phragmites australis* roots. *Physiol Plant.* 121: 66-74, 2004.

Ghnaya, T., Nouairi, I., Slama, I., Messedi, D., Grignon, C., Abdelly, C., Ghorbel, M.H.: Cadmium effects on growth and mineral nutrition of two halophytes: *Sesuvium portulacastrum* and *Mesembryanthemum crystallinum. J. Plant Physiol.* 162: 1133-1140, 2005.

Ghnaya, T., Nouairi, I., Messedi, D., Grignon, C., Abdelly, C., Ghorbel, M.H.: Evaluation de l'aptitude de deux halophytes à extraire les métaux lourds. Actes du Meeting International Aridoculture et Cultures Oasiennes: voies pour un développement durable en zones arides, Jerba Tunisia, Pp185-191, 2004.

Ghosh, M., Singh, S.P.: A comparative study of cadmium phytoextraction by accumulator and weed species. *Environmental pollution* 133: 365-371, 2005.

Hazardous Waste Consultant. Remediating soil land sediment contaminated with heavy metals, Nov/Dec. Elsevier Science, Netherlands 1996.

Herren, T., Feller, U.: Effect of locally increased zinc contents on zinc transport from the flag leaf lamina to the maturing grains of wheat. *J. Plant Nutr.* 19: 379-387, 1996.

Herren, T., Feller, U.: Transport of cadmium via xylem and phloem in maturing wheat shoots: comparison with the translocation of zinc, strontium and rubidium. *Ann. Bot.* 80: 623-628, 1997.

Hewitt, E.J.: Sand and water culture methods used in the study of plant nutrition. *Commonwealth Bureau of Horticulture Tech. Com.*, Pp. 22. 1966.

Hunt R.: Basic growth analysis. *Plant Growth Analysis for beginners.* UNWIN HYMAN. Pp.112, London 1990.

Jacobson, L.: Maintenance of iron supply in nutrient solutions by a single addition of ferric-potassium-ethylene-diamine-tetraacetate. *Plant Physiol.* 26: 411-413, 1951

Kim, C-G., Bell, J.N.B., Power, S.A.: Effects of soil cadmium on *Pinus sylvestris* L. seedling. *Plant Soil.* 257: 443-449, 2003.

Kim, Y.Y., Yang, Y.Y., Lee, Y.: Pb and Cd uptake in rice roots. *Physiol Plant.* 116: 368-372, 2002.

Krämer U, Talke I N., Hanikenne M, 2007: *Transition metal transport Minireview*, FEBS Letters: 12, 2263-72

Larbi, A., Morales, F., Abadía, A., Gogorcena L.J., Abadía, J.: Effects of Cd and Pb in sugar beet plants grown in nutrient solution: induced Fe deficiency and growth inhibition. *Funct Plant Biol.* 29: 1453-1464, 2002.

Lombi, E., Tearall, K.L., Howarth, J.R., Zhao, F.J., Hawkesford, M.J., McGrath, S.P.: Influence of iron status on cadmium and zinc uptake by different ecotypes of the hyperaccumulator Thlaspi caerulescens. *Plant Physiol* .128: 1359–1367, 2002

Lu, X., Kruatrachue, M., Pokethitiyookb, P., Homyok, K.: Removal of cadmium and zinc by water hyacinth, *Eichhornia crassipes*. *Sci. Asia.* 30: 93-103, 2004

Nouari, I., Ghnaya, T., Messeddi, D., Zarrouk, M., Abdelly, C., Ghorbel, M.H.: Etude de la croissance et de la capacité d'accumulation des métaux lourds chez deux halophytes: *Mesembryanthemum crystallinum* et *Sesuvium portulacastrum. XIIIèmes J. Biol l'Assoc Tun Sci Biol.* Pp.112 2002.

Olivi, L., Bressler, J.: Maitotoxin stimulates Cd^{2+} influx in Madin-Darby Kidny cells by activating Ca-permeable cation channels. Cell Calcium. 27: 187-193, 2000.

Olmos, E.O., Martínez-Solano, J.R., Piqueras, A., Hellín E.: Early steps in the oxidative burst induced by cadmium in cultured tobacco cell (BY-2 line). *J. Exp. Bot.* 54:291-301, 2003

Perfus-Barbeoch, L., Leonhardt, N., Vavasseur, A., Forestier, C.: Heavy metal toxicity: cadmium permeates through calcium channels and disturbs the plant water status. *Plant J.* 32: 539-548, 2002

Robinson, B.H., Millis, T.M., Petit, D., Fung, L.E., Gren, S.R., Clotheir, B.E.: Natural and induced cadmium accumulation in poplar and willow: implications for phytoremediation. *Plant Soil.* 227: 301-306, 2000

Roosens, N., Verbruggen, N., Meerts, P., Ximenez-Embun, P., Smith, J.A.C.: Natural variation in cadmium tolerance and its relationship to metal hyperaccumulation for seven populations of Thlaspi caerulescens from western Europe. *Plant Cell Environ* 26: 1657-1672, 2003

Sandalio, L.M., Dalurzo, H.C., Gómez, M., Romero-Puertas, M.C., del Rio, L.: Cadmium-induced changes in the growth and oxidative metabolism of pea plants, *J. Exp. Bot.* 52: 2115-2126, 2001

Sanità di Toppi, L., Gabbrielli, R.: Responses to cadmium in higher plants. *Environ. Exp. Bot.* 41: 105-130, 1999

Shen, J.B., Jiang, B., Pappano, A.J.: Comparison of L-type calcium channel blockade by nifedipine and/or cadmium in guinea pig ventricular myocytes. *J. Pharmacol Exp Ther.* 294: 562-570, 2000

Suzuki, N.: Alleviation by calcium of cadmium-induced root growth inhibition in *Arabidopsis* seedlings. *Plant Biotechnology* 22: 19–25, 2005

Thomine, S., Wang, R., Ward, J.M., Growford, N.M., Schroeder, J.I.: Cadmium and iron transport by members of a plant metal transporter family in *Arabidopsis* with homology to *Nramp* genes. Proc. Nat. Acad .Sci .USA. 97: 4991-4996, 2000

Wallace, A., Wallace, G.A., Cha, J.W.: Some modifications in trace metal toxicities and deficiencies in plants resulting from interactions with other elements and chelating agents. The special case of iron. *J. Plant Nutr.* 15: 1589-1598, 1992

White, F.P.: Calcium channels in higher plants. Biochim Biophys Acta. 1465: 171-189, 2000

Wojas. S., Ruszczyńska. A., Bulska E., Wojciechowski M., Antosiewicz D.M. Ca^{2+}-dependent plant response to Pb^{2+} is regulated by LCT1, *Environ. Pollution:* 147: 584-592, 2007

Zornoza, P., Vázquez, S., Esteban, E., Fernández-Pascual, M., Carpena, R.: Cadmium-stress in nodulated white lupin: strategies to avoid toxicity. *Plant Physiol. Biochem.* 40: 1003-1009, 2002.

Zhao, F.J., Hamon, R.E., Lombi, E., McLaughlin, M.J., McGrath, S.P.: Characteristics of cadmium uptake in two contrasting ecotypes of the hyperaccumulator *Thlaspi caerulescens*. *J. Exp. Bot.* 53: 535–543, 2002

In: Soil Ecology Research Developments
Editors: Tian-Xiao Liu, pp. 197-216

ISBN: 978-1-60021-971-9

Chapter 8

A COLD SITE LOCATED BELOW THE LIMIT OF DISCONTINUOUS ALPINE PERMAFROST: PLANT CHARACTERISTICS AND SOIL NUTRIENT DYNAMICS DURING THE WINTER SEASON

Michele Freppaz[a,1]***, Luisella Celi***[a]***, Fulvia Rosso***[a]***, Veronika Stöckli***[b]***, Marcia Phillips***[b] ***and Ermanno Zanini***[a]

[a] University of Torino, Di.Va.P.R.A. – Laboratorio Neve e Suoli Alpini, 44 Via Leonardo da Vinci, 10095 Grugliasco (TO), Italy

[b] Swiss Federal Institute for Snow and Avalanche Research, 7260 Davos Dorf, Switzerland

ABSTRACT

Exceptional occurrences of permafrost exist in the forest belt well below the limit of discontinuous alpine permafrost. There are several such sites in the Alps, all steep scree slopes located at the foot of high limestone cliffs. A particularly interesting and important characteristic is that they are vegetated with subalpine vegetation and that some plants display signs of severely limited growth. Research carried out in the last years has demonstrated that the ground is not frozen as a result of a particularly cold microclimate at these locations, but that permafrost is present due to a particular air circulation phenomenon. Under these conditions, soil organic matter accumulates due to the slowing down in degradative processes of plant residues. Consequently, the nutrient availability to plants may be limited and contribute to the reduced plant growth, although scant attention has been devoted to this topic.

To determine vegetation and soil characteristics, tree height, tree ring width measurements and soil analysis were performed in two contiguous areas differently affected by frost conditions in Western Switzerland and characterized by different growth of trees: (I) dwarf trees; (II) reference trees. Near surface soil temperature in the two sites

[1] Corresponding author, Di.Va.P.R.A. – Chimica Agraria e Pedologia – Laboratorio Neve e Suoli Alpini 44 Via Leonardo da Vinci, 10095 Grugliasco (TO). Tel: 0039 011 6708514, Fax: 0039 011 4031819,e-mail: michele.freppaz@unito.it.

has been collected from a whole winter (November 2001-May 2002). In the same period the N and C dynamics in the organic OH horizon were measured by the buried bag technique and in the undisturbed soils.

Tree growth was significantly lower in the permafrost affected sites, where the winter soil temperature was significantly lower than the reference forest.

During winter an increase of nitrate concentration was recorded in all sites, providing an inorganic N pool ready available for plant growth under the reference forest, but not under the dwarf trees, due to the lower soil temperature which inhibit plant nutrient absorption. Moreover, under the dwarf forest the microbial N immobilization, with a corresponding DON and NH_4^+ decrease, was more evident than in the reference site. The critical conditions under the dwarf trees could have selected a microbial community particularly tolerating cold temperatures, and then more resistant to moderate freeze/thaw events.

Thus, an inorganic N pool, constituted mainly by the leachable NO_3^-, is available in the early growing season in the cold site and in the reference site, but the lower soil temperature under the dwarf trees may inhibit soil nutrient adsorption by plants. Therefore in the cold site there is as asynchronies between the availability of nutrients and their utilization, which may affect the plant growth.

INTRODUCTION

Permafrost distribution ranges from continuous coverage in the northern continuous permafrost zone to discontinuous ice lenses in the localized permafrost zone. Discontinuous alpine permafrost exists roughly above 2400 m ASL.

Below the timberline, permafrost is assumed to exist only at scree slopes located at foot of high cliffs, at very shaded sites as reported by Kneisel et al. (2000). Slopes composed of large blocky materials are reported to determine a mean annual ground temperature colder than in fine materials (Harris and Pedersen, 1998). Unexpectedly low ground temperatures and ground ice during the summer were reported in talus at low altitudes (200-1000 m ASL) throughout the Daisetsu Mountains (Hokkaido Island) (Sawada et al., 2003) and in the mountains of Central Asia (Altai, Northern and Inner Tien Shan) (Gorbunov et al., 2004).

The difference in air temperature between the open space of the block slope and the atmosphere is thought to create an air circulation phenomenon which is supposed to be the predominant factor for the preservation of permafrost on mountain belts (Wakonigg, 1996; Tanaka et al., 2000). In winter, the warmer air in the blocks tends to rise and escape in the upper part of the slope, displaced by cold air, entering through holes in the snow cover. In the absence of snow cover in summer, the cold air trapped between the blocks sinks downslope through the blocks and escapes into the air at the bottom of the deposit, permitting warm air to replace it. These processes are assumed to operate in conjunction with other factors, such as slope and aspect which modify the potential radiation at a given site (Harris and Pedersen, 1998).

Due to the occurrence of permafrost conditions in the mountain belt these sites are vegetated with plants typical of subalpine and alpine areas and some species display signs of severely limited growth (Wegmann, 1995; Kneisel et al., 2000; Gorbunov et al., 2004). The ground may be covered by a thick layer of mosses (e.g. *Sphagnum*), due to the high soil humidity linked to heavy precipitation and air condensation. The vegetation composition and the specific thermal conditions lead to a great accumulation of organic material deposited on the block substrate (Kneisel et al., 2000; Freppaz et al., 2003; Gobat et al., 2004).

In cold regions, such as arctic ecosystems, the reduced organic matter decomposition rate is a key factor for the control of plant productivity, because plant biomass production generally is limited by low availability of N and P (Schmidt et al., 1999). Sveinbjornsson (1993) for example found that in *Betula pubescens* N availability was more important than soil temperature, in large part due to slow organic matter mineralization. The decomposition of organic matter and the nutrient dynamics in those environments are strongly affected by seasonal changes. In the winter time the lack of plant activity and the proliferation of a cold adapted microbial community may cause a relatively great nutrient immobilization (Schmidt and Lipson, 2004) by changing the quality itself of the organic matter (Dalias et al., 2001). During the snow melting and the starting of the plant growing season the nutrients become available through microbial lysis and mineralization of the easily degradable material and are rapidly up taken and consumed by plants during the short summer season (Lipson et al., 1999). Soil N mineralization during the nongrowing season represents an important portion of annual N cycle, and the accumulation of inorganic N in winter supply a large pulse of mobile and available N during the early growing season (Schimel at al., 2004).

In low elevation permafrost affected sites, where soil frost conditions are not directly connected to air temperature, the link between seasonal fluctuation of nutrient availability and plant growth may be more complex and is still unknown.

In this work we aimed to explore the soil nutrient dynamics and plant characteristics in a well-known cold site located below the limit of discontinuous alpine permafrost in the Jura Range, in north-western Switzerland (Lesquereux, 1844, Duchafour, 1976, 1983). In this site peculiarly dwarfed Norway spruce trees inside a semicircular escarpment are present. The limited plant growth has been associated for many years with permafrost (Richard, 1961) and recently Delaloye et al. (2003) revealed that ground temperatures are low and are clearly associated with permafrost. Previous works have attributed the reduced plant growth mainly to plant physiological stresses, *since roots are cold but not shoots* (Korner and Hoch, 2006), but they invoked soil N and P limitation as potential and additional causes.

This paper is aimed to describe and estimate the contribution of soil nutrient availability to the limited plant growth by evaluating the seasonal cycles of nutrients during the winter and the early spring season.

MATERIALS AND METHODS

Study Site

This research was conducted in Creux du Van, Swiss Jura (46°56'N, 6°44'E). The talus slope is located at the bottom of a cirque facing east-northeast between 1170 and 1300 m ASL (Figure 1). The cirque is closed from its southern to western rim by a 100-150 m high cliff of hard limestone. The fallen blocks build up a well-sorted talus slope at the base of the cliff. The mean annual air temperature (MAAT) measured over the period 1998-2001 is +5.5°C (Delaloye and Reynard, 2001). The annual precipitation is estimated to be 1600 mm.

Figure 1. Study area.

The very steep and unstable uppermost part of the scree consists of gravels without soil formation. Metric blocks are numerous in the stable and less inclined lower part and are covered by an organic soil (Gobat et al., 2004). The bedrock under the scree consists mainly of marls.

The presence of a permafrost table reported by Delaloye et al. (2003) is referred to the advective energy fluxes due to air density contrasts. The distribution of vegetal associations on the talus slope also indicates different ground thermal conditions, with growth of dwarf red spruces in several patches of the lower part of the talus slope. The areas situated downwards are covered by a mixed forest of beeches and spruces.

To compare soil characteristics and tree growth (*Picea abies*) in areas with different evidence of permafrost occurrence the site was divided into two strata, corresponding to the prevailing tree growth form: I) stunted forest and II) reference forest (Figure 1).

Tree Growth, Age and Size

Ring width is the most commonly studied parameter. Large, thin-walled cells are laid down early in the growing season when conditions are favourable for growth. These are followed by smaller, thick-walled and denser cells late in the growing season which form due to the onset of cooler temperatures, lack of soil moisture, and shorter days. The production of latewood cells terminates abruptly, followed by larger cells the following growing season. This abrupt termination at the end of one year and the beginning of the next year marks a ring boundary. Ring width is measured between two successive ring boundaries.

Around each soil profile in every stratum, three trees exhibiting representative sizes were selected. The height of the trees was estimated and the tree stems were cored perpendicular to their axes at heights of 30 cm above the ground. In the laboratory, the wood cores were processed with the usual dendrochronological methods (Schweingruber 1988).

Soil Chemical and Physical Characterization

The soil under the dwarf trees was classified as a Histic Organosol Insaturé and under the reference trees as an Organosol Insaturé (Gobat et al, 2004).

The soil samples were collected from the different soil horizons, at depth between 5 and 75 cm. The electrical conductivity was measured by a conductivity cell, measuring the resistance of a 1:20 soil/water suspension. The bulk density was determined weighing an intact soil core (500 mL). The pH was determined in a 1:20 soil-water suspension. Total soil carbon and nitrogen were measured using a THERMOQUEST NC 2005 combustion analyzer. $CaCO_3$ content was determined by a Scheibler's calcimeter. The total P was determined colorimetrically after solfo-perchloric digestion (Martin et al. 1999).

To determine ammonium and nitrate concentrations, 10 g of soil was added to 50 mL of 0.5M K_2SO_4 and shaken for 1 h, centrifuged and filtered. NH_4^+ in 0.5M K_2SO_4 extracts was diffused into 0.01M H_2SO_4 after treatment with magnesium oxide (Bremner 1965) and the trapped NH_4^+ was determined colorimetrically (Crooke and Simpson 1971). NO_3^- in the same extracts was determined colorimetrically as NH_4^+ after the reduction with Devarda Alloy. The organic nitrogen (N_{org}) was determined as the difference between total N and the inorganic N species.

For dissolved organic compounds, sub-samples (10 g fresh weight) were shaken with 100 mL 0.5M K_2SO_4 for 1 h and the suspension filtered at 0.45 μm under suction. Total dissolved N (TDN) in the extracts was measured as NH_4^+ after oxidation of aliquots of extracts with alkaline persulphate and subsequent reduction with Devarda Alloy (Williams et al., 1995). Dissolved organic N was calculated as DON = TDN – ($NH_4^+ + NO_3^-$).

Dissolved inorganic phosphorus (DIP) in the extracts was measured using the malachite green method (Ohno and Zibilske, 1991). Total dissolved P (TDP) concentrations in the extracts were measured as inorganic P in the persulphate oxidized extracts used to determine TDN (Williams et al. 1995). The difference between TDP and DIP was considered to be organic P (DOP).

Dissolved organic carbon (DOC) in the 0.5M K_2SO_4 suspension was measured by the Walkley and Black wet oxidation method (Nelson and Sommers, 1982).

To measure microbial biomass C and N a set of samples was fumigated overnight with chloroform and extracted with 0.5 M K_2SO_4 in parallel with a set of unfumigated samples (Williams and Sparling 1988). Biomass C and N values were calculated from the flushes of extractable C and N using the recovery factors of 0.45 for C (Sarathchandra et al. 1989) and 0.54 for N (Brookes et al. 1985). The mineralizable N was determined by the anaerobic incubation of soil samples in closed containers at 40°C for 10 days, as reported by Waring and Bremner (1964).

Temperature Measurements

To record soil temperature, thermistors combined with dataloggers (UTL-1) were buried in pairs in the two soil profiles at a depth of about 10 cm (OH horizon). The loggers recorded the temperature throughout the period November 2001-May 2002. Mean soil temperature was calculated in each stratum.

Soil Incubation

At the same time of temperature measurement C and N dynamics in the OH horizon were evaluated through field incubation experiments (November 2001- May 2002). Inorganic N variations, microbial uptake of mineralised nutrients and dissolved organic forms variation were measured in the field by the buried bag technique (Eno, 1960; Pastor et al., 1984; Adams et al., 1989), which enables measurements that account for the differences and fluctuations in soil temperature to be made while the water content is kept constant (Schmidt et al., 1999). The technique prevents plant uptake and leaching of mineralised nutrients but allows air diffusion and uptake by the microbial biomass.

A 100 g of soil was placed in a polyethylene bag, returned to the soil and buried to a depth of approximately 20 cm. Three bags have been incubated in each area. An another aliquot of soil was returned to the laboratory and immediately treated with 0.5M K_2SO_4 for NH_4^+ and NO_3^- analysis as previously described (Williams et al., 1995; Schmidt et al., 1999).

Net N mineralization was calculated as the difference between the sum of NH_4^+ and NO_3^- concentrations in the incubated and initial samples. Microbial N immobilization was calculated by differences in the microbial N before and after incubation. This is the potential amount of nutrients that microbes can immobilize inside the bags when the uptake by plant roots is eliminated (Schmidt et al., 1999). Furthermore, changes in dissolved organic nitrogen (DON) and carbon (DOC) during incubation were quantified because nutrients may be released as dissolved organic compounds during microbial dieback, or microbes may assimilate dissolved organic compounds during re-growth (Schmidt et al., 2002).

Statistical Analysis

All results were expressed as mg N, or P, or C kg^{-1} oven dry soil. Statistical analysis of the data was carried out using SPSS (SPSS Inc., Chicago, IL) software.

RESULTS

Vegetation Characteristics

The analysis of tree ring pattern revealed that the smallest trees of stratum I exhibited ages around 90 years, despite their small size (Table 1). Trees in stratum II at the edge of the site show distinctly taller size, higher ages and faster growth.

Soil Characteristics

The two soil profiles exhibited unaltered (OL) and emialtered (OF) organic horizons above a OH horizon which filled the interstices between blocks of hard limestone. The permafrost body was 15-20 m thick. The dwarf spruce stratum was restricted to the area where the active layer remained thin reaching the maximal thickness of about 2 m in October (Delaloye et al., 2003). The organic soil consisted of a hydromor with sphagnum moss rests on the hard limestone blocks. In the dwarf tree stands, raw humus was overlaying blockfields, with pockets of sphagnum peat accumulating in the gaps between boulders.

Soil characteristics recorded in the late fall are reported in Table 2. Moisture content was equal to 75 and 60% in the OH horizon of stratum I and II, respectively. The pH of the OL and OF horizons was around 4.2 in both sites due to the acidifying effect of spruce litter and of *Sphagnum* in stratum I. The higher pH (6.2-7.2) in the OH horizon was related to the limited weathering of limestone blocks (Gobat et al., 2004) and consequent incorporation of $CaCO_3$. Electrical conductivity was generally low and strongly decreased with depth in the stratum I while it remained constant in the reference site. The organic C content decreased from the OL to the OH horizons in both sites (Table 2). The total nitrogen increased along the profile in stratum I from 12.8 to 16.0 g kg^{-1} causing a C/N ratio reduction in the deeper horizons. In stratum II the total N decreased as the C trend resulting in a constant C/N ratio.

The different forms of N, C and P are reported only for the OF and OH horizons (Table 3). The ammonium content was lower in the stratum I than in stratum II with a great depletion from the OF to the OH horizons. This trend was observed also for nitrate in the Stratum II, whereas low and comparable values were found in the two horizons of stratum I. The microbial C was lower in stratum I and generally higher in the upper horizons. The microbial C/N ratio ranged from 28 in the OF of stratum I to 4 in the OH of stratum II. The highest DON content was found in the OH of stratum I, but was represented by a low amount of mineralizable material (NPot). The DOC content as well was more present in the stratum I, in both horizons.

Table 1 .Vegetation and site characteristics of two stratum in Creux du Van. Radial growth was analysed as the sum of tree ring growth over the past 50 years, 1952 to 2001. The standard deviation is indicated in brackets

Creux du Van Stratum	I	II
Altitude (m a.s.l.)	1220	1190
Slope (°)	29	25
Vegetation	"Dwarf" spruce *Vaccinium myrtillus* *Sphagnum quinquefarium* *Sphagnum acutifolium*	"Reference spruce" *Sorbus aucuparia* *Vaccinium myrtillus*
Mean high (spruce) (m)	3	25
Radial growth (spruce) (cm 50 $year^{-1}$)	3.0 (0.9)	17.2 (28.5)
Age (spruce) (year)	87.6 (24.5)	124.3 (5.2)

Table 2. Main chemical-physical characteristics of the soil

Profile	I			II		
Horizon	OL	OF	OH	OL	OF	OH
Humidity %	nd	nd	75	nd	nd	60
Depth cm	5	10	25	10	30	75
pH_{H2O}	4.3	4.3	7.2	4.2	4.2	6.4
Bulk density gcm^{-3}	nd	0.1	0.2	nd	0.2	0.2
Electrical conductivity μScm^{-1}	279	59	74.6	102.9	102.9	101.5
$CaCO_3$ %	0.0	0.0	2.6	0.0	0.0	3.3
C-$CaCO_3$ %	0.0	0.0	0.3	0.0	0.0	0.4
C_{org} g kg^{-1}	450	442	379	458	423	347
N_{tot} g kg^{-1}	12.8	12.7	16.0	16.0	14.2	12.0
C/N	35	35	24	29	30	29
Ptot $mgkg^{-1}$	nd	160	170	nd	250	223

Total phosphorus content ranged from 160 to 250 mg kg^{-1} from stratum I to II, respectively. The dissolved inorganic P (DIP) was significantly higher in the OF horizon of stratum II.

Table 3. Concentration of C (mg kg^{-1}), N (mg N kg^{-1}) and P (mg P kg^{-1}) forms in the soil horizons (OF, OH). In parentheses the percentages of total C, N or P

	I		II	
Horizon	OF	OH	OF	OH
$N\text{-}NH_4^+$ mg kg^{-1}	119.9(0.9)	15.0(0.1)	146.3(1.0)	38.6(0.3)
$N\text{-}NO_3^-$ mg kg^{-1}	10.2(0.1)	8.5(0.05)	50.2(0.4)	10.2(0.1)
N_{micr} mg kg^{-1}	450(3.5)	120(0.7)	266(8.7)	716(5.9)
DON mg kg^{-1}	nd	443(2.8)	nd	193(1.6)
NPot (anaerobic) mg kg^{-1}	30.1(0.2)	14.8(0.1)	0.0	39.1(0.3)
DOC mg kg^{-1}	27000(6.1)	16900(4.5)	18000(4.2)	15000(4.3)
C_{micr} mg kg^{-1}	13000(3.0)	1137(0.3)	23000(5.4)	2796(0.8)
$C/N_{(microbic)}$	28	9	8	4
DIP mg kg^{-1}	0.6(0.38)	0.4(0.24)	1.0(0.40)	0.2(0.09)
DOP mg kg^{-1}	0.7(0.44)	0.8(0.47)	0.4(0.16)	0.7(0.31)

Soil Temperature

The mean soil temperature during the winter season was equal to –0.45 and +0.13°C in stratum I and II, respectively (Figure 2). The snow cover during winter 2001-2002 was less than 10 cm thick until December 26th and then reached 50 cm for the remaining winter season (Delaloye et al., 2003). The minimum soil temperature in stratum I (–3.6°C) was registered after the snow fall (January 9-12, 2002), whereas in the other stratum the minimum value (-1.6°C) occurred on December 24, 2001. The presence of a deep snow cover was able to increase the soil temperature to 0°C only from January 30th to May the 8th in stratum I, while for the other stratum the snow insulation effect started previously, from the 28th of December.

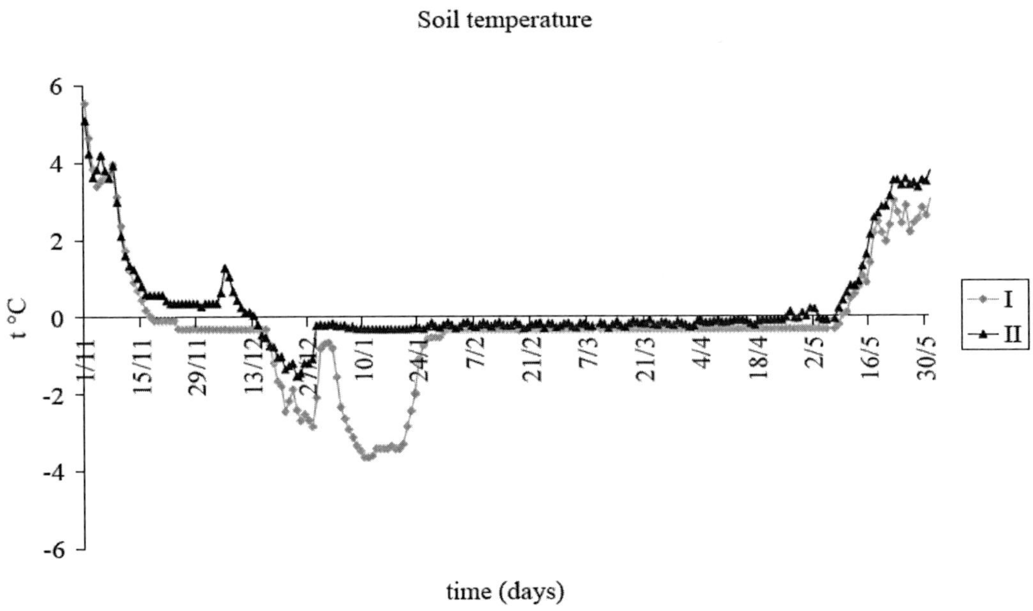

Figure 2. Soil temperature (average daily values) recorded in the three soils at 10 cm depth from November 2001 to May 2002.

Seasonal N and C Dynamics in the Buried Bags

The various forms of N and C were evaluated during the period November-May. The comparison inside and outside bags allow to estimate the amounts of nutrients that can be up taken by plants and mosses or leached along the profile.

During the winter months a significant reduction of ammonium concentration was recorded in both strata (Figure 3) and no statistical differences were observed between the incubated and the undisturbed soil (Table 4). Conversely, net nitrification (the difference of NO_3^- concentration before and after the incubation) was positive in both strata (Figure 3), with a higher nitrate production in stratum II and in the incubated soil (Table 4).

A positive net N mineralization was found in both strata, equal to 12.3 and 6.1 mg N kg^{-1} in stratum I and II, respectively.

The microbial N and C increased in stratum I (Figure 4), and, for microbial C, more in the buried bag than in the not incubated soil (Table 4). In the stratum II no statistical differences in the microbial N were observed along the winter season (Figure 4), while microbial C increased and slightly more in the buried bag (Table 4). The microbial N immobilization was equal to 560.7 mg N kg^{-1} and –147 mg N kg^{-1} respectively in site I and II.

DON and DOC decreased in both sites (Figure 5), with no influence of incubation in stratum II (Table 4).

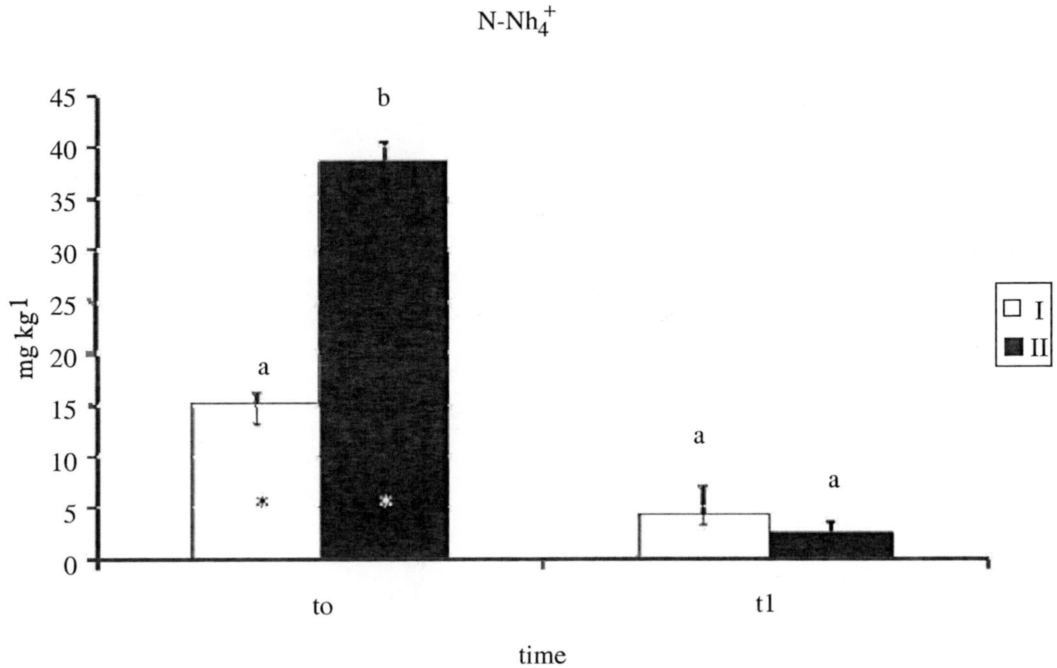

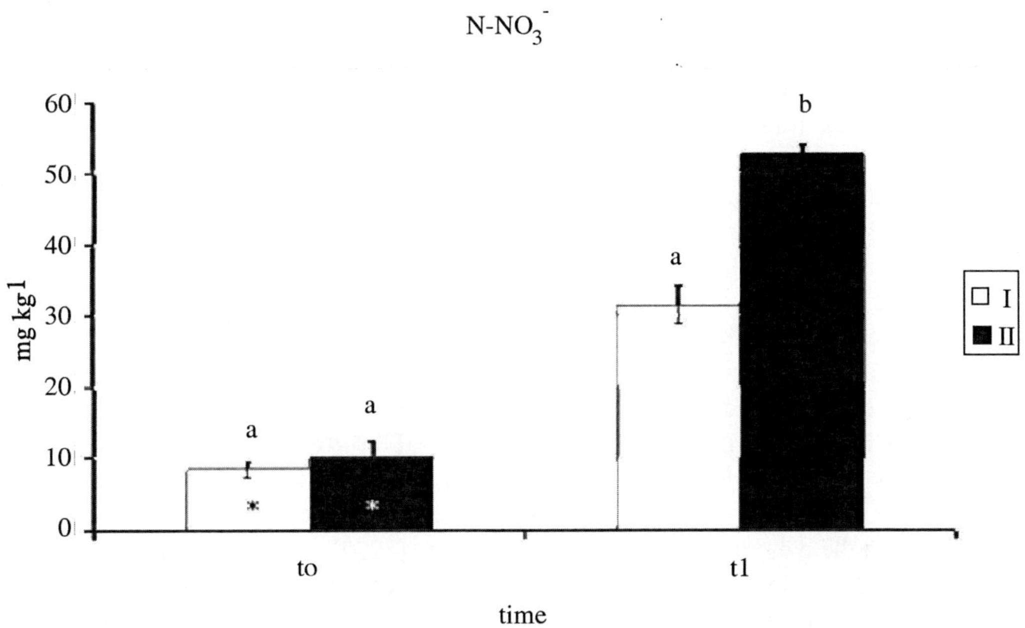

Figure 3. Ammonium and nitrate concentration before (to, November 2001) and after the incubation (t1, May 2002). Different letters indicate significant differences between stratum I and stratum II ($p<0.05$). Asterisks indicate significant differences between to and t1 in the same stratum.

Table 4. Effect of soil incubation in the buried bags (t1b: C and N forms (mg kg^{-1}) in the buried bags in May 2002; t1notb: C and N forms in the undisturbed soil in May 2002). Asteriscs indicate significant differences between means (p<0.05). ns: not significant differences between means

Stratum I	t1b	t1notb	
$N\text{-}NH_4^+$	4.4	4.8	ns
$N\text{-}NO_3^-$	31.4	6.8	*
Nmicr	680.7	580.3	ns
Cmicr	7243	3702	*
DON	11.9	40.60	*
DOC	320	438	ns
Stratum II	t1b	t1notb	
$N\text{-}NH_4^+$	2.4	9.4	ns
$N\text{-}NO_3^-$	52.5	10.9	*
Nmicr	569	763.5	ns
Cmicr	5002	4039	*
DON	13.2	15.5	ns
DOC	293	355	ns

DISCUSSION

The tree growth was significantly lower in stratum I, where the permafrost active layer remains thin. Although the size of the dwarf trees was hardly more than 3 m, they were of similar age to the adjacent reference trees, ranging from 90 to 130 years, with values similar to those reported by Korner and Hoch (2006).

The winter soil temperature decreased to a mean value of –0.45 °C at the stunted forest, with the minimum peaks registered during the December-February period, well below the values recorded during the same months in a forested (*Picea mariana*) peatland in North Central Minnesota (mean annual air temperature of +3.3°C) (Nichols, 1998). The presence of the permafrost table seems to reduce the insulation effect due to snow cover and keep the soil at temperature lower than 0°C, with possible stress or even damages to the root system (Edwards and Cresser, 1992). In the reference site, even if the greater presence of trees shades the ground from solar radiation and intercepts some of the snowfall in winter (French, 1996; Hardy et al., 2001), the soil temperature in the cold season resulted significantly higher. This is in accord with the data reported by Korner and Hoch (2006), who in the same site, during the previous winter season (2000-2001), recorded a soil temperature 2-3 °C colder than under the reference forest canopy, although the air temperature in the two sites was very similar.

The low temperature at the site I determined a greater soil moisture due to the H_2O vapour condensation following the cooling of the air during circulation inside the permafrost lens. This has allowed the development of a particular flora, dominated by a dense layer of *Sphagnum*. Nevertheless the high moisture, the stratum I appeared to be well oxygenated due to the air flux through the soil.

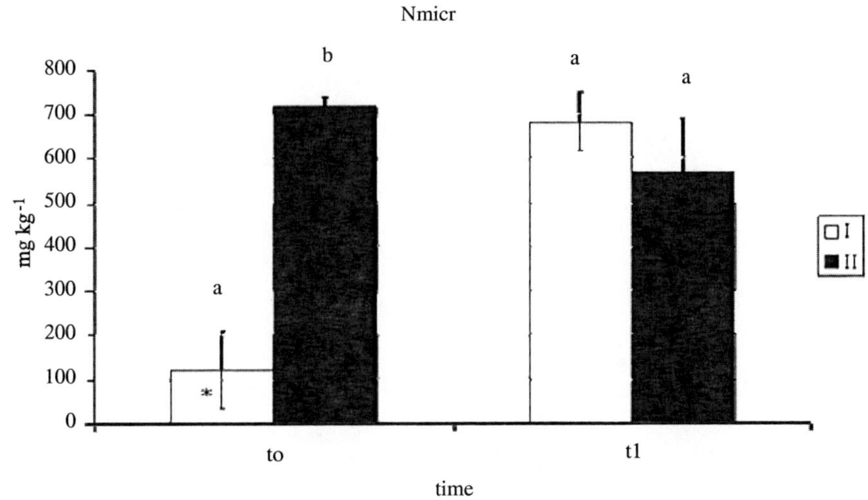

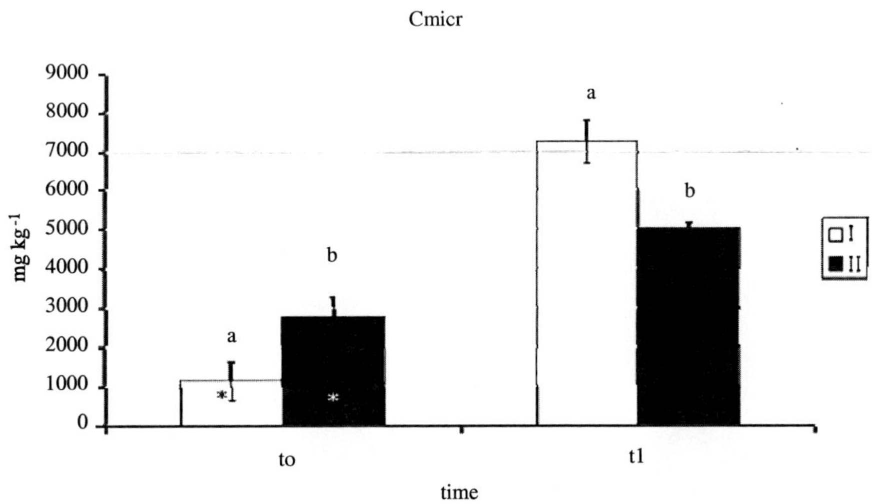

Figure 4. Nmicr and Cmicr concentration before (to, November 2001) and after the incubation (t1, May 2002). Different letters indicate significant differences between stratum I and stratum II ($p<0.05$). Asterisks indicate significant differences between to and t1 in the same stratum.

The microclimatic conditions at the site I have caused a great accumulation of soil organic matter leading to formation of an organosol with histic characteristics (Gobat et al., 2004). The large presence of *Sphagnum spp.* has contributed to C accumulation. These species are reported to decompose more slowly than vascular plant species (Clymo and Hayward, 1982; Ovenden, 1990) probably because of the high resistance to degradation of their lignin-like compounds and lipid-derived polymers (Johnson and Damman, 1991). The high C/N ratios are index of the low organic matter decomposition (Anderson 1973; Edwards 1975; Nadelhoffer et al. 1991; Reichstein et al. 2000). Surprisingly, a low C/N ratio was found in the OH horizon of the stunted forest, but it cannot be attributed to higher mineralization, nevertheless the increased pH which should be more favourable to the bacterial growth.

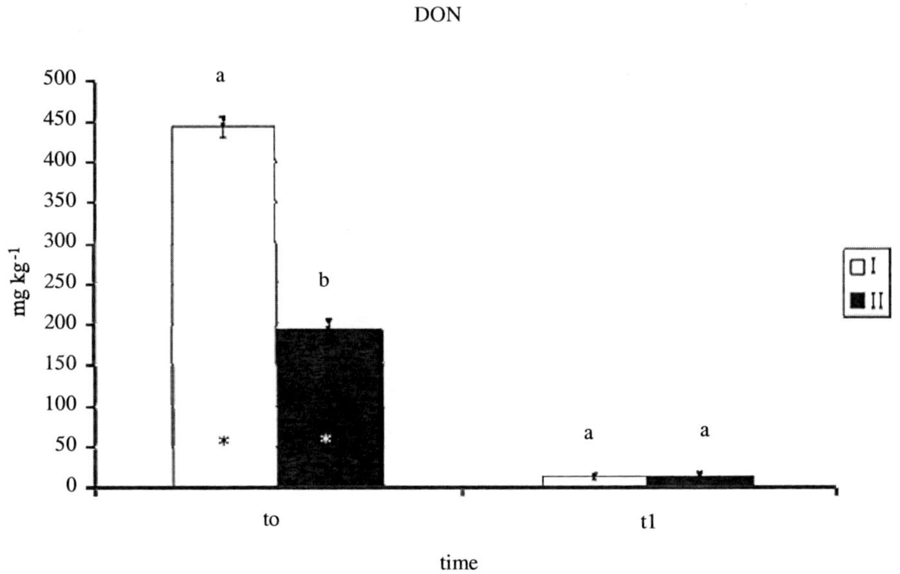

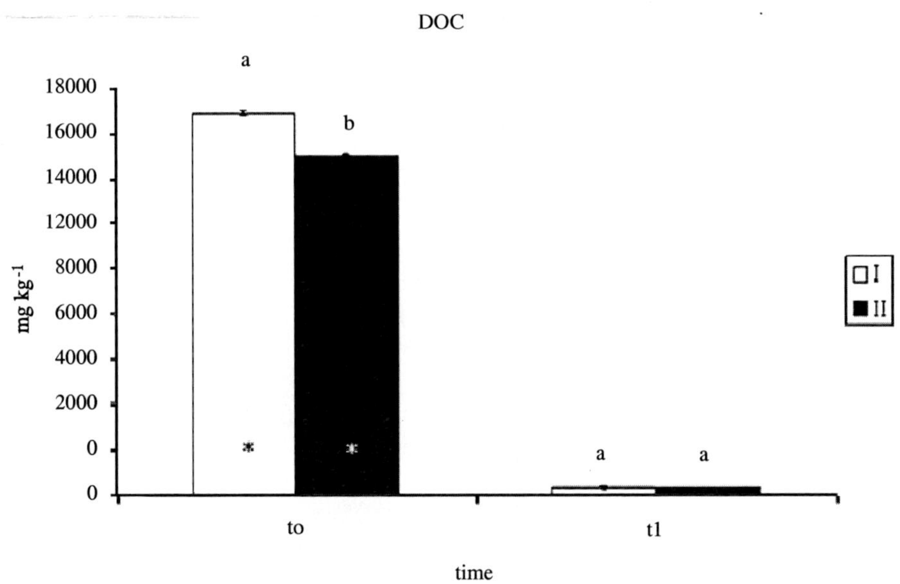

Figure 5. DON and DOC concentration before (to, November 2001) and after the incubation (t1, May 2002). Different letters indicate significant differences between stratum I and stratum II ($p<0.05$). Asterisks indicate significant differences between to and t1 in the same stratum.

The microbial N and C in the OH horizon in the late fall revealed in fact a limited biomass content compared to the reference site and even lower than that reported in Arctic soils (Schmidt et al., 1999). The high N accumulation could be due rather to a particular adaptation of spruce to the cold environment, which results in a lower growth with a greater photosynthesis and a higher N incorporation in the needles (Oleksyn, 1998). Moreover, the N accumulation in the soil followed by a decrease of C could be attributed to a selective degradation of only easily decomposable O-containing compounds (Saiz-Jimenez, 1996). This would affect the degradation of N-containing compounds and the production of dissolved inorganic and organic forms.

The inorganic N content in both sites was significantly higher in the OF horizon than in the OH horizon, evidencing how the upper organic horizons have a high N absorption capability. Low quantities of N were found into the OH horizon, where the major part of roots developed.

Ammonium in the late fall was present in very lower amounts in the dwarf forest and prevailed to the nitrate in all the OF and OH horizons, probably in relation to a stronger action of *Sphagnum* that can assimilate even short NO_3^- pulses through a re-adsorption from senescing lower stem parts (Jauhiainen et al., 1998).

During the winter and early spring season, NH_4^+ was reduced while nitrate increased in the buried bags in both sites but the changes were always more expressed in the reference site. The NH_4^+ consumed in the buried bags was likely used by microbes and recycled, ruling out a possible uptake by plants, since no differences between the incubated and the undisturbed soil were found. The increase of nitrate concentration during winter and early spring was in agreement with several studies (for instance Williams et al., 1996). In more temperate sites the increasing nitrification in early spring is related to a higher activity of soil microorganisms when air temperature is still low and plants remained dormant. The opposite occurs in the permafrost zone: a protraction of negative soil temperatures also when the air temperature started to increase and reached values of +15°C in April (Delaloye et al., 2003; Korner and Hoch, 2006) could cause an unbalance between the nutrient availability in the rizosphere and the spring plant request. In the dwarf forest the production of nitrate might be not exploited by spruce since soil temperatures until the end of May were still below that threshold temperature of +3.2 °C in soil indicated for the beginning of the growing season (Korner and Hoch, 2006). By contrast in the reference site the soil temperature was above this threshold. The major amount of NO_3^- inside the bag compared to outside suggested a large nutrient consumption during the early spring. It can be inferred that while in the stratum II the higher amount of NO_3^- produced during winter and early spring was likely consumed through plant uptake and leaching, in the dwarf forest the lower amount of NO_3^- produced was lost only trough leaching and/or translocation into depth, highly favoured by voids between boulders.

This suggests that the major part of nitrate was no more available when roots started later their activity. Therefore there is as asynchronies between the availability of critical resources and their utilization in the dwarf forest. To confirm this hypothesis further studies with different incubation techniques are currently being investigated.

Although net N mineralization followed the same trend in the two sites, the major soil N immobilization in the microorganisms of the site I compared to the other site suggested that the critical conditions under the dwarf trees could have selected a different microbial community particularly tolerating cold temperatures and then more active and more resistant to the winter season (Brooks et al., 1996) and to moderate freeze/thaw events (Lipson and Monson, 1998; Freppaz et al., 2007). Such tolerance has been observed in Antarctic peat microbes (Wynn-Williams, 1982) and in pure cultures of cold-adapted bacteria (Panoff et al., 1995). The microbial activity seems to continue down to –5°C, although around theses values the free water availability starts to be a limiting factor (Coxson and Parkinson, 1987). The microbial C increase is linked to a significant reduction of DOC concentration, both outside and outside the bags, revealing how the microbial access to labile C was equal.

The greater immobilisation of N in the microbial biomass and the comparable behaviour inside and outside the bags revealed a higher competition of microorganisms with plants for nutrients under the dwarf forest than in the reference site. However, the microbial immobilisation was not so competitive to determine a net negative N mineralization, as found by Bauhus and Barthel (1995) and Raubuch and Joergensn (2002). The winter N mineralization created a significant pool of N still available for plant. In essence, spruce could survive on the exceeding inorganic N discarded by microbes (Foster et al., 1989; Lipson and Monson, 1998) if other phenomena such as leaching would not cause a large depletion of the few nutrients produced under the dwarf forest.

Conclusion

Tree growth was limited in the permafrost affected site, where the winter and early spring soil temperature was significantly lower than in the reference forest. A smaller amount of ammonium was available in the dwarf forest and was consumed during winter while nitrate seemed not a limiting factor and was produced during winter and early spring. However, the low soil temperatures did not allow plant uptake and the nutrient was lost trough leaching and/or translocation into depth before roots started their activity. Moreover under the dwarf forest the greater microbial N immobilization, with a corresponding DON and NH_4^+ decrease revealed a significantly higher competition of microorganisms with plants for nutrients. The critical conditions under the dwarf trees could have selected a microbial community particularly tolerating cold temperatures, and then more active and more resistant to the winter season and to moderate freeze/thaw events.

In conclusion, an inorganic N pool, constituted mainly by the leachable NO_3^-, was available in the early growing season in both sites, but the lower soil temperature under the dwarf trees may inhibit soil nutrient absorption by plants. Therefore in the cold site there is asynchrony in the early spring season between the availability of critical resources and their utilization by plants, with potential loss of nutrients.

Acknowledgements

We are grateful to the Swiss Federal Institute for Snow and Avalanche Research for the logistic and technical support.

REFERENCES

Adams, MA; Polglase, PJ; Attiwill, PM; Weston, CJ. (1989). In situ studies of nitrogen mineralization and uptake in forest soils; some comments on methodology. *Soil Biology and Biochemistry* 21: 423-429.

Anderson, JM. (1973). The breakdown and decomposition of sweet chestnut (Castanea sativa Mill.) and beech (Fagus sylvatica L.) leaf litter in two deciduous woodland soils. II Changes in carbon, hydrogen, nitrogen and polyphenol content. *Oecologia* 12: 275-288.

Bauhus, J and Barthel, R (1995). Mechanisms for carbon and nutrient release and retention in beech forest gaps. II the role of microbial biomass. *Plant and Soil* 168-169: 585-592.

Bremner, JM (1965). Inorganic forms of nitrogen. In: Black Ca (eds), *Methods of soil analysis*, part 2: 1179-1206. American Society of Agronomy, Madison, Wisconsin.

Brookes, PC; Landman, A; Pruden, G; Jenkinson, DS. (1985). Chloroform fumigation and the release of soil nitrogen: a rapid direct extraction method to measure microbial biomass nitrogen in soil. *Soil Biology and Biochemistry* 17: 837-842.

Brooks, PD; Williams, MW; Schmidt, SK. (1996). Microbial activity under alpine snowpacks, Niwot Ridge, Colorado. *Biogeochemistry* 32: 93-113.

Clymo, RS and Hayward PM. (1982). The Ecology of Sphagnum. In: AJE Smith (eds), *Bryophyte Ecology*. Pp. 229-289. Chapman and Hall, London, UK.

Coxson DS and Parkinson D (1987). Winter respiratory activity in aspen woodland forest floor litter and soils. *Soil Biology and Biochemistry* 19: 49-59.

Crooke, WM and Simpson, WE. (1971). Determination of ammonium in Kjeldahl digests of crops by an automated procedure. *Journal of the Science of Food and Agriculture* 22: 9-10.

Dalias, P; Anderson, JM; Botner, P; Couteaux, MM. (2001). Temperature responses of carbon mineralization in conifer forest soils from different regional climates incubated under standard laboratory conditions. *Global Change Biology* vol. 7: 181-192.

Delaloye, R and Reynard, E. (2001). Les éboulis gelés du Creux du Van (Chaîne du Jura, Suisse). *Environnements périglaciaires* 8 : 105-113.

Delaloye, R; Reynard, E; Lambiel, C; Marescot, L; Monnet, R. (2003). Thermal anomaly in a cold scree slope (Creux du Van, Switzerland). 8th International Conference on Permafrost, Zurich, 21-25th July 2003. Wilfried Haeberli and Dagmar Brandova (Ed.) pp. 175-180.

Duchaufour, P. (1976). *Atlas écologique des sols du monde*. Masson, Paris.

Duchaufour, P. (1983). *Pédologie 1 : Pédogénèse et classification*. Masson, Paris.

Edwards, AC and Cresser MS. (1992). Freezing and its effect on chemical and biological properties of soil. *Advances in Soil Science* 18: 59–79.

Edwards, NT. (1975). Effects of temperature and moisture on carbon dioxide evolution in a mixed deciduous forest floor. *Soil Science Society of America Journal* 39: 361-365.

Eno, CF. (1960). Nitrate production in the field by incubating the soil in polyethylene bags. *Soil Science Society of America Proceedings* 24: 277-279.

Foster, NW; Nicolson, JA; Hazlett, PW. (1989). Temporal variation in nitrate and nutrient cations in drainage waters from a deciduous forest. *Journal of Environmental Quality* 18 : 238-244.

French, HM. (1996). Permafrost In: Longmans (eds), *The Periglacial Environment.*

Freppaz, M; Celi, L; Stockli, V; Phillips, M. (2003). Results of a first study on tree growth and soil characteristics at a cold site located below the limit of discontinuous alpine permafrost. 8th International Conference on Permafrost, Zurich, 21-25th July 2003. Wilfried Haeberli and Dagmar Brandove (Ed.) pp 39-40.

Freppaz, M; Williams, BL; Edwards, AC; Scalenghe, R; Zanini, E. (2007). Simulating soil freeze/thaw cycles typical of winter alpine conditions: implications for N and P availability. *Applied Soil Ecology* 35: 247-255.

Gobat, JM; Aragno, M; Matthey, W. (2004). The Living Soil: Fundamentals of Soil Science and Soil Biology. Science Publishers (USA). Pp. 626.

Gorbunov, AP; Marchenko, SS; Seversky, EV. (2004). The thermal environment of blocky materials in the mountains of Central Asia. *Permafrost and Periglacial Processes* 15: 95-98.

Hardy, JP; Groffman, PM; Fitzhugh, RD; Henry, K; Welman, AT; Demers, JD; Fahey, TJ; Driscoll, CT; Tierney, GL; Nolan, S. (2001). Snow depth manipulation and its influence on soil frost and water dynamics in northern hardwood forest. *Biogeochemistry* 56: 151-174.

Harris, SA and Pedersen, DE. (1998). Thermal regimes beneath coarse blocky materials. *Permafrost and Periglacial Processes* 9: 107-120.

Jauhiainen, J; Vasander, H; Silvola, J. (1998). Nutrient concentration in shape Sphagna at increased N-deposition rates and raised atmospheric CO_2 concentrations. *Plant Ecology* vol. 138, n.2: 149-160.

Johnson, LC and Damman, WH. (1991). Species controlled Sphagnum decay on a South Swedish raised bog. *Oikos* 61: 234-242.

Kneisel, C; Hauck, C; Vonder Muhll, D. (2000). Permafrost below the timberline confirmed and characterized by geoelectrical resistivity measurements, Bever valley, eastern Swiss Alps. *Permafrost and Periglacial Processes* 11: 295-304.

Korner, C and Hoch, G. (2006). A test of treeline theory on a montane permafrost island. *Arctic Antarctic and Alpine Research* 38: 113-119.

Lesquereux, L. (1844). Quelques recherches sur les marais tourbeaux en general. H. Wolfrath, Neuchatel, Switzerland.

Lipson, DA and Monson, RK. (1998). Plant-microbe competition for soil amino acids in the alpine tundra: effects of freeze-thaw and dry-rewet events. *Oecologia* 113: 406-414.

Lipson, DA; Schmidt, SK; Monson, RK. (1999). Links between microbial population dynamycs and nitrogen availability in an alpine ecosystem. *Ecology* 80: 1623.1631.

Martin, M; Celi, L; Barberis, E. (1999). Determination of low concentrations of organic phosphorus in soil solution. *Communication Soil Science Plant Analysis* 30 : 1909-1917.

Nadelhoffer, K.J; Giblin, AE; Shaver, GR; Laundre, JA. (1991). Effects of temperature and substrate quality on element mineralization in six arctic soils. *Ecology* 72: 242-253.

Nelson, DW and Sommers, LE. (1982). Total carbon, organic carbon and organic matter. Page, A.L. (eds.) In *Methods of Soil Analysis*. Part 2 pp. 199. Madison, WI: American Society of Agronomy.

Nichols, DS. (1998). Temperature of upland and peatland soils in a north central Minnesota forest. *Canadian Journal of Soil Science* 78, n.3: 493-509.

Ohno, T and Zibilske, LM. (1991). Determination of low concentrations of phosphorus in soil extracts using malachite green. *Soil Science Society of America Journal* 55:892–895.

Oleksyn, J; Modrzynski, J; Tjoelker, MG; Zytkowiak, R; Reich, PB; Karolewski, P. (1998). Growth and physiology of Picea abies populations from elevational transects: common garden evidence for altitudinal ecotypes and cold adaptation. *Functional Ecology* 4: 573-590.

Ovenden, L. (1990). Peat accumulation in northern peatlands. *Quaternary Research* 33: 377-386.

Panoff, JM; Thammavongs, B; Laplace, JM; Hartke, A; Boutibonne, P; Auffray, Y. (1995). Cryotolerance and cold adaptation in Lactococcus lactis. *Cryobiology* 32: 516-520.

Pastor, J; Aber, JD; McClaugherty, CA; Melillo, JM. (1984). Aboveground production and N and P cycling along a nitrogen mineralization gradient on Blackhawk Island, Wisconsin. *Ecology* 65: 256-268.

Raubuch, M and Joergensen, RG. (2002). C and net N mineralisation in a coniferous forest soil: the contribution of the temporal variability of microbial biomass C and N. *Soil Biology and Biochemistry* 34: 841-849.

Richard, JL. (1961). Les forêts acidophiles du Jura. Hans Huber Berne (ed.) p. 101.

Saiz-Jimenez, C. (1996). The chemical structure of humic substances: recent advances, pp. 1-44. Piccolo A (eds.) *Humic substances in terrestrial ecosystems*. Elsevier Science B. V..

Sarathchandra, SU; Perrott, KW; Littler, RA. (1989). Soil microbial biomass: influence of simulated temperature changes on size, activity and nutrient-content. *Soil Biology and Biochemistry*. 21, 8: 987-993.

Sawada, Y; Ishikawa, M; Ono, Y. (2003). Thermal regime of sporadic permafrost in a block slope on Mt. Nishi-Nupukaushinupuri, Hokkaido Island, Northern Japan. *Geomorphology* 52: 121-130.

Schimel, JP; Bilbrough, C; Welker, JA. (2004). Increased snow depth affects microbial activity and nitrogen mineralization in two Arctic tundra communities. *Soil Biology and Biochemistry* 36: 217-227.

Schmidt, IK; Jonasson, S; Michelsen, A. (1999). Mineralization and microbial immobilization of N and P in arctic soils in relation to season, temperature and nutrient amendment. *Applied Soil Ecology* 11: 147-160.

Schmidt, IK; Jonasson, S; Shaver, GR; Michelsen, A; Nordin. A. (2002). Mineralization and distribution of nutrients in plants and microbes in four arctic ecosystems: responses to warming. *Plant and Soil* 242: 93-106.

Schmidt, SK and Lipson DA. (2004). Microbial growth under the snow: Implications for nutrient and alleochemical availability in temperate soils. *Plant and Soil* 259: 1-7.

Schweingruber, FH. (1988). Tree rings: basics and applications of dendrochronology. Kluwer Academic Publishers, Dordrecht (NE), pp. 276.

Sveinbjornsson, B. (1993). Climate and growth of mountain birch near the treeline on northern Sweden and Icreland: In Frenzel B. (eds) *Oscillations of the alpine and polar tree limits in the Holocene*. Stuttgart: Gustav Fischer Verlag: 57-67.

Tanaka, HL; Nohara, D; Yokoi, M. (2000). Numerical simulation of wind hole circulation and summertime ice formation at Ice valley in Korea and Nakayama in Fukushima, Japan. *Journal of the Meteorological Society of Japan*. 78: 611-630.

Wakonigg, H. (1996). Unterkühlte Schutthalden. Arbeiten aus dem Institut für Geographie der Karl-Franzens-Universität Graz 33: 209-223.

Waring , SA and Bremner, JM. (1964). Ammonium Production in Soil under Waterlogged Conditions as an Index of Nitrogen Availability. *Nature* 201: 951 – 952.

Wegmann, G. (1995). Permafrostvorkommen auf geringer Meereshohe, eine Fallstudie im Brueltobel. Unpublished diploma thesis, Univeristy of Zurich, Switzerland.

Wynn-Williams, DD. (1982). Simulation of seasonal changes in microbial activity of maritime Antarctic peat. *Soil Biology and Biochemistry* 14: 1-12.

Williams, BL and Sparling GP. (1988). Microbial biomass carbon and readily mineralized nitrogen in peat and forest humus. *Soil Biology and Biochemistry* 20: 579-581.

Williams, BL; Shand, CA; Hill, M; O'Hara, C; Smith, S; Young, ME. (1995). A procedure for the simultaneous oxidation of total soluble nitrogen and phosphorus in extracts of fresh and fumigated soils and litters. *Communication Soil Science Plant Analysis* 26: 91-106.

Williams, MW; Brooks, PD; Mosier, A; Tonnessen, KA. (1996). Mineral nitrogen transformations in and under seasonal snow in a high-elevation catchment in the Rocky Mountains, United States. *Water Resource Research* vol. 32: 3161-3171.

In: Soil Ecology Research Developments
Editors: Tian-Xiao Liu, pp. 217-231

ISBN: 978-1-60021-971-9

Chapter 9

EFFECT OF MANURE AND CROP RESIDUES CONTINUOUS APPLICATION ON HUMIC C IN A CULTIVATED SOIL

O. Francioso[1,]*, P. Gioacchini*, V. Tugnoli**, D. Montecchio* and C. Ciavatta*

*Dipartimento di Scienze e Tecnologie Agroambientali, Università degli Studi di Bologna, V.le Fanin 40, Bologna 40127, Italy

**Dipartimento di Biochimica, Università degli Studi di Bologna, via Belmeloro 8/2, Bologna 40127, Italy

ABSTRACT

The chemical characteristics of soil organic matter (SOM) can be influenced by management and amendment practices which effects can be measured only after long-term experiment. In this long-term study of over 30 years, with a rotation wheat-corn, we compared the effects of adding cattle manure (CM) and crop residues (CR) wheat straw or corn-stalks after each crop, on humic substances (HS). Potentiometric titration, thermal analysis (TG-DTA), and spectroscopic methods such as diffuse reflectance infrared Fourier (DRIFT) and liquid nuclear magnetic resonance (^{13}C NMR) spectroscopies were used in order to investigate humic acid (HA) structure. The amendment practices clearly influenced the humic C and the COOH groups content that only increased in CM treatment. The quality of this humic fraction was affected by the different agricultural practices, so that when the soil did not receive any amendment, the aromatic and carboxylic C decreased, whereas the aliphatic C increased as an effect of the crop rotation. With the amendments, in contrast, the aromatic C generally increased, this increase was mainly due to the incorporation of aromatic groups in the structure of HA, arising from the phenolic groups present in the lignin of the crop residues.

Keywords: *δ^{13} (IRMS), DRIFT, ^{13}C NMR, humic acids, TG-DTA.*

1 Corresponding author: Tel.:+39-051-2096205; Fax:+ 39-051-2096203, e-mail address: ornella.francioso@unibo.it

INTRODUCTION

Long-term field experiments have demonstrated how the chemical characteristics of soil organic matter (SOM) can be influenced by management and amendment practices [1,2] Usually, intensive cultivations characterized by frequent soil tillage cause a mineralization of soil organic carbon (SOC) while organic fertilisation (manure and crop-residues) improves the quantity and quality of SOC [3]. However, the chemical changes in SOM are complex because no well defined chemical structure has been discovered. The key is to find reliable analytical methods to determine the qualitative and quantitative changes in SOM and finally to identify a category of molecules that are representative of those changes that taking place. Recently, emphasis has been placed on the refractory fraction of SOC, in the dynamics of organic C and C sequestration. This fraction is an heterogeneous mixture of organic compounds, consisting of aromatic, aliphatic, phenolic, and quinonic functional groups with varying molecular sizes and molecular weights, called "humic" substances (HS) [4,5]. The main characteristic of these macromolecules is their extreme stability to biochemical or chemical degradation over time, whether in soil [6] or in coals [7] or composts [8]. In spite of the indefinite nature of HS structure, some materials, such as cellulose or lignin components, may be accumulated and preserved during the humification process [9]. Their presence in HS is considered as an early humification process [10]. In particular the early humic fraction is that more important for agronomic practices [11] with respect to that associated to mineral complexes (insoluble in alkali). Although numerous studies have investigated the dynamic fraction of SOC [12,13] few have examined how the molecular structure of humic acids (HA), the main fraction of the HS, is influenced by long-term field amendment experiments. A better understanding of the changes occurring in HS might give additional information on the effects of management and fertilisation practices. Since the molecular structure of HA is complex, different analytical techniques are an essential precondition for more accurate structural interpretations to predict the ability of HS to sequester C in different cropping and soil management systems. Carbon isotopic composition has been successfully used in studies of SOM dynamics [14]. The $\delta\ ^{13}C$ value of both total organic C and HS depends on the $\delta\ ^{13}C$ value of soil vegetation [15-16]. Due to the different isotopic fractionation in C_3 and C_4 plants during the photosynthetic pathways, $\delta\ ^{13}C$ can be used as a natural tracer to follow the dynamics of C_3 and C_4-derived C in soil [14].

A rapid characterisation of SOM can be obtained by using spectroscopic techniques and thermal analysis. In particular among spectroscopic techniques the diffuse reflectance infrared Fourier transform (DRIFT) has often been applied to investigate HS from different origins [17-20], to follow organic matter degradation in soil [21] and in forest [22], and to study as management influence the SOM characteristics [3, 23-25]. Another spectroscopic technique which has arguably provided the most interpretable information on the nature and relative abundance of chemical groups present in HS is liquid-state ^{13}C NMR spectroscopy [26-30]. The application of liquid-state ^{13}C NMR and DRIFT spectroscopies resulted useful tools to analyze the influence of agricultural management on structural changes of SOM [23-25].

In addition thermal analysis (TA) represents a rapid and accurate approach in the investigation of HS [16, 20, 31-33]. This technique involves a continuous and simultaneous measurement of weight loss (TG) and energy change (DSC) during heating of the sample. During heating of HS a first mass loss ($\approx$300°C) is produced by decomposition of the labile C fraction while the more refractory C is decomposed at higher temperatures (≈450°C) [31, 33]. The mass loss that occurs during heating can be used to compare the relative abundance of more and less labile C whilst the position of peaks reflects the structure and chemical composition of the sample.

Moreover, combining thermal analysis with the spectroscopic techniques can be obtained important structural information on soil organic matter composition.

In this study, advanced analytical techniques such as δ^{13} isotopic ratio-mass spectrometry (IRMS), DRIFT, and ^{13}C NMR spectroscopies and thermal analysis (TG-DTA) were exploited to characterise humic acids isolated from a soil amended with crop residues and cow manure over a 30-year period. The objective was to verify how the amendment practices may influence the HA structure.

Materials and Methods

The soil samples analyzed (horizon Ap, 0-40 cm depth) were taken from plots of a field experiment conducted for over 30-years at the agricultural farm of the University of Bologna's Agricultural Faculty (Cadriano, Italy). This soil is classified (Soil Survey Staff - USDA SCS 1989) as a *Typic Udochrept.* The main physical-chemical characteristics of the soil studied were the following: pH (in water) 6.9; texture: sand 56%; silt 16%; clay 28%; total calcium carbonate ($CaCO_3$) < 1%; total organic C (TOC) 7.5 g kg^{-1}; total N 1.1 g kg^{-1}; cation exchange capacity 16.5 $cmol_c$ kg^{-1}. A more detailed description of the field experiments have previously been reported [34].

However, briefly, the experimental design included plots amended over a 30-year period with cattle manure (CM) and crop residues (CR) constituted by wheat straw or corn-stalks (each biomass followed in succession), using un-amended soil as the control (C). CM was added to the plots at the following rate: 6 t ha^{-1} of dry matter after wheat tillage and 7.5 t ha^{-1} of dry matter after corn tillage. All the plots examined, including the control C, were fertilized with 100 kg N ha^{-1} yr^{-1} added as ammonium nitrate [34]. Each treatment (CM, CR and C) was conducted in triplicate using a randomized block design.

Analyses were carried out on soil samples taken from: i) un-amended soil in 1972 (C_0), and ii) after 30-years (C_{30}) and iii) after a 30-year amendment period with CM_{30} and CR_{30}. All analytical determinations were conducted in triplicate.

Humic Acid Isolation

10 g of air-dried soil previously crushed and sieved at 2 mm, were put into a 250 mL flask and extracted under N_2 with 100 mL of 0.1 M NaOH plus 0.1 M $Na_4P_2O_7$ at 65 °C using an end-over-end shaker at 100 rpm for 24 hours. The suspension was centrifuged at 5,000 x *g* for 30 min and then filtered through a 0.45 μm filter (Millipore, Bedford MA - USA). The solution was acidified with 5 M HCl to pH < 2 to precipitate the HA and afterwards was centrifuged at 5,000 x *g* for 20 min in order to eliminate the supernatant. The HA solutions were dialyzed against d-H_2O until a neutral pH was achieved and finally freeze-dried and

stored under vacuum over silica gel. HAs were extracted from individual replicate plots and were then combined before analyses.

Total Organic C and $\delta^{13}C$

Total organic C (TOC) in soils and HA samples were measured with an elemental analyzer (Thermo Finnigan mod. EA 1110). The percentage of C in the sample was calculated using acetanilide as a certified standard containing 71.09 % of carbon. The HA samples were weighed in tin capsules and analyzed directly, whereas the soil samples were pretreated with 6 M HCl in order to eliminate the carbonate-bound C. About 10 mg of finely ground dry soil were weighed in silver capsules placed on a cast iron plate. A few drops of acid solution were added directly to the capsules and the plate was kept at 80 °C until the soil was dry. The samples were measured using CF-IRMS (Continuous Flow-Isotope Ratio Mass Spectrometry) by entering the CO_2 from the elemental analyzer into an IRMS (Isotope Ratio Mass Spectrometer mod. Delta Plus Thermo Electron, Bremen, Germany). The C isotope values were expressed in delta notation (δ), where: $\delta(‰) = [(R_{sample} / R_{standard}) - 1 \times 1000$, and R_{sample} and $R_{standard}$ are $^{13}C/^{12}C$ of the sample and standard, respectively. Delta ^{13}C is referred to the Vienna Pee Dee Belemnite (VPDB) standard [35]. All measurements were made in triplicate for each sample. The reproducibility of the delta ^{13}C values of the samples was, in 90 % of the cases, better than 0.1 ‰. The variation coefficient was < 0.1%.

Potentiometric titration. Samples were prepared by dissolving 12 mg of the freeze-dried HA in 20 mL of Milli-Q Millipore water containing 0.05 M NaCl to keep the ionic strength constant. The pH was adjusted to pH 3 by adding about 1 mL of 0.05 M HCl. The solutions were titrated to pH 10.5 with 0.05 M NaOH using a VIT 90 Titrator Radiometer (Radiometer, Copenhagen, DK). The potentiometric titrations were carried out in triplicate at 25 °C, under N_2 flow and the delivery range was 10 μL min^{-1} (± 0.01). The first derivative method was performed to determine the concentration of COOH groups.

DRIFT Spectroscopy

The spectra were recorded with a Nicolet Impact 400 FT-IR Spectrophotometer (Madison, WI) equipped with an apparatus for diffuse reflectance (Spectra-Tech. Inc., Stamford, CT). Spectra were collected as Kubelka-Munk units using KBr (Aldrich Chemical Co.) as the background reference. The sample chamber was continuously flushed with CO_2 free dry air. DRIFT spectra were collected from 4000 to 400 cm^{-1} and averaged over 200 scans (resolution ± 4 cm^{-1}). Analyses of spectral data were performed with Grams/386 spectral software (Galactic Industries, Salem, NH). Spectral sections from 1850 to 800 cm^{-1} were baseline-corrected to an absorbance value of 0.00 at 1850 cm^{-1}.

NMR Spectroscopy

The NMR spectra were obtained by dissolving the HA (70-80 mg) in 0.5 mL of 0.5 M NaOD. The spectra were recorded with a Bruker ACF 250 spectrometer (Karslruhe, Germany), using a 5 mm multinuclear probe. Proton decoupled 62.890 MHZ ^{13}C spectra were accumulated with a 32 K data point, gated decoupling sequence, 30° pulse angle, 12.8 KHz spectral width, an acquisition time of 1.3 s and a relaxation delay of 3 s. Generally 100,000 scans were accumulated. A line broadening (20-30 Hz) was used prior to Fourier transformation. Chemical shifts are relative to tetramethylsilane (TMS).

The spectra were divided into the following four chemical regions [28]: to the aliphatic carbons (0-40 ppm), to the peptidic and oxidized carbons (40-110 ppm), to the aromatic and phenolic carbons (110-160 ppm) and to the carboxylic (160-190 ppm) carbons [26, 36]. In particular we considered an additional subdivision in the region at 115-99 ppm that might be assigned to anomeric C in polysaccharides, acetal or ketal.

TG-DTA analysis. Thermogravimetric analysis (TG) and differential thermal analysis (DTA) were carried out simultaneously using a TG-DTA92 instrument (SETARAM, France) as described in previous paper [20].

The Statgraphics version 5 plus (Statistical graphics system by statistical graphics corporation) was used for statistical calculations.

RESULTS AND DISCUSSION

The amount of SOC after 30-years farming is shown in Table 1. Compared to the beginning of the experiment C_0 (1972), a significant decrease ($P<0.05$) in TOC of about 23% was observed in the control after 30-yrs (C_{30}), whereas with the CR treatment the C loss was around 10%. The soil treated with manure, on the contrary, did not show any significant change with respect to the control (C_0).

The $\delta\ ^{13}C$ of the TOC over the 30-years of farming became progressively less negative (Tab. 1) in all the treatments but the greatest variation was seen in the CR_{30} plot. The changes in all treatments are due to the presence of corn in the crop-rotation, while additionally in CR, to the amendment practice of adding corn residues after the corn harvest. Comparing the isotopic signature of TOC to that of humic C after 30-years, a much smaller variation is seen and this would suggest a preferential utilization of C_3-derived C in the humification process. The $\delta\ ^{13}C$ change of TOC, therefore, would mainly be due to the presence of C_4-derived C in the non-humic fraction. In contrast to soil TOC, the percentage of humic C significantly ($P<0.05$) increased in both CM_{30} and CR_{30} treatments. These changes in C content reflect the effect of long-term application of manure and crop-residues on HA. In spite of the substantial decrease in TOC, even the control after 30 yrs (C_{30}) measured a small increase in humic C (approximately 6%). Since the control did not receive any addition of organic material, the C sources used to maintain the level of humic C throughout the 30 year period, were from both rhyzodeposition during plant growth and C derived from mineralization of the root residues remaining in the soil after the harvest [37]. The fact that the two treatments (CM and CR) both show an increase in humic C and a decrease in TOC indicates that HA are able to store C over time and that they are resistant to C mineralization even when the conditions promoting C loss are prevalent.

Table 1. Effect of soil amended with crop residues (CR), and cattle manure (CM) on total organic carbon (TOC), δ^{13}C of TOC, and on C, δ^{13}C and COOH groups of humic acids with respect to the control unamended (C_0) at the start of the experiment and after 30 yrs. of trials

	Treatments			
Soil	C_0	C_{30}	CR_{30}	CM_{30}
TOC (%)	0.82 ± 0.01a	0.63 ± 0.005	0.74 ± 0.005	0.85 ± 0.003a
Δ(%)†	-	- 23.1	- 9.8	+ 3.7
δ^{13}C (‰)	-24.1 ± 0.2	-22.0 ± 0.5a	-22.0 ± 0.1a	-22.4 ± 0.2a
Humic acid				
C (% d.m)	38.0 ± 0.05	40.2 ± 0.02	44.0 ± 0.03	45.7 ± 0.03
Δ(%)†	-	+ 6	+ 16	+ 20
δ^{13}C (‰)	-24.6 ± 0.1a	-24.5 ± 0.07a	-23.4 ± 0.3	-24.3 ± 0.1a
COOH meq/100g	560 ± 2.0	468 ± 2.5	520 ± 3.2	590 ± 1.6
Amendants				
	Wheat straw		Manure	
C (% d.m)	42.0 ± 0.17		23.02 ± 0.35	
δ^{13}C (‰)	-24.0 ± 0.10		-20.75 ± 0.30	

Means followed in the same row by the same letter are not statistically different at $P< 0.05$ according to Duncan's test.

†Δ (%) percentage change = $[(TOC_{1972} - TOC_{2002})/TOC_{1972}] \times 100$

± standard error

The quantification of HA carboxylic acids (COOH) is shown in Table 1. Our data shows a statistically significant increase ($P<0.05$) in COOH groups in plots amended with CM and a net loss of these groups of about 16% and 7%, in C_{30} and CR_{30}, respectively. These results are generally in agreement with previous findings on HS from several amendment trials [38]. Moreover, it is generally accepted that the carboxylic groups increase in organic matter during humification [5], therefore, the higher the amount of COOH groups in CM_{30}, the greater the chemical reactivity towards xenobiotic compounds and the better the soil buffer activity.

DRIFT Spectroscopy

The DRIFT spectra of HA from control and amended plots (Figure 1) were attributed in accordance with previous papers [17-20, 39].

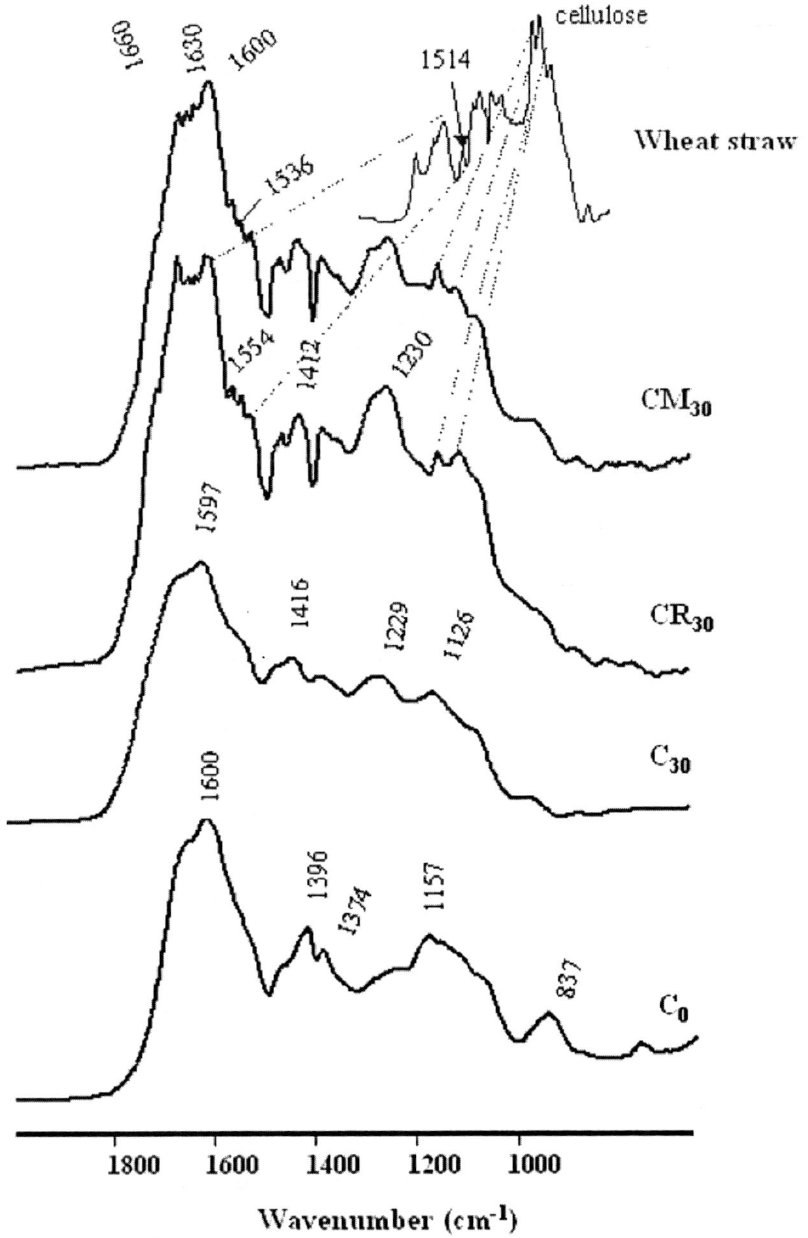

Figure 1. DRIFT spectra of humic acids extracted from unamended soil (C_0) at the start of the experiment and after 30 yrs. (C_{30}) and amended consecutively for 30 yrs with cow manure (CM) and crop-residues (CR). On the top of the spectra is reported the wheat straw spectrum .

In detail, the band around 1600 cm^{-1}, due to ν_{as} (-COO^-) and ν(C=C) of aromatic rings vibrations [17-18], increased in the CM_{30} sample compared to CR_{30} and C_{30}. We suggest that the carboxylate groups provide a dominant contribution to this vibration in accordance with the potentiometric data titration and ^{13}C NMR analysis (see below). Another important modification appeared around 1660 cm^{-1}, assigned to C=O stretching in aromatic compounds or in amide I [40]. It appeared as a shoulder in C_{30} and its relative intensity increased in CR_{30} compared to the CM_{30} spectrum. The variation of the relative intensity of this band observed in CM_{30} seems likely to be attributable to the presence of polar substituents. Moreover, the CM_{30} spectrum was characterized by two well resolved bands at 1640 and 1630 cm^{-1}. The region around 1500 cm^{-1}, in both CR_{30} and CM_{30}, showed an important structural change compared to the control (C_0). The peaks at 1550 cm^{-1}, 1535 cm^{-1} and 1514 cm^{-1} indicate a large aromatic content from lignin derivates in HA extracted from the soil amended with CR_{30} (data supported by NMR). In the C_{30} sample this region was not well resolved. These components are chiefly identified by the presence of the peak at 1514 cm^{-1}, assigned to the lignin component absorption produced by the stretching of *ortho, para* and 1,3,4-tri-substituted aromatic rings of guaiacyl [41]. This is in agreement with the presence of bands around 1660 cm^{-1}, 1235 cm^{-1} and 1080 cm^{-1}, which can be attributed to the vibration of 2-methoxy phenol originated from lignin degradation that appeared more intense in the CR_{30} spectrum.

On the basis of these results we may infer that lignin residues from crop rotation (wheat/corn) do not undergo a complete degradation in soil, but they can persist over time and they can be included in the molecular structure of humic substances during the humification process. However, considering that the $\delta^{13}C$ value (Table 1) of HA from soil amended with CR_{30} is similar to that of HA from C_0 and is also closer to that of wheat, the relative contribution of wheat-derived lignin to HA seems to be greater than that of corn-derived lignin.

The relative intensity of bands in the 1460-1300 cm^{-1} region increased particularly in CM_{30} and CR_{30} than to the controls. Moreover over 30 years in all spectra appeared a broad band at around 1230 cm^{-1} assigned to ν(C-O) stretching motion of phenolic groups, and acid dimers with electron-withdrawing groups and C-C (in condensed aromatic rings) [42]. This band noteworthy increased in CR_{30} whereas no vibration appeared in the C_0 spectrum. Furthermore, the region between 1120 -1084 cm^{-1} is characteristic of ν_s(C-OH) and C-O-C in sugar and ether group vibrations, respectively.

^{13}C NMR Spectroscopy

The spectra of samples after 30 yrs (Figure 2) significantly differed in all regions if compared to the control (C_0). The region corresponding to the alkyl-C (50-0 ppm) showed a decreasing trend from C_{30} to CM_{30} and CR_{30}; compared to C_0, this region increased in C_{30}, decreased in CR_{30}, and remained unchanged in the CM_{30} treatment. The increase in alkyl-C in the HA, therefore, was observed only when the soil did not receive any amendment and it was probably a result of crop rotation. In C_{30}, this C fraction seems to be the main cause of the small increase in humic C observed, in spite of the large decrease in TOC (Tab.1). The alkyl-C component is actually involved in the long-term stabilization of SOM [23].

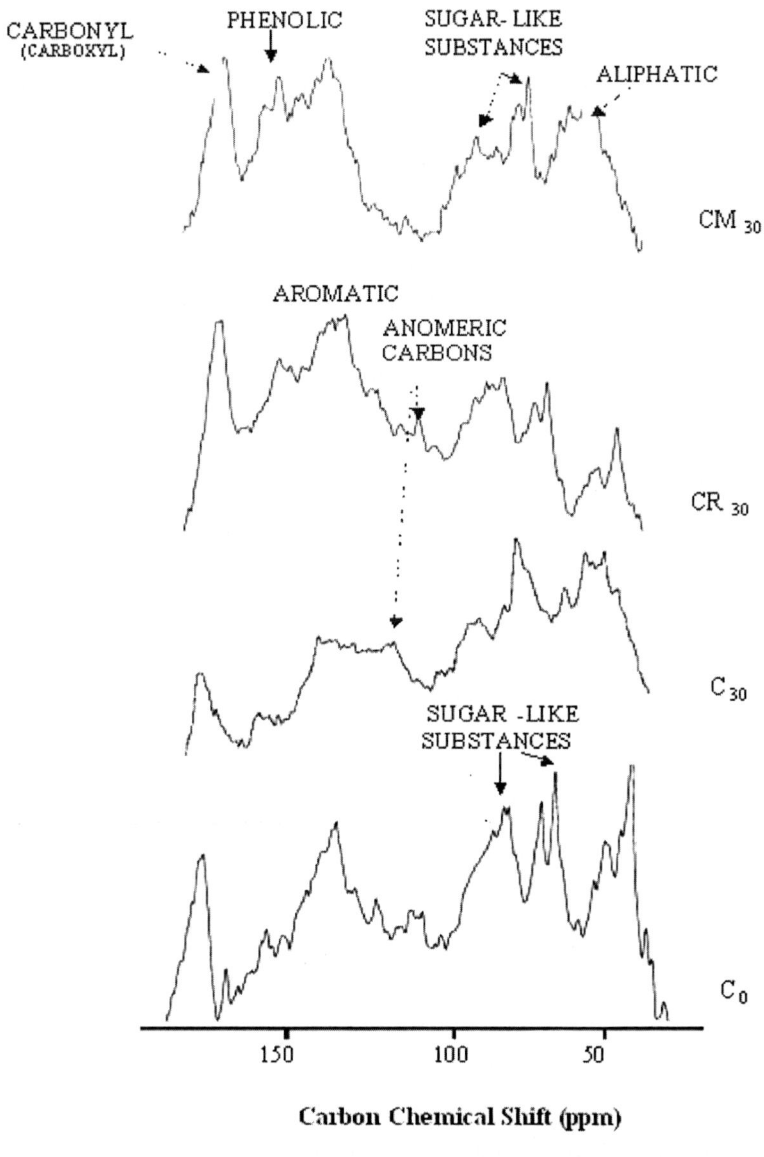

Figure 2. 1-D liquid-state [13] C NMR spectra of humic acids in NaOD extracted from unamended soil (C_0) at the start of the experiment and after 30 yrs. (C_{30}) and amended consecutively for 30 yrs with cow manure (CM) and crop-residues (CR). Tentative chemical shift assignements for these samples correspond to representative major functional groups of humic substances.

The amendments, on the contrary, did not favour any accumulation of alkyl-C in the HA, thus modifying the positive effect of the crop rotation alone. In CR_{30}, in particular, the observed decrease in alkyl-C may be due to: i) the relatively low alkyl-C content in crop residues (wheat/corn), ii) the modest microbial production of alkyl-C during soil crop-residue decomposition, and iii) the small amount of transformation of mobile alkyl-C into rigid alkyl-C by cross-linking or association with soil mineral particles [43]. Both CR_{30} and CM_{30} treatments showed, after 30-years, a C increase in the aromatic region (165-130 ppm), thus with the amendments, the increase in humic C content (Table 1) seems to be mainly due to this kind of C component. However, differences can be found between the two treatments. In CR_{30} the high anomeric C signal, associated with carbohydrates, is related to the presence of polymers such as hemi-cellulose or cellulose often encrusted with lignin. In this condition, these substances are not readily accessible to microorganisms and therefore, their decomposition slowly occurs, allowing their preservation in soil. In addition, the resonance at 105 ppm assigned to unsubstituted aromatic rings is likely to derive from non-degraded lignin, indicating a crop-residue preservation with little or no alteration in soil. This is again supported by DRIFT which showed stronger signal in the 1500-1570 cm^{-1} region due to lignin moieties. The aromatic C content, also increased in CM_{30}, but no increase in anomeric C signal was observed. This would probably suggest a microbiologically-mediated breakdown of resistant material favoured by the amendment and synthesis of new organic molecules. This hypothesis is further supported by the more intense resonance corresponding to N or O substituted aromatic rings observed in CM_{30} as compared to CR_{30} and C_{30}. This was particularly evident in the phenolic C (150-170 ppm) that showed a significant increase in CM_{30}. The considerable amount of aromatic components substituted, (i.e. hydroxtyl, methoxyl or carboxyl groups) suggests that residue transformation in soil is aimed towards in situ-synthesis of new HS. Additional support is given by the region between 70-60 ppm, characteristic of compounds with hydroxyl groups on adjacent carbon atoms [44]. Among these groups, inositols and sugar alcohols might be considered the most common natural compounds present in this sample. The intensities of the signals at 73 ppm (C_α in units with β-O-linkages) and 53 ppm (methoxyl groups) also provide evidence of the different effects of the treatments: O-CH_3 groups are more intense in CM_{30} than CR_{30} spectra, whereas these groups disappear in C_{30}. Finally, the resonance areas of the carboxylic C (195-165 ppm) showed an increase in carboxylic groups only in CM_{30}, consistent with the results obtained using the other techniques. Manure originally containing components with more oxygenate groups, such as carboxylic groups [45], of course, influences the genesis of HA towards an enrichment of oxygenate functional groups in their structures.

Thermal Analysis

The TG-DTA curves of HA from control and amended plots are shown in Figure 3. Each curve was characterized by two intense exothermic reactions at 300°C and at 460°C, respectively. The strong thermal combustion of the first peak might be influenced by the accumulation or formation of recent organic carbon. The increase in carboxylic groups in the plot amended with CM, as supported by titration and NMR analyses, might mainly be responsible of the strong exothermic reaction observed in this sample. The second exothermic peak remarkabily changed in CM_{30} whereas no significant loss weight appeared in CR_{30} with respect to the control (C_0). In addition the position of the second exothermic peak is shifted towards higher temperature in both treatments (Figure 3). This suggests that high resistant temperature substances such as aromatic structures and cleavage of C-C bonds [16, 31] are involved in exothermic reaction. The different thermal behaviour of CR_{30} and CM_{30} might be the consequence of structural and chemical modification of HA undergone to continuous application with manure and crop residues.

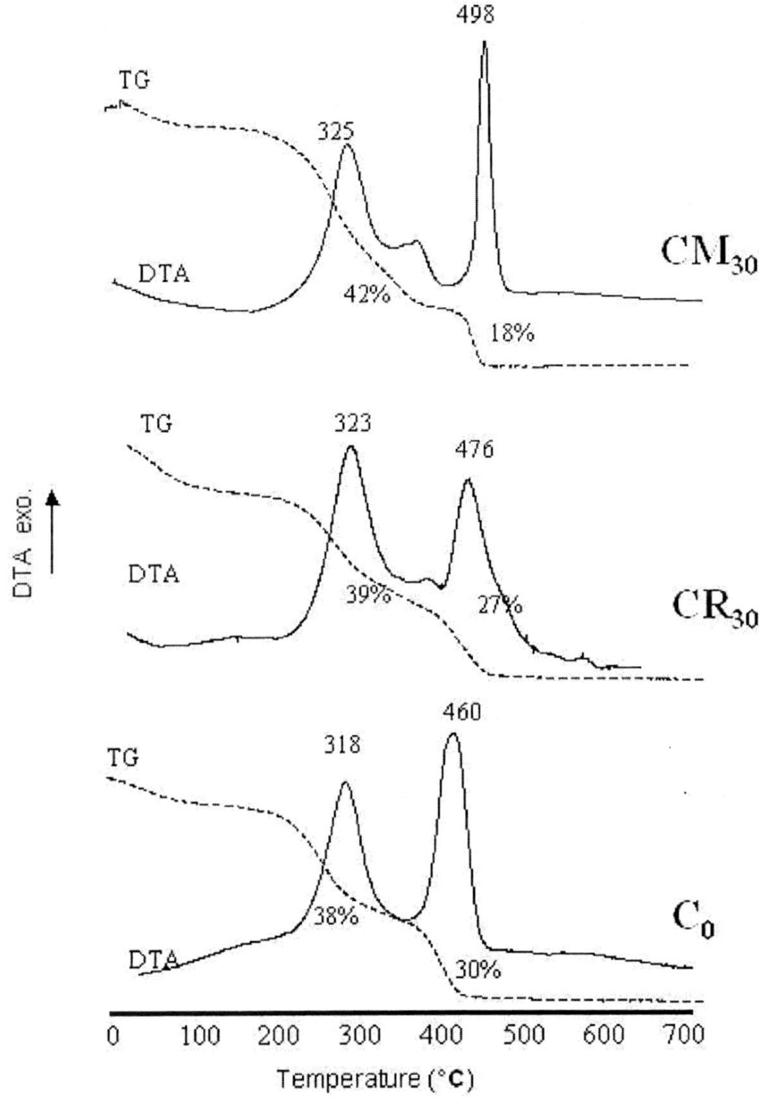

Figure 3. TG-DTA curves of humic acids extracted from unamended soil (C_0) at the start of the experiment and amended consecutively for 30 yrs with cow manure (CM) and crop-residues (CR).

Conclusion

This long-term field experiment highlighted how both the amount and quality of organic matter of agricultural soils can be affected by management and amendment practices such as crop-rotation and amendment with different kinds of organic materials (manure and crop-residues). Conventional farming caused, in the long run, a loss of soil organic C. The addition of crop residues, on the contrary, significantly reduced this loss, whereas the addition of manure preserved the initial level of soil organic C. However, the C loss mainly involved the non humic fraction, since the amount of humic C increased in all the treatments including the control after 30 yrs (+ 6%). This behaviour further confirms the positive role of the humic fraction in C sequestration and storage. The quality of this humic fraction was also affected by the different agricultural practices, so that when the soil did not receive any amendment, the aromatic and carboxylic C decreased, whereas the aliphatic C increased as an effect of the crop rotation. With the amendments, in contrast, the aromatic C generally increased, but with CR this increase was mainly due to preservation of undecomposed lignin components derived from the crop-residues added to the soil, whereas with CM, this kind of C was the result of a new synthesis of HS. The presence of a high amount of carboxylic groups, highly substituted aromatic and alkyl-C components further supports this idea. The study of the effects of different agricultural practices on SOC dynamics in the long run is very important in order to achieve a sustainable agriculture. This research provided additional evidence of the important role played by amendment practices on soil HA formation and on the ability of HS to sequester C.

ACKNOWLEDGEMENTS

This research was supported with funds provided by i) the "Ispettorato Centrale delle Repressioni Frodi" of the Italian Ministry of Agricultural and Forestry Politics, ii) University of Bologna (ex quota 60% - 2006) Thanks are due, in particular, to Prof. G. Toderi and Dr. G. Giordani for supplying the samples

REFERENCES

[1] Venteris, E.R., McCarty, G.W., Ritchie, J.C., Gish, T. (2004). Influence of management history and landscape variables on soil organic carbon and soil redistribution. *Soil Sci.* 169, 787-795

[2] Kaiser, M., Ellerbrock, R.L. (2005). Functional characterization of soil organic matter fractions different in solubility originating from a long-term field experiment. *Geoderma* 127, 196-206.

[3] Francioso, O., Ciavatta, C., Sánchez- Cortés, S., Tugnoli, V., Sitti, L., and Gessa, C. (2000). Spectroscopic characterization of soil organic matter in long-term amendment trials. *Soil Sci.* 165, 495-504.

[4] Aiken, G.R. Isolation and concentration techniques for aquatic humic substances. In: Aiken, G.R., McKnight, D.M., Wershaw, R.L., MacCarthy, P. (Eds.). Humic substances in soil, sediment and water: geochemistry and isolation. New York, *Wiley-Interscience*, 1985.

[5] Stevenson, F.J. Humus Chemistry-Genesis, Composition, Reactions. 2nd edition. New York, Wiley, 1994.

[6] Campbell, C.A., Paul, E.A., Rennie, D.A., and McCallum, K.J. (1967). Factors affecting the accuracy of the carbon-dating method in soil humus studies. *Soil Sci.* 104, 81-85

[7] Lawson, G.J., Steward, D. (1989). Coal Humic Acids. In: Hayes, M.H.B., MacCarthy, P., Malcom, R.L., and Swift, R.S., (Eds.), *Humic Substances II,* John Wiley and Sons Ltd, pp. 642-677.

[8] Senesi N., Plaza, C., Brunetti, G., Polo, A. (2007). A comparative survey of recent results on humic-like fractions in organic amendments and effects on native soil humic substances. *Soil Biol. Biochem.* 39, 1244-1262.

[9] Hatcher, P.G., Clifford, D.J. (1997). The organic geochemistry of coal: from plant materials to coal. *Org. Geochem.* 27, 251-274.

[10] Bates, A. L., and Hatcher, P.G. (1989). Solid-state 13C NMR studies of a large fossil gymnosperm from the Yallourn Open Cut, Latrobe Valley, Australia. *Org. Geochem.* 14, 609-617.

[11] Olk, D.C., Cassman, K.G. and Fan, T.W.M. (1995). Characterization of two humic acid fractions from a calcareous vermiculitic soil: implications for the humification process. *Geoderma* 65, 195-208.

[12] Sohi, S.P., Mahieu, N., Powlson, D.S., Madari, B., Smittenberg, R.H., and Gaunt, J.L. (2005). Investigating the chemical characteristics of soil organic matter fractions suitable for modeling. *Soil Sci. Soc. Am. J.* 69, 1248-1255.

[13] Hooker, B.A., Morris, T.F., Peters, R., and Cardon, Z.G. (2005). Long-term Effects of Tillage and Corn Stalk Return on Soil Carbon Dynamics. *Soil Sci. Soc. Am. J.* 69, 188-196.

[14] Balesdent, J., Mariotti, A. Measurement of soil organic matter turnover using 13C natural abundance. In: Boutton, T.W., Yamasaki, S. (Eds.). *Mass Spectrometry of Soils.* New York, Marcel Dekker, 1996, pp. 83-111.

[15] Roscoe, R., Buurmann, P., Velthorst, E.J., Vasconcellos, C.A. (2001). Soil organic matter dynamics in density and particle size fractions as revealed by the $^{13}C/^{12}C$ isotopic ratio in a Cerrado's oxisol. *Geoderma* 104, 185-202.

[16] Francioso, O., Montecchio, D., Gioacchini, P., and Ciavatta, C. (2005). Thermal analysis (TG-DTA) and isotopic characterization (13C and 15N) of humic acids from different origins. *Appl. Geochem.* 20/3, 537-544.

[17] Baes, A.U., and Bloom P.R. (1989). Diffuse reflectance and transmission Fourier transform infrared (DRIFT) spectroscopy of humic and fulvic acids. *Soil Sci. Soc. Am. J.* 53, 695-700.

[18] Niemeyer, J., Chen, Y., and Bollag, J.M. (1992). Characterization of humic acids, composts, and peat by diffuse reflectance Fourier-transform infrared spectroscopy. *Soil Sci. Soc. Am. J.* 56, 135-140.

[19] Francioso, O., Sànchez-Cortés, S., Tugnoli, V., Marzadori, C., and Ciavatta, C. (2001). Spectroscopic study (DRIFT, SERS and ^{1}H NMR) of peat, leonardite and lignite humic substances. *J. Mol. Struct.* 565, 481-485.

[20] Montecchio, D., Francioso, O., Carletti, P., Pizzeghello, D., Chersich, S., Previstali, F., Nardi S. (2006). Thermal analysis (TG-DTA) and DRIFT spectroscopy applied to investigate the evolution of humic acids in forest soil at different vegetation stages, *J. Thermal. Anal. Calorim.* 83, 393-399.

[21] Spaccini, R., Piccolo, A., Haberhauer, G., Stemmer, M., Gerzabek, M.H. (2001). Decomposition of maize straw in three European soils as revealed by DRIFT spectra of soil particle fractions. *Geoderma.* 99, 245-260.

[22] Haberhauer, G., and Gerzabek, M.H. (1999). DRIFT and transmission FT-IR spectroscopy of forest soils: an approach to determine decomposition processes of forest litter. Vibrational Spectr. 19, 413-417.

[23] Capriel, P. (1997). Hydrophobicity of organic matter in arable soils: influence of management. *Eur. J. Soil Sci.* 48, 457-462.

[24] Ding, G., Novak, J.M., Amarasiriwardena, D., Hunt, P.G., Xing, B. (2002). Soil organic matter characteristics as affected by tillage management. *Soil Sci. Soc. Am. J.* 66, 421-429.

[25] Ding, G., Liu X., Herbert, S., Novak, J.M., Amarasiriwardena, D., Xing, B. (2006). Effect of cover crop management on soil organic matter. Geoderma 130, 229-239.

[26] Wilson, M.A. NMR techniques and applications in geochemistry and soil chemistry. Oxford UK, Pergamon Press, 1987.

[27] Preston, C. (1996). Application of NMR to soil organic matter analysis, history and prospects. *Soil Sci.* 161, 144-166.

[28] Conte, P., Piccolo, A., van Lagen, B., Buurman, P., de Jager, P. A. (1997). Quantitative differences in evaluating soil humic substances by liquid and solid-state 13C-NMR spectroscopy. *Geoderma* 80, 339-352.

[29] Kingery, W.L., Simpson, A.J., Hayes, M.H.B., Locke, M.A., and Hicks, R.P. (2000). The application of multidimensional NMR to the study of soil humic substances. *Soil Sci.* 165, 483-494.

[30] Mathers N.J., Mao, X.A., Xu, Z.H., Saffigna, P.G., Berners-Price, S.J., and Perera, M.C.S. (2000). Recent advances in the application of ^{13}C and ^{15}N NMR spectroscopy to soil organic matter studies. *Aust. J. Soil Res.* 38, 796-87.

[31] Provenzano, M.R., and Senesi, N., (1999). Thermal Properties of Standard and Reference Humic Substances by Differential Scanning Calorimetry. *J. Thermal Anal. Calorim.* 57, 517-526.

[32] Rosa, A.H., de Oliveira, L.C., Bellin, I. C., Rocha, J.C., Cruz Romão, L.P., and Dias Filho, N. L., (2005). Influence of alkaline extraction on the characteristics of humic substances in Brazilian soils.*Thermochimica Acta* 433, 77-82.

[33] Kučerík, J., Kamenářová, D., Válková, M., Pekař, M., and Kislinger, J. (2006). The role of various compounds in humic acids stability studied by TG and DTA. *J. Thermal. Anal., Cal.* 84, 715-720

[34] [34] Toderi, G., and Giordani, G. (1982). Effetti dell'apporto di materia organica al terreno, sotto diverse forme, sulle colture di mais e di frumento in rotazione biennale. I. Effetti sulle rese di frumento. *Riv. Agron.* 16, 197-202.

[35] Coplen, T.B., (1994). Reporting of stable hydrogen, carbon and oxygen isotopic abundances. *Pure Appl. Chem.* 66, 272-276.

[36] Almendros, G., Tinoco, P., Gonzalez-Vila, F.J., Ludemann, H.D., Sanz, J. Velasco, F. (2001). ^{13}C-NMR of forest soil lipids. *Soil Sci.* 166, 186-196.

[37] Wander, M.M., and Yang, X. (2000). Influence of tillage on the dynamics of loose and occluded particulate and humified organic matter fractions. *Soil Biol. Biochem.* 32, 1151-1160.

[38] Plaza, C., Senesi, N., García-Gil, J.C., Brunetti, G., D'Orazio, V., and Polo, A. (2002). Effects of pig slurry application on soils and soil humic acids. *J. Agric. Food Chem.* 50, 4867-4874.

[39] Francioso, O., Sánchez-Cortés, S., Casarini, D., Garcia-Ramos, J.V., Ciavatta, C., Gessa, C. (2002). Spectroscopic study of humic acids fractionated by means of tangential ultrafiltration. *J. Mol. Struct.* 609, 137-147.
[40] [40] Zaccheo, P., Cabassi, G., Ricca, G., Crippa, L. (2002). Decomposition of organic residues in soil: experimental technique and spectroscopic approach. *Org. Geochem.* 33, 327-345.
[41] Amalfitano, C., Pignalosa, V., and Ramunni, A. Chemical and physical properties of humic acids extracted from an andisol manured with green horse bean. In: Senesi, N. and Miano, T.M. (Eds.). Humic substances in the global environment and implications on human health. Amsterdam, Elsevier, 1994, pp. 219-226.
[42] Rao, C.N.R., *Chemical applications of infrared spectroscopy*. New York and London, Academic Press, 1963.
[43] Baldock, J.A., and Preston, C.M. Chemistry of carbon decomposition processes in forests as revealed by solid-state carbon-13 NMR. In: Kelly, J.M. and McFee, W.W. (Eds.). *Carbon forms and functions in forest soils.* Madison, WI, SSSA, 1995, pp. 89–117.
[44] Breitmaier, E., Voelter, W. 13CNMR Spectroscopy. In: Ebel H.F. (Ed.) Monographs in Modern Chemistry, Vol. 15, 2nd Edition. Weinheim, Germany, *Verlag Chemie,* 1978, pp. 247-260.
[45] Ellerbrock, R.H., Höhn, A., Rogasik, J. (1999). Functional analysis of soil organic matter as affected by long-term manurial treatment. *Eur. J. Soil Sci.* 59, 65-71.

In: Soil Ecology Research Developments
Editors: Tian-Xiao Liu, pp. 233-245

ISBN: 978-1-60021-971-9

Chapter 10

DYNAMICS OF TOTAL AND HUMIC CARBON IN A LONG-TERM FIELD EXPERIMENT DETERMINED BY ^{13}C NATURAL ABUNDANCE

Paola Gioacchini*[1]*, Daniela Montecchio, Ornella Francioso and Claudio Ciavatta

Dipartimento di Scienze e Tecnologie Agroambientali, *Alma Mater Studiorum* Università di Bologna, viale Fanin 40, Bologna I-40127, Italy

ABSTRACT

Soil organic carbon (C) preservation in agroecosystems is crucial point to maintain soil fertility and productivity, and to reduce losses of CO_2 in the atmosphere. Agricultural management practices can differently affect the level of soil organic C (SOC). In the present paper or chapter the results of a long-term field experiment (30 years) were investigated to evaluate the effect of mineral fertilization and organic amendments on soil organic C content and on the humic acids (HA) that represent the most important and stable reservoir of soil organic C. The effect of the plant species was also evaluated by comparing wheat and corn monocultures. The amount of corn-derived C in soil and HA at the end of the experiment was calculated by ^{13}C natural abundance measurements. After 30 years of cultivation, the SOC significantly decreased in both unfertilized (Control) cropping systems, especially with continuous corn. Mineral fertilization (Min) and organic amendments (Org) always caused an increase in SOC, especially with Org treatment on continuous corn. The C always increased in HA, except in the unfertilized plots of corn monoculture. Again the highest increase was observed with Org treatment. The amount of corn-derived C in total organic C (TOC) increased in the following order: unfertilized < Min < Org treatments, ranging from 19 to 29%. The turnover time of the older C_3-derived C increased in the same order ranging between 55 and 86 years in the Control and Org treatment, respectively. In the HA the proportion of corn-derived C was similar in the Control and Org treatment (26.4%), lower in the Min treatment (23.7%). Nevertheless if we consider the total amount of corn-derived C in soil and the proportion recovered in the HA, the highest was measured in the unfertilized control. In general a

[1] Corresponding authore: mail address: paola.gioacchini@unibo.it.

proportion ranging from 35% (Org) to 40% (control) of the total corn-derived C in soil was recovered as humic C, confirming the important role of this pool as a C reservoir in soil.

Keywords: Soil orgnic C, Humic Acids, $\delta^{13}C$, manure, mineral fertilization

INTRODUCTION

Preservation of soil organic C (SOC) in agricultural lands is of primary importance for maintaining soil fertility and productivity as well as environmental quality. Intensive cultivation often causes reductions in C content, contributing to the increase in atmospheric CO_2 concentration. This negative effect can be mitigated by sustainable management practices such as crop residues input, organic amendments or reduced tillage, that can increase organic matter input to the soil, or decrease decomposition of soil organic matter and oxidation of SOC (Follett, 2001; Paustian et al., 2000).

To evaluate the effects of these practices on SOC storage, however, requires a better knowledge of SOC dynamics and long-term field experiments. The measure of the transformations of SOC and its pools from a quantitative point of view can only be obtained in experiments where a change in vegetation from C_3 to C_4 plants or vice versa has occurred. In this case the SOC turnover can be followed by measuring the isotopic composition of C in soil after the vegetation change (Balesdent et al., 1988; Flessa et al., 2000; Clapp et al., 2000; Wilts et al., 2004). The different photosynthetic pathway of C_3 and C_4 plant species is responsible for their different isotopic composition that is reflected in that of the soil. This natural label allows to quantify the new C input. When a C_4 plant, usually, in agroecosystems, corn or sorghum, is grown on a soil that has always carried C_3 plants, the $\delta^{13}C$ of SOC progressively changes, and the shift is directly related to the proportion of new C_4-derived C (Balesdent et al., 1987). The longer is the period of cultivation of the C_4 plant, the more the $\delta^{13}C$ value of the SOC approaches that of the plant community (Nadelhoffer and Fry, 1988) because only slight isotope fractionation may occur during early stages of soil organic matter decomposition in well-drained, mineral soils (Boutton, 1996). Since these unique $\delta^{13}C$ values persist during decomposition and soil organic matter formation, the SOC turnover rate can be determined by the rate at which its $\delta^{13}C$ value changes to approach that of the new plant community (Balesdent et al., 1987; Boutton et al., 1998).

Based on the same assumption, we can calculate the turnover rate for particular pools of SOC by measuring the change of their $\delta^{13}C$ values. One of the most important pools of SOC is that of the humic substances and in particular of humic acids (HA) that are complex macromolecules modified from plant compounds or newly synthesized during decomposition, that accumulate in soil (Stevenson, 1994). Because of their resistance to microbial degradation (Quails, 2004), they represent a crucial component of SOC where C tends to be stored for long periods and an important reservoir of nutrients.

In the long-term field experiment in Cadriano, at the University of Bologna, Italy, in 1966, there was a change of vegetation from C_3 to C_4 species, since a monosuccession of corn started on a soil previously cropped with C_3 plants. Plots with continuous wheat cropping were also included in the experiment and for both monocultures different treatments were compared. This experimental setup allows the use of the ^{13}C abundance technique in order to study the C transformations and dynamics since the wheat monoculture represents the reference site that has the same soil properties, land use and environmental conditions of the site with corn. In particular, in this study we could evaluate the effects of the two plant species cropped in monosuccession for a long time period, on the level of SOC and of its stable pool of humic acids (HA). Moreover, we could quantify the amount of corn-derived C in total organic C and in the humic fraction after 30 years of continuous corn cropping, and compare the effects of different treatments such as mineral fertilization and organic inputs, on the amount of corn-derived C accumulated in soil.

MATERIALS AND METHODS

The long-term field experiment located in Cadriano, (University of Bologna, Italy, Department of Agricultural Sciences and Technologies) initiated in 1966, but soil samples from 1966 to 1972 are no longer available. The soil is a *Typic Udochrept* with the following chemical characteristics: pH (in water) 6.9; sand 56%; silt 16%; clay 28%; total organic C (TOC) 8.5 g kg^{-1}; total N 1.1 g kg^{-1}; cation exchange capacity 16.5 $cmol_C$ kg^{-1}.

The experiment compares two continuous cropping system, wheat (*Triticum durum*) and corn (*Zea maize*) that did not receive any mineral or organic fertilization (Control), or treated with mineral fertilizers (Min) or with organic amendments (Org). In the Min treatment, wheat received 200 kg N ha^{-1} and 150 kg P ha^{-1}, whereas corn received 300 kg N ha^{-1} and 150 kg P ha^{-1}, in the Org treatment both monocultures received 20 q ha^{-1} of manure applied to soil at the end of the summer at the ploughing. The crop yields and the whole above-ground biomass were removed at harvest, whereas the below-ground biomass remained in soil.

Each treatment was conducted in triplicate using a randomized block design. Soil samples from 1973 and from 2002 were taken at the depth of 25 cm, air dried and grinded, then analyzed for soil total organic C (TOC), humic acids C (HA) and their isotopic ratio. All analytical determinations were conducted in six replicates.

Humic acid isolation. Thirty g of dry soil finely grinded, were extracted with 300 mL of 0.1 M NaOH plus 0.1 M $Na_4P_2O_7$ at 65 °C by shaking for 24 hours. The suspension was then centrifuged at 5,000 x *g* for 30 min and filtered through a 0.45 μm filter (Millipore, Bedford MA - USA). Twenty-five ml were separated from the rest of the filtered solution, acidified with H_2SO_4 to precipitate HA and analyzed according to Ciavatta and Govi, 1993 to determine the amount of C from HA expressed on a dry soil basis. The rest of the filtered solution was acidified with 5 M HCl to pH < 2 to precipitate the HA and centrifuged at 5,000 x *g* for 20 min. The precipitated HA were re-solubilized in 0.5 M NaOH and then dialyzed against H_2O until neutral pH before being freeze – dried. The measurement of the humic C isotopic ratio was carried out on the freeze-dried HA samples.

Total organic C and $\delta^{13}C$. Soil total organic C was measured with an elemental analyzer (Thermo Electron mod. EA 1110), since this soil does not contain inorganic C the pretreatment with HCl was not necessary. The percentage of C in the sample was calculated using acetanilide as a certified standard containing 71.09% of carbon. The C isotopic ratio of soil total C and HA-C was measured with CF-IRMS (Continuous Flow-Isotope Ratio Mass Spectrometry mod. Delta Plus Thermo Electron, Bremen, Germany). The C isotope values were expressed in delta notation (δ), where: $\delta(‰) = [(R_{sample} / R_{standard}) - 1 \times 1000$, and R_{sample} and $R_{standard}$ are $^{13}C/^{12}C$ of the sample and standard, respectively. Delta ^{13}C is relative to the Vienna Pee Dee Belemnite (VPDB) standard. All measurements were conducted in triplicate. The reproducibility of the delta ^{13}C values of the samples was, in 90 % of the cases, better than 0.1 ‰. The coefficient variation was < 0.1%.

The proportion of corn-derived C, *X%*, after the 30 years period in soil total organic C and in the HA was calculated as:

$$X\% = (\delta_{cs} - \delta_{ws})/(\delta_c - \delta_w) \times 100$$

where $\delta_{cs} = \delta^{13}C$ of soil or HA after 30 years of continuous corn, $\delta_{ws} = \delta^{13}C$ of soil or HA after 30 years of continuous wheat, $\delta_c = \delta^{13}C$ value of corn (-13,4‰), $\delta_w = \delta^{13}C$ value of wheat (-27,5‰).

The relative turnover times of the relic C_3-C were also calculated for the soil TOC and HA-C by assuming that the initial organic C (C0) decayed exponentially with time (t) according to this equation: $C_t = C_0e^{-kt}$.

C_0 was the value measured in the soil cultivated with wheat in 1972, the C_3-C at C_t was calculated using the $\delta^{13}C$ value and the mass balance equation. The calculated k for each treatment was then used to calculate a half-life of C_3-C by setting C_t in the above equation to $C_0/2$.

Statistical Analysis

The Student-Newman-Keuls test was used to compare significance differences within the treatments by using Statgraphics version 5 plus (Statistical graphics system by statistical graphics corporation). Differences were declared at 0.05 level of significance.

Results

Soil Organic C and Humic C

Table 1 gives an overview of TOC content in control and plots differently treated with Min and Org. In both monocultures the TOC content significantly differ ($P<0.05$) in the treatments. Compared with the TOC of 1973 (8.5 g TOC kg^{-1} in soil cropped with wheat) the lost of C was approximately of 5% for the control with wheat, and 15% and 3% for control and Min cropped with mais, respectively. The treatment with Org, in both monocultures, not only reduced the C loss, but induced an increase about 7% for wheat and 9% for corn.

The humic-C level underwent the same influence by the two plant species, in the presence of wheat the Humic-C content was always significantly ($P<0.05$) higher than with corn (Tab. 1). After 30 years, some humic-C was lost from the control plots of corn, 2.5 g kg^{-1} was the value in 1973, 2.1 g kg^{-1} statistically different ($P<0.05$), whereas no variation was detected in the control plots of wheat. On the contrary, with the treatments, we found an increase in HA-C in both monocultures that was again more pronounced with the Org one that showed the highest content of HA.

Table 1. Soil total organic C (TOC) and humic C (HA-C) content in the control and in the plots of both monocoltures treated consecutively over 30 years with mineral fertilizers (Min) and manure (Org)

Treatments	Wheat	Corn
	TOC g kg^{-1}	
Control	8.03 ± 0.06[†]	7.23 ± 0.29
Min	8.80 ± 0.17[a]	8.25 ± 0.04
Org	9.12 ± 0.07[a]	9.65 ± 0.13
	HA-C g kg^{-1}	
Control	3.03 ± 0.05	2.08 ± 0.03
Min	4.00 ± 0.03	3.35 ± 0.03
Org	4.35 ± 0.05	3.76 ± 0.02

[†] Mean ± standard error, n=6 replicates

[a] Within column, means followed by the same letter are not statistically different at 0.05 level of probability.

Corn-Derived C in Soil and HA

After 30 years of continuous corn cropping the $\delta^{13}C$ of the TOC became significantly ($P<0.05$) less negative as a result of the corn-derived C entered into the soil (Figure 1a). This shift in isotopic signature was significantly greater in the treated plots than in the control. The mass balance equation indicated that, whereas in the control about 20% of corn-derived C was measured in 2002, both treatments induced an increase in the proportion of corn-derived C in SOC, so the C_4-C input at the end of the experiment was about 26% and 30% with Min and Org treatment respectively (Table 2). The amount of new corn-derived C, on a dry soil basis, present in SOC after 30 years is shown in Figure 2a, the highest input was measured with the Org treatment, then with the Min treatment whereas in the control we found the lowest amount of corn-derived C.

The $\delta^{13}C$ of the HA-C in the corn plots also showed a shift towards higher values due to new C_4-C that entered into this C pool (Figure 2a). However, unlike what observed in SOC, the difference in $\delta^{13}C$ between the wheat reference plot and the corn plot, was similar for all the treatments and of about 4‰ δ units (Figure 2b). As a consequence the proportion of corn-derived C in HA turned out to be similar for all the treatments, ranging from 24% in the Min, to 26% in both control and Org treatment (Table 2). The amount of corn-derived C present in the humic fraction, and expressed on a dry soil basis, however was different and, in the treatments significantly greater than in the control (Figure 1b).

By comparing the proportion of corn-derived C in HA and SOC, we observed that in the control the value in HA (26%) was higher than that found in the SOC (20%), whereas in the Org treatment it was lower than the 30% measured in SOC. It seems that, without any treatment, the new corn-derived C entering the soil, would be preferentially immobilized.

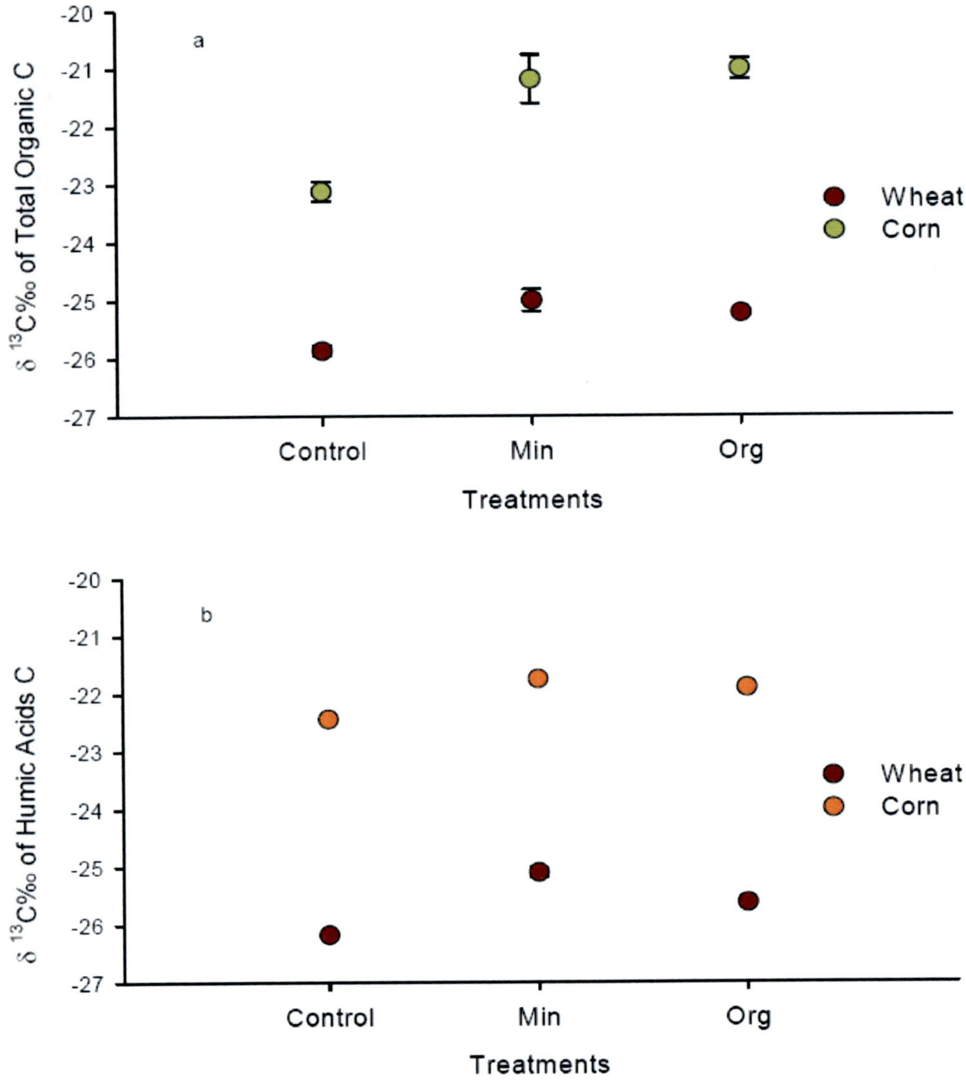

Figure 1. $\delta^{13}C$ ‰ values of soil total organic C (a) and of humic acids C (b) of control and plots of both monocultures treated consecutively over 30 years with mineral fertilizers (Min) and manure (Org).

With the two treatments, on the contrary, and especially with the Org one, the lower proportion of corn-derived C in HA compared to that of SOC suggests that no preferential storage of new C input through immobilization in HA was induced by these treatments. This different behaviour of the new C_4-C depending on the treatments is further confirmed by relating the amount of corn-derived C measured in HA to the total corn-derived C in SOC. This relationships gives evidence that in the control about 40% of the total new C_4-derived C in SOC was in the HA fraction, 38% was measured with the Min treatment, whereas 35% was found in the Org one.

Table 2. Proportion of new corn-derived C in SOC and HA and half-life of old C_3-C in SOC and HA measured in the control and in the plots treated consecutively over 30 years with mineral fertilizers (Min) and manure (Org)

	Soil		HA	
Treatment	Proportion of corn-derived C	Half-life of old C_3-C	Proportion of corn-derived C	Half-life of old C_3- C
	%	yr	%	yr
Control	19.4 ± 0.56	55	26.4 ± 0.08	29
Min	23.9 ± 0.16	65	23.7 ± 0.62	100
Org	29.8 ± 0.88	86	26.4 ± 0.37	156

Mean ± standard error, n=6 replicates

Relic C

The turnover of the old C_3-C was significantly affected by the treatments (Table 2). The half-life of the previous C_3-C in the total SOC pool increased in the following order: control, Min, Org treatment, meaning that both treatments, but especially the Org one, were able to slow down the turnover of the relic C.

The half-life of HA C_3-C was shorter than that of SOC in the control, but significantly longer than SOC with both treatments. The C_3-C turnover time in the control was about 30 years, whereas the treatments increased the half-life of C_3-C by three fold with repeated mineral fertilization and by five fold with the addition of manure.

DISCUSSION

Total Organic C and Humic C

The absence of nutrient addition in the control emphasized the effect of the plant species on the SOC content and dynamics. In the long-term field experiment at Halle, Germany, that had rye as a reference plot, Flessa et al., 2000, measured a higher value of SOC concentration in the rye plot compared to the corn plot although in this experiment the plots got mineral fertilization. Our results showed that the addition of mineral fertilizer had a positive effect on the TOC of the control with both plant species. By comparing the values of TOC concentration of the treated and no-treated plots in 2002, we found that the mineral fertilization produced the greatest positive effect on C storage in the corn plots than in the wheat plots (Table1). Our results are not in agreement with those of literature demonstrating that this effect on C storage is a complex phenomenon not well understood. In fact . in a long-term field experiment with two levels of mineral fertilization, Wilts et al., 2004 observed an overall decrease in SOC during continuous corn production with moldboard tillage. Similar results were obtained by Reicosky et al., 2002 and by Clapp et al., 2000.

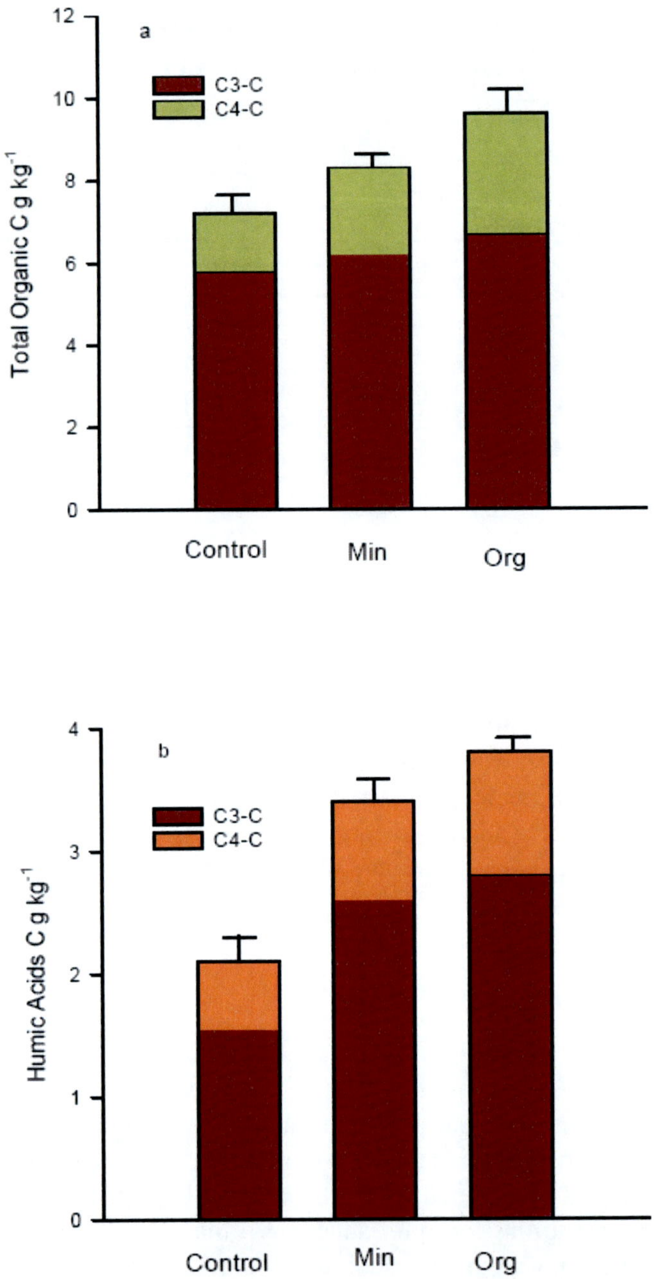

Figure 2. C_4-C from corn and C_3-C in soil total organic C (a) and in humic acids (b) of control and plots of corn monoculture treated consecutively over 30 years with mineral fertilizers (Min) and manure (Org). Bars are standard errors, n = 6 replicates.

Other experiments have also shown either no change or increase in SOC concentration with N fertilization applied to continuous corn production with moldboard tillage (Huggins et al., 1998; Gregorich et al., 1996).

The effect of manure addition in the Org treatment was even more pronounced than the mineral fertilization alone and also in this case the better effect on SOC concentration was observed in the corn plots. Other long-term studies highlighted the positive effect of manure addition on C sequestration as a result of increased root yields, higher humification rate constant and the direct application of organic matter through manure (Kundu et al., 2007; Bhattacharyya et al. 2007). In particular the authors measured in a 30 years rotation with wheat and soybean an increase in above and below-ground biomass, with the application of farmyard manure together with mineral N alone or NPK that was probably associated, apart from N, P, and K supply, to other benefits induced by manure addition, such as improvements in microbial activity and soil physical conditions.

The HA-C content measured in the corn plots at the first sampling in 1973 was significantly lower than the values found in the wheat plots. This difference was probably due to the previous soil exploitation by the corn that had already been cropped for 7 years. Such exploitation continued with corn over the 30 years of the experiment since in the control the level of HA-C in 2002 was significantly lower than in 1973. This trend was not observed with wheat that at the end of the experiment did not show any variation in the HA-C level.

The treatments significantly increased ($P<0.05$) the HA-C compared to the control and this positive effect was more evident in the corn plots. The greater increase in HA-C induced by the Org treatment was probably related to the direct application of organic matter through manure and to a higher humification rate constant (Bhattacharyya et al., 2007). The repeated addition of mineral fertilizers over years also maintained a high HA-C content. Similar results were obtained by Galantini and Rosell, 2006 that after 8 years of continuous wheat found a significant higher content of HA-C when the wheat was annually fertilized compared to the control without fertilization.

The capability of soil to maintain a high content of HA-C or even to increase this C pool is dependent on the balance between humification rate and degradation. The humification rate is positively affected by C and nutrients availability that both are factors stimulating the microbial activity. Degradation is the counterbalanced mechanism that breaks down the formed humic fraction and that is itself mediated by microorganisms. It has been reported that HA-C from long-term fertilized soils are more resistant to degradation than those from non fertilized soils and that the HA's were better utilized by microorganisms when they serve as sole source of C or N than under more favourable nutrients conditions (Filip and Kubàt, 2001).

Corn-Derived C in TOC and Humic Acids

In SOC after 30 years, a proportion ranging from 20 to 30% was C originating from corn. In the Halle experiment Flessa et al., 2000, found that after 37 years of continuous corn, 15% of SOC accounted for corn-derived C. In other studies higher proportions of corn-derived C are reported (Puget et al., 1995; Balesdent et al., 1990), but these differences are mainly related to different management practices. The amount of corn-derived C in soil usually increases with no-till management compared to conventional tillage and with the return of above-ground biomass to soil compared to the removal. In our experiment the above-ground biomass was removed so the contribution of corn-derived C to the SOC only originated from roots system and rhyzodeposition. Soil characteristics also affect the amount of new C_4-C in soil because in sandy soil the organic matter is hardly protected by the formation of stable aggregates or through adsorption to mineral surfaces (Sollins et al., 1996). Moreover the depth of soil sampling must be considered when results from different experiments are compared. Hooker et al., 2005, for instance, in a 28 years experiment comparing no-till or conventional tillage with residues returned or residues removed from soil, measured the contribution of corn-derived C at two different soil depths: 0-5 and 5-15 cm. The values they found for the conventional tillage with mineral fertilization and residues removal that was the same management practice of our experiment, were similar to ours if the whole depth of sampling (0-15 cm) is considered.

The mineral fertilization and the manure addition caused an increase in the proportion of corn-derived C in SOC as a consequence of the increase in biomass production both above and below-ground. The higher crop yield obtained with mineral fertilization and manure addition, also imply a higher production of roots and rhyzodeposition that can explain the greater amount of corn-derived C in the treated plots compared to the control.

The treatments positively influenced the ability of soil to sequester C not only by increasing the proportion of new C_4-C entering the SOC, but also by slowing down the turnover rate of the relic C_3-C. In our experiment the values of C half-life we found ranged between 55 and 86 years that fall within ranges presented by other authors (Balesdent et al., 1990; Gregorich et al., 1994; Angers et al., 1995; Clapp et al., 2000). Lower values are also reported in literature but these are usually referred to surface soil layers. Hooker et al., 2005 calculated for the surface layer of 0-5 cm shorter C_3-C half-life ranging between 14 and 19 years, compared to the deeper layer of 5-15 cm. In our experiment the reason for the higher values is that they were calculated for the whole sample without distinguish between surface and deeper horizon.

The treatments had a different influence on the proportion of corn-derived C found in the HA-C compared to what observed in SOC. The values ranged between 24% and 26% of the HA-C and this implies a different turnover rate compared to SOC. In the control, in spite of the net loss of SOC and HA-C over the experimental period, we observed an increase in the proportion of new C_4-C in this pool (26%) compared to the proportion measured for the SOC (20%). Without addition of mineral nutrients over long time period, this pool was actually able to sequester organic C, by including in its structure a great amount of new C_4-C input entering into the soil. We calculated that 40% of the total corn-derived C in soil was recovered as HA-C. On the other hand, in the control, the HA fraction was less efficient in storing the old C_3-derived C compared to SOC, with a turnover time 25 years shorter than that measured for SOC. It must be considered that in the control the only source of nutrients derived from mineralization of soil organic matter and in this particular condition the humic fraction could have been preferentially mineralized than other fractions such as the humin that is reported to be more recalcitrant to degradation (Stevenson,1994). Moreover, as already mentioned, the HA from non fertilized long-term fields seems to be less resistant to microbial degradation than those from fertilized fields.

In the HA extracted from the Min treated plots we did not observe any preferential immobilization of corn-derived C since the proportions measured in HA and SOC were similar, meaning that the amount of corn-derived C entering the HA increased parallely to the increase in HA content. With mineral fertilization, however, the HA were able to store a greater amount of C compared to the control, mainly the old C_3-C that with this treatment had a much slower turnover rate. This, again, was probably related to the quality of HA because those formed in long-term fields annually supplied with mineral fertilization are reported to be more resistant to degradation (Filip and Kubàt, 2001).

The addition of manure, did not induce a preferential accumulation of corn-derived C in HA in spite of the increase in HA-C content and although with this treatment the greatest proportion of corn-derived C was found in SOC. Nevertheless this treatment caused the highest C sequestration in the HA fraction that was not only due to increased yields of roots residues, but also to higher humification rate and the direct application of organic matter through the manure itself. Moreover we measured a very slow turnover rate of the old C_3-C in the plots treated with manure, that might be due to slower breakdown rate.

CONCLUSION

Long-term field experiments where a change in vegetation from C_3 to C_4 plant species has occurred and where plots of C_3 species are maintained have a great worth for the study of C dynamics. The agricultural management practices are crucial for the maintenance of the SOC level over long time periods and our results indicate that the mineral fertilization and the addition of manure are both able to reduce the C loss or even to increase the SOC over years. The addition of manure had the most positive effect on C sequestration as a result of several factors, in particular the supply of organic matter through the manure itself, but also other more general benefits related to the microbial activity and physical conditions of soil (Kundu et al., 2007). Both treatments, however, positively affected soil C sequestration, by inducing an increase in the amount of corn-derived C in SOC compared to the control and by slowing down the turnover rate of the old C_3-C.

When we considered the humic fraction, being one of the most important pools of SOC involved in C sequestration, we found that both treatments were able to significantly increase this pool compared to the unfertilized control, but the mechanism involved was mainly a preservation of the old C_3-C from degradation process, not an increase in the proportion of new C_4-C in the humic pool. This proportion was actually similar for the control and the two treatments, but the turnover rate of the C_3-C from the treated plots was threefold and fivefold slower for Min and Org treatments respectively suggesting that the annual supply of mineral fertilizers and even more of manure, can significantly increase the preservation of old C in the humic C pool.

AKNOWLEDGEMENTS

The authors would like to thank Profs. Giovanni Toderi, and Guido Baldoni and Dr. Gianni Giordani for kindly providing the historical soil samples, as well as soil data and management information. This research has been funded by University of Bologna ex quota 60%.

REFERENCES

Angers D.A., R.P. Voroney, and D. Cote. 1995. Dynamics of soil organic matter and corn residues affected by tillage practices. *Soil Sci. Soc. Am. J.* 59:1311-1315

Balesdent J., D. Boisgontier, and A. Mariotti. 1990. Effect of tillage on soil organic carbon mineralization estimated from ^{13}C abundance in maize fields. *J. Siol Sci.* 41:587-596

Balesdent, J., A. Mariotti, and B. Guillet. 1987. Natural 13C abundance as a tracer for studies of soil organic matter dynamics. *Soil Biol. Biochem.* 19:25-30

Balesdent, J., G.H. Wgner, and A. Mariotti. 1988. Soil organic matter turnover in long-term field experiments as revealed by carbon-13 natural abundance. *Soil Sci. Soc. Am. J.* 52:118-124

Bhattacharyya, R., S. Chandra, R.D. Singh, S. Kundu, A.R. Srivastva, and H.S. Gupta. 2007. Long-term farmyard manure application effects on properties of a silty clay loam soil under irrigated wheat-soybean rotation. *Soil Till. Res.* 94:386-396

Boutton, T.W. 1996. Stable carbon isotope ratios of soil organic matter and their use as indicators of vegetation and climate change. P. 47-82. *In* T.W. Boutton and S. Yamasaki (ed.) Mass spectrometry of soils. Marcel Dekker, New York.

Boutton, T.W., S.R. Archer, A.J. Midwood, S.F. Zitzer, and R. Bol. 1998. $\delta^{13}C$ values of soil organic carbon and their use in documenting vegetation change in a subtropical savanna ecosystem. *Geoderma* 82:5-41

Ciavatta C., and M. Govi. 1993. Use of insoluble polyvinylpyrrolidone and isoelectric focusing in the study of humic substances in soils and organic wastes: a review. *J. Chromatogr.* A. 643:261-270

Clapp, C.E., R.R. Allmaras, M.F. Layese, D.R. Linden, and R.H. Dowdy. 2000. Soil organic C and C-13 abundance as related to tillage, crop residues, and nitrogen fertilization under continuous corn management in Minnesota. *Soil Till. Res.* 55:127-142

Filip Z., and J. Kubàt. 2001. Microbial utilization and transformation of humic substances extracted from soils of long-term field experiments. *Eur. J. Soil Biol.* 37:167-174

Flessa, H., L. Bernard, H. Balint, and W. Merbach. 2000. The origin of soil organic C and respiration in a long-term maize experiment in Halle, Germany, determined by 13C natural abundance. *J. Plant Nutr. Soil Sci.* 163:157-163

Follett, R.F. 2001. Soil management concepts and carbon sequestration in cropland soils. *Soil Tillage Res.* 61:77-92

Gregorich, E.G., B.H. Ellert, and C.M. Monreal. 1994. Turnover of soil organic matter and storage of corn residues carbon estimated from natural ^{13}C abundance. *Can. J. Soil Sci.* 75:161-167

Gregorich, E.G., B.H. Ellert, C.F. Drury, and B.C. Liang. 1996. Fertilization effects on soil organic matter turnover and corn residue placement. *Soil Sci. Soc. Am.J.* 60:472-476

Hooker B.A., T.F. Morris, R. Peters, and Z.G. Cardon. 2005. Long-term effects of tillage and corn stalk return on soil carbon dynamics. *Soil Sci. Soc. Am. J.* 69:188-196

Huggins, D.R., C.E. Clapp, R.R. Allmaras, J.A. Lamb, and M.F. Layese. 1998. Carbon dynamics in corn-soybean sequences as estimated from natural carbon-13 abundance. *Soil Sci. Soc. Am. J.* 62:195-203

Kundu, S., R. Bhattacharyya, V. Prakash, B.N. Ghosh, and H.S. Gupta. 2007. Carbon sequestration and relationship between carbon addition and storage under rainfed soybean-wheat rotation in a sandy loam soil of the Indian Himalayas. *Soil Till. Res.* 92:87-95

Nadelhoffer, K.J., and B. Fry. 1988. Controls on natural nitrogen-15 and carbon-13 abundances in forest soil organic matter. *Soil Sci. Soc. Am. J.* 52:1633-1640

Paustian, K., J. Six, E.T. Elliott, and H.W. Hunt. 2000. Management options for reducing CO2 emissions from agricultural soils. B*iogeochemistry* 48:147-163

Puget P., C. Chenu, and J. Balesdent. 1995. Total young organic matter distribution in aggregates of silty cultivated soils. *Eur. J. Soil Sci.* 46:449-459

Qualls R. G. 2004. Biodegradability of humic substances and other fractions of decomposing leaf litter. *Soil Sci. Soc. Am J.* 68:1705-1712

Reicosky, D.C., S.D. Evans, C.A. Cambardella, R.R. Armaras, A.R. Wilts, and D.R. Huggins. 2002. Continuous corn with moldboard tillage: Residue and fertility effects on soil carbon. *J. Soil Water Conserv.* 57:277-284

Sollins P., P. Homann, and B.A. Caldwell. 1996. Stabilization and destabilization of organic matter: mechanisms and controls. *Geoderma* 74:65-105

Stevenson F. J. 1994. Humus chemistry – Genesis, composition, reactions. 2nd ed. Wiley, New York

Wilts, A.R., D.C. Reicosky, R.R. Allmaras, and C.E. Clapp. 2004. Long-term corn residues effects: harvest alternatives, soil carbon turnover, and root-derived carbon. *Soil Sci. Soc. Am. J.* 68:1342-1351

In: Soil Ecology Research Developments
Editors: Tian-Xiao Liu, pp. 247-263
ISBN: 978-1-60021-971-9

Chapter 11

RHIZOBIUM LEGUMINOSARUM ISOLATED FROM AGRICULTURAL ECOSYSTEMS SUBJECTED TO DIFFERENT CLIMATIC INFLUENCES: THE RELATION BETWEEN GENETIC DIVERSITY, SALT TOLERANCE AND NODULATION EFFICIENCY

S.I.A. Pereira*[1]*, A.I.G. Lima and E.M.A.P. Figueira
[1]Centre for Cell Biology, Department of Biology, Campus Santiago, 3810-193 AVEIRO, Portugal

ABSTRACT

Salinity is a major environmental constraint of crop production, and with the climate changes that are being announced for the next decades, like global warming or local reduction of rainfall, this problem will be amplified. Salinity stress negatively impacts agricultural yield throughout the world affecting production whether it is for subsistence or economic gain. *Rhizobium* has considerable scientific, economical and ecological interest because of their ability to establish nitrogen-fixing nodules on leguminous hosts. This feature enables plants to grow in soils with low nitrogen levels, to achieve good crop yields without massive nitrogen fertilization, and, as a consequence, to decrease the contamination of water reservoirs by inorganic nitrogen compounds. Salinity not only affects free-living rhizobia but also considerably restrains the nodulation process and symbiotic nitrogen fixation. To fix nitrogen in saline environments leguminous plants require both free-living rhizobia and hosts tolerant to salt. Therefore, the selection of tolerant phenotypes, which can withstand the negative impact of saline soils, can be of great use to improve nitrogen fixation and productivity in salt-affected soils.

In this work *Rhizobium* was isolated from several locations in Portugal with different environmental conditions. *Rhizobium* isolates were screened for their tolerance to salinity as free-living organisms and for their efficiency to fix N_2 under salt conditions in

[1] Corresponding author address: Pereira, Sofia Isabel Almeida ,Centro de Biologia Celular, Departamento de Biologia, Campus de Santiago, Universidade de Aveiro, 3810-196 Aveiro, Portugal,E-mail: siapereira@portugalmail.com, Phone: 00351 234 370 782, Fax: 00351 234 865008.

symbiosis with a legume. Furthermore, *Rhizobium* protein and plasmid profiles were evaluated, in order to identify variability and to relate it with salt tolerance.

To accomplish these goals isolates were grown in YEM supplemented with NaCl (25 to 1800 mM). According to their growth responses isolates were classified in three groups: sensitive (0-50 mM NaCl), tolerant (100-500 mM NaCl) and extremely tolerant (600-1800 mM NaCl). Salt tolerance was a reflex of the conditions experienced in their natural habitats. Forty-one polypeptides were separated by SDS-PAGE. Salt conditions induced differences in protein profiles when compared with controls. Extremely tolerant strains were the less affected, suggesting the putative presence of constitutive mechanisms conferring tolerance to NaCl. Strains displayed different plasmid profiles and classification analysis suggests that the presence of certain plasmids (828, 734, 147 and 82 MDa) can be correlated to salt tolerance. Finally, tolerant strains were demonstrated to efficiently nodulate *Pisum sativum* plants, in the absence of salt, and more importantly in the presence of moderate levels of sodium chloride, which, due to the announced climate changes, could be important in a near future to maintain the present areas of legume cultivation. This is of particular importance because it points out the need of further studies to predict the influence of climate alterations on soil microbial populations. Under this context, because *Rhizobium* is important to several natural and agricultural communities, it may be potential use as an indicator and enabler of agricultural sustainability in affected soils.

1. INTRODUCTION

The impact of climate changes on biota has recently gained prominence, given the significant concern towards global warming and impending climate changes. According to the Third Assessment Report of Working Group I of the Intergovernmental Panel on Climate Change (IPCC, 2007) the global average surface temperature has increased over the 20th Century by about 0.6°C and for 2100 estimates predict increases between 1.4 and 5.8 °C, depending on natural changes and human activities. With the predicted climate alterations of increasing temperature and rainfall distribution is expected that this problem will not only be intensified but also be spread worldwide in a very near future. The resulting land degradation is a major constraint of crop yield worldwide, with erosion, salinization, and desertification as important consequences (Rozelle et al., 1997). About one-third of the world's irrigated land is salt affected (Shannon et al., 1984), and up to 40% of world's land surface shows potential salinity problems (Bouhmouch et al., 2005). Being a Mediterranean country, Portugal is strongly affected by high temperature (which frequently surpass 40°C) and reduced rainfall, specially in late Spring and Summer, constraining agriculture production, particularly in the southern areas.

In many developing countries, the effective management of nitrogen (N) in the environment is an essential element for agricultural sustainability (Rehman and Nautiyal, 2002). The 6 billion people on earth consume an average of nearly 11 g of N per person per day (Fink et al., 1999). Plant sources satisfy up to 80% of dietary needs in much of the tropics and sub-tropics. With the earth's population increasing 1.4% annually, and expected to reach 8.3 billion in 2025 (Mannion, 1998), unprecedented increases in crop production will be needed if the current levels of dietary proteins and caloric intake are to be maintained. Furthermore, leguminous plants are frequently used for degraded soil sites in arid and semiarid regions because they can grow in barren soils that are unsuitable for most crops.

Biological nitrogen fixation in legumes is often used to improve infertile agricultural soils (Rehman and Nautiyal, 2002). The input of mineral nitrogen in the soil mainly consists of symbiotic nitrogen fixation, transformation of organic matter and fertilization. Biological activity is of great importance for the nitrogen balance in the soil because nitrogen fixation and organic matter transformation both depend on biological activity. Fixation of atmospheric nitrogen results from symbiosis between leguminous crops and rhizobia (van Hoorn et al., 2001). This symbiotic association is by far the most important contributor to the world's supply of biologically fixed N_2. Rhizobial symbioses with more than 100 agriculturally important legumes contribute nearly half the annual quantity of biological nitrogen fixation entering soil ecosystems (Graham and Vance, 2000; Somasegaran and Hoben, 1994). For this reason, taking into consideration the importance of legumes in animal and human consumption, some attention has been given to the effects that environmental stresses exert on *Rhizobium* populations (Ibekwe et al., 1995).

Leguminous plants growing in saline environments require both free-living rhizobia and host salt tolerant phenotypes. Therefore, it becomes important to find symbiotic partners that can fix nitrogen under stress conditions (Abdelmoumen et al., 1999; Singleton et al., 1982). The ability of rhizobia to establish nitrogen-fixing nodules on leguminous hosts enables plant growth in soils with low nitrogen levels and consequently decreases in the contamination of water reservoirs by inorganic nitrogen compounds (Hynes and O' Connell, 1990).

Many biotic and abiotic factors affect the growth and survival of rhizobia in soil. *Rhizobium* spp. strains are very sensitive to soil environment abiotic factors such as high salt, pH and temperature stresses that affect their dinitrogen fixation and hence the productivity of legumes (AbdelGadir and Alexander, 1997; Athar and Johnson, 1997).

Most rhizobial strains, which nodulate important crops, are also sensitive to soil desiccation. Therefore, for the good growth of legumes in arid and semiarid regions of the world where fertilizers are unavailable or expensive, it seems deemed necessary to plants, being nodulated by an effective strain of *Rhizobium* that tolerate these adverse environmental conditions (Rehman and Nautiyal, 2002).

Salinity not only affects free-living rhizobia but also considerably restrains the nodulation process and symbiotic nitrogen fixation (Abd-Alla et al., 1998; Elsheikh and Wood, 1995). Salt may affect symbiosis by its effects on the growth and survival of rhizobia in soil, restrictions on root colonization, inhibition of processes of infection and nodule development, or impairment of active nodule functioning (Rehman and Nautiyal, 2002). Some authors reported *Rhizobium* and *Bradyrhizobium* strains that can survive and grow as free living organisms under salt concentrations that inhibit the growth of most legumes (Bordeleau and Prévost, 1994; Figueira 2000; Zahran et al., 1994). According to Brewing et al. (1993), the genetic variability in *Rhizobium* is wide enough to ensure the survival of strains selected under extreme conditions. Zahran et al. (1994) also obtained different salt responses among several rhizobia isolates. Thus, host and symbiotic processes are generally thought to be the factors limiting N_2 fixation under salt conditions (Abd-Alla et al., 1998; Cordovilla et al., 1999; Delgado et al., 1993).

Genes conferring specific adaptation to adverse conditions, including salt stress are often plasmid-borne (Silver and Misra, 1988). Plasmids are important genetic components for the divergence and adaptation of microbial populations because they contribute to genomic plasticity (Zhang, 2001). These plasmids might be lost or regained in populations, rapidly change copy number and undergo higher mutation rates because of the common occurrence of reiterated DNA (Fawzy and Kuykendall, 1994). The correlation between the selection pressure caused by stress and the existence of the same plasmids suggests that plasmids play a major role in the adaptation of bacteria to environmental stresses (Lakzian et al., 2002). In addition to symbiotic plasmids, *Rhizobium* strains may carry 1-10 other plasmids, which range in size from about 30 to more than 1000 MDa (Fawzy and Kuykendall, 1994) and may contribute to saprophytic competence of rhizobia (Chen et al., 1993).

Bacteria living under salt stress also experience a number of physiological alterations. Salinity imposes both ionic and osmotic stress, which can be extremely detrimental for microflora survival (Saxena et al., 1996). For example, according to Saxena et al. (1996) and Unni and Rao (2001) the imposition of any stress to bacteria results in adaptative responses, which lead to changes in the regular metabolic processes that are then reflected in protein profiles. For this reason, one approach for understanding *Rhizobium*'s ability to tolerate salt has been to identify stress induced protein alterations. Several authors (Saxena et al., 1996; Shamseldin et al., 2006; Soussi et al., 2001) reported that changes in protein profiles in response to salt stress are sufficient to substantiate the presumption that proteins have a role to play in salt tolerance. Völker et al. (1992) also suggested that stress protein induction is an important cellular adaptations to growth-limiting conditions, such as heat and salt-stress.

This chapter aimed to study the halotolerance of rhizobial populations from different agricultural ecosystems in order to evaluate *Rhizobium*'s vulnerability to salinity, a growth constraint that can affect many soils, even non saline, due to secondary salinization, which is expected to increase due to climate alterations. To fix nitrogen in saline environments leguminous plants require both free-living rhizobia and hosts tolerant to salt. Therefore, the selection of tolerant phenotypes, which can withstand the negative impact of saline soils, can be of great use to improve nitrogen fixation and productivity in salt-affected soils. For these reasons, *Rhizobium* isolates were also screened for their efficiency to fix N_2 under salt conditions in symbiosis with a legume. *Pisum sativum* plants were the host chosen, since they are reported as one of the most halotolerant cultivated legumes (Subbarao and Johansen, 1994). Since plasmids code for genes that are important for bacterial adaptation and survival to environmental constraints, plasmid profile analysis can offer a basic genetic tool to elucidate if different populations have the same or different genetic features to cope with salt stress. The comparative study of protein pool alterations between halotolerant strains under salt stress shad some light the mechanisms underlying salt tolerance.

2. Salt Tolerance of Rhizobium Isolated from Different Agricultural Ecosystems

Six soils were obtained from arable fields in different locations of Portugal subjected to different climates. S. Bernardo (SB) soil is a silt loam soil with low salinity (0.29 g Na^+ Kg^{-1} dry soil) and water availability throughout the year (69.3% field capacity in June); Vagos (V) soil is a sandy loam soil with low salinity (0.23 g Na^+ Kg^{-1} dry soil) and high water availability throughout the year (86.6% field capacity in June); Costa Nova (CN) soil is a sandy soil with sea influence (1.12 g Na^+ Kg^{-1} dry soil) and a low water content during part of the year (53.6% field capacity in June); Alentejo (A) soils are a clay loam soils with low water availability and affected by high temperatures during part of the year (84.4% field capacity in January, 27.8% field capacity in June). AS soil was collected in late Spring (0.36 g Na^{+} Kg^{-1} dry soil) while AWA and AWB soils were collected in early Winter (0.32 g Na^{+} Kg^{-1} dry soil).

Considerable genetic variability in salt tolerance among legumes species and cultivars has been reported. Some tree legumes such as *Prosopis* and *Acacia* spp. are highly tolerant to salinity, while grain legumes have long been recognized as either sensitive or only moderately tolerant to salinity. However, according to Subbarao and Johansen (1994) *Pisum sativum* is one of the most halotolerant-cultivated legumes. For this reason, in this work, pea plants were used as hosts to obtain *Rhizobium* isolates from soils.

To evaluate the influence of salt stress in the soil microflora, thirty-nine isolates of *Rhizobium leguminosarum* bv. *viciae* were obtained from root nodules of pea (*Pisum sativum* L.) plants grown in SB, V, CN, AWA, AWB and AS soils.

According to Hirsch et al. (1993) and Ibekwe et al. (1997) the distribution of *Rhizobium* strains in agricultural soils is affected both by current and previous cultivation: the presence of the host plant generally leads to a population increase which persists for some years. The difficulties of establishment and persistence of these bacteria in soils in the absence of their specific leguminous hosts have emphasised the need for legume inoculation and have prompted studies of their survival in soil under adverse conditions. In this chapter, *Pisum sativum* plants were grown in all soils in the moment of sampling or in the last three years.

USDA strains were provided by Peter vanBerkum (United States Department of Agriculture), and GRA19 strain was a kind gift by Carmen Lluch (Granada University).

2.1. Characterization of NaCl Tolerance

Salt tolerance of *Rhizobium* isolates and strains were screened in YEM medium supplemented with increasing NaCl concentrations (0 to 1800 mM). Salt tolerance was defined experimentally as the maximum NaCl concentration where cell growth was inhibited less than 75%. Isolates showed distinct responses to salt, being classified in three groups, according to their salt tolerance: sensitive (0-50 mM NaCl); tolerant (100-500 mM NaCl); and extremely tolerant (600-1800 mM NaCl). Figure 1 shows the percentage of sensitive, tolerant and extremely tolerant isolates in each location, and Table 1 shows the growth response of *Rhizobium* isolates and strains to salinity. Isolates from Vagos and S. Bernardo evidenced high vulnerability to NaCl, being all sensitive. Massive growth inhibitions were observed at concentrations of 25 mM for Vagos and 50 mM for S. Bernardo (Table 1). These responses can probably be related to the large availability of water present in those locations throughout the year.

Costa Nova isolates showed high salt susceptibility since 86% of them were sensitive (Figure 1). The halotolerance of *Rhizobium* CN isolates (Table 1 and Figure 1) did not correlate to salt levels normally present near the sea. It might be expected that salt stress would be a constant condition at this site, especially in late Spring and Summer, when the soil is influenced by reduced water content and concomitant increase in ion concentration. According to Bååth (1992) and Bååth et al. (1998) environmental stresses exert a selective pressure in soil microflora. The soil microbial population presents variability, which in a non-stressed environment would not bring any advantage.

Table 1. Growth (10^7 cells ml^{-1}) of *Rhizobium leguminosarum* isolates in YEM supplemented with NaCl (0 to 1800 mM). Values are means of three replicates

	NaCl (mM)											
	0	25	50	100	200	300	500	600	700	1000	1400	1800
SB 2	89.41	59.00	24.75	11.30	10.92	10.53						
SB 4	81.66	87.11	37.41	11.42	10.18	8.93						
SB 6	81.74	16.64	19.92	11.88	90.49	7.09						
SB 8	85.54	18.69	15.16	9.32	8.66	7.99						
SB 10	77.02	48.26	17.15	8.12	7.57	7.01						
SB 13	84.47	68.47	17.44	6.18	6.74	11.37						
SB 16	87.35	57.75	23.81	14.85	12.24	9.62						
V 2	45.52	14.34	13.72	11.37	5.69							
V 7	50.28	7.16	7.87	5.94	2.20							
V 10	50.35	3.58										
V 13	57.60	13.29										
V 16	31.70	3.64										
V 17	54.16	7.18	9.28									
V 26	50.23	11.54										
V 30	55.35	5.54	5.41									
CN 7	52.76	42.75	12.14	13.3								
CN 9	56.75	41.32	8.21	8.93	4.15	0.63						
CN 11	35.18	21.44	6.76	6.28	5.24	2.12						
CN 13	48.35	26.74	3.52	1.11								
CN 15	60.07	54.75	16.06	4.62								
CN 17	59.80	32.66	10.63	7.95								
CN 20	49.94	49.42	10.26	2.24	3.18	2.67						
AWA 1-2	23.85	25.33	30.09	33.66	41.64	39.63	5.58	5.04				
AWA 1-3	27.86	23.91	27.79	27.93	46.94	38.03	7.61	7.61				
AWA 2-1	74.16	74.46	24.94	8.78	3.33							
AWA 2-2	118.00	128.7	113.51	115.4	4.21							
AWA 2-5	45.15	9.15	7.48	2.54	0.81							
AWB 1-1	47.03	23.02	34.06	2.81	3.76							
AWB 2-1	36.15	15.25	10.34	8.02	1.11							
AWB 2-4	46.15	40.12	8.88	2.61	2.48							
AWB 2-7	89.67		120.4	120.8	97.34	54.04	2.22	1.88				
AWB 5-1	47.25	56.61	62.28	72.45	50.92	37.11	12.1	10.04	7.97			
AS 1-2	55.00	53.76	56.99	50.27	58.37	59.25	42.40	49.01	49.71	53.88	5.53	1.94
AS 2-1	52.08	52.17	51.93	56.84	53.84	52.39	46.67	46.40	52.49	39.71	39.36	12.37
AS 4-2	67.83	68.12	69.63	61.93	45.56	45.56	60.76	52.90	50.75	37.18	4.22	1.76
AS 20-4	89.58	52.39	5.75									
AS 20-7	140.26	145.25	149.24	133.36	99.67	22.01	7.03	5.59	5.58			
AS 20-8	155.89	151.97	149.62	140.26	123.18	33.59	7.25	6.19	5.16			
AS 24-1	68.09	64.93	63.04	65.02	59.65	56.97	46.23	42.66	45.78	44.58	7.18	2.37
GRA 19	58.45	6.92	6.36	4.97								
USDA 2455	49.58	18.82	10.66	3.72								
USDA 2335	52.72	45.73	18.24	8.73								
USDA 2343	30.92	33.04	33.38	26.37	7.95							
USDA 2353	26.64	18.89	20.20	24.53	32.31	33.01	50.18	52.25	37.60	3.85	2.85	
USDA 2433	27.32	19.18	19.18	20.82	27.33	30.57	48.11	50.01	43.46	3.72	2.36	
USDA 2479	38.55	6.98	26.14	34.23	36.79	45.71	48.9	38.71	24.62	2.85	3.92	

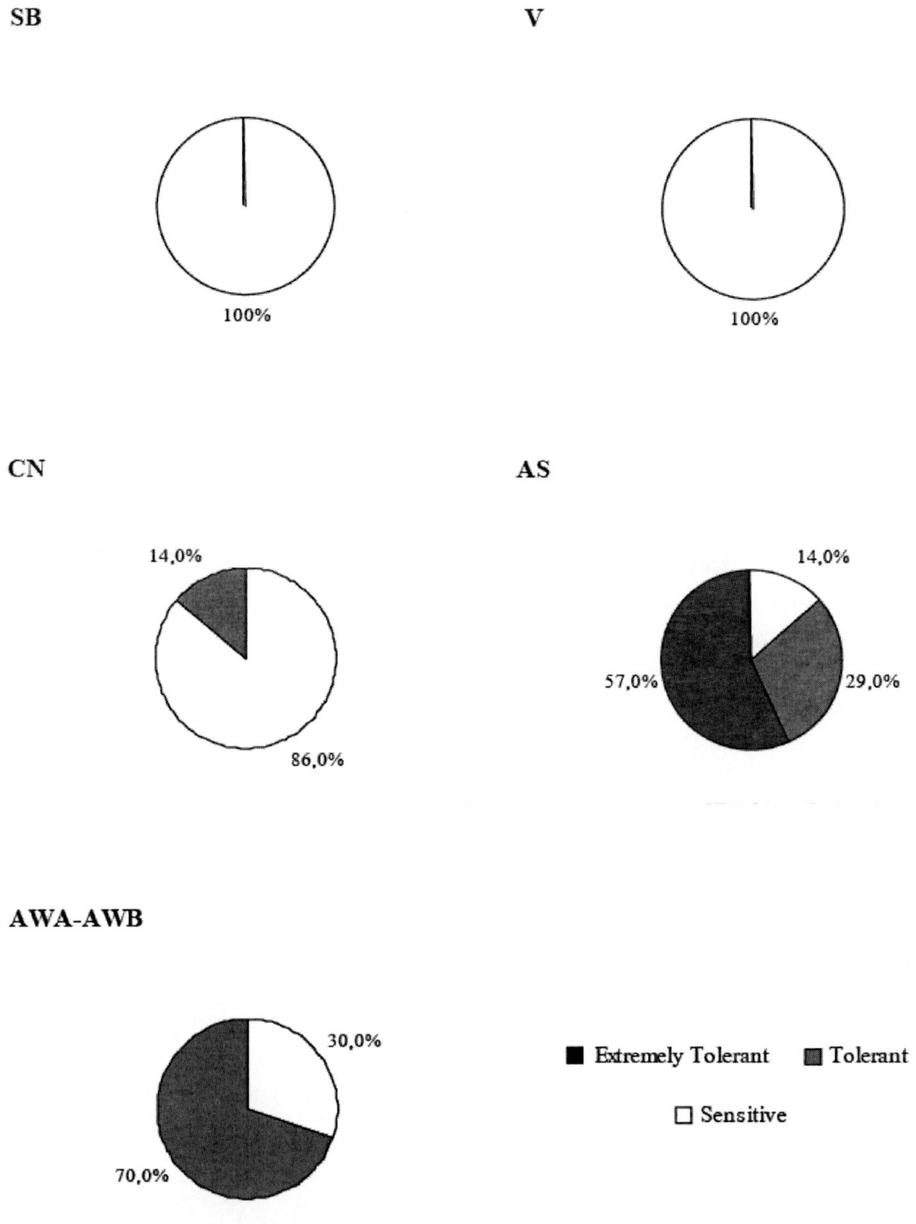

Figure 1.Percentage of extremely tolerant, tolerant and sensitive Rhizobium isolates obtained from each location. CN - Costa Nova isolates; V - Vagos isolates; SB - S. Bernardo isolates; AS - Alentejo Spring isolates; and AWA- and AWB- Alentejo Winter isolates.

Nevertheless, when the conditions are changed and this population is submitted to a selective pressure, the genotypes presenting tolerance ability can survive, hence yielding a tolerant population, with lower genetic variability. However, all CN isolates showed a strong growth inhibition at 50 mM NaCl (Table 1). This finding lead us to conclude that *Rhizobium* population was not well adapted to the environmental conditions and the isolates obtained from this location probably survive due to microsites where the soil moisture was maintained at levels that assure their survival, as suggested by Sprent and Zahran (1988) and Zerhari et al. (2000).

Alentejo isolates showed higher growth rates under NaCl stress, tolerating salinity levels up to 1600 mM (Table 1). Interestingly, Winter (AWA and AWB) isolates showed lower salt tolerance than those isolated in Spring (AS). Actually, 57% of AS isolates were extremely tolerant and only 14% of them were sensitive (Figure 1). The high salt tolerance observed in the isolates from non-saline soils (AS) of Alentejo could be explained by the climatic conditions observed in their habitat, considering the shortage of water experienced by *Rhizobium* populations in the Alentejo region, at least during part of the year. Drought and salinity tolerance are often associated (Mohammad et al., 1989; Yeo, 1994) probably because some of the mechanisms involved in water stress tolerance, as osmotic adjustment, can also contribute for their tolerance to salinity. Bordeleau and Prévost (1994) corroborate these findings, reporting that *Rhizobium* isolated from arid soils were capable of effective legume nodulation under drought and saline conditions. Cordovilla et al. (1999) and Mohammad et al. (1989) have also used strains of salt tolerant *Rhizobium* to nodulate plants of *Medicago sativa, Vicia faba* and *Pisum sativum* under water stress. However, there were differences in salt tolerance of strains isolated in different seasons of the same location. These results may be explained by the different water availability during the year. In June, the *Rhizobium* population was experiencing low water availability and high osmotic stress, since in this season, extreme water shortage is frequent in the Alentejo region. The drought conditions could select water stress tolerant strains, which would be more abundant in this time of the year. In winter, water stress is a feature not so selective as in late spring. The high soil moisture would therefore provide an increase in the number of individuals more adapted to the new environmental conditions, which do not include water shortage, thus yielding a less NaCl tolerant population.

In this chapter a marked variation in *Rhizobium* salt tolerance was observed, which is in agreement with other reports (Zahran et al., 1994). Our results also suggest that salt tolerance of *Rhizobium* populations was strongly influenced by their origin, which may represent evidence of the population's adaptation to the climate conditions experienced in the habitat they colonize. Hence, Alentejo populations seem to be well adjusted to extreme conditions of water shortage and high temperatures, which yielded high salt tolerance. On the other hand, SB and V populations are much more sensitive to salt stress because they come from a mild environment.

2.2. Screening Rhizobia for Nitrogen Fixation Potential with *Pisum Sativum*

Nodulation and nitrogen fixation in legume and rhizobial associations are commonly limited by soil infertility conditions, including salinity and drought. According to Somasegaran and Hoben (1994) the N content correlates well with the shoot dry weight, thus presenting an acceptable basis for the comparison of the strains nodulation efficiency. Relative nodulation and effectiveness of symbiotic dinitrogen fixation of *Rhizobium* isolates was evaluated by shoot dry weight comparison between plants under 0 and 90 mM NaCl, inoculated with different *Rhizobium* genotypes and controls (plants not inoculated and not receiving inorganic nitrogen, (-N) control; and plants not inoculated and supplied with 5 mM NO_3^-, (+N) control), as described by Somasegaran and Hoben (1994).

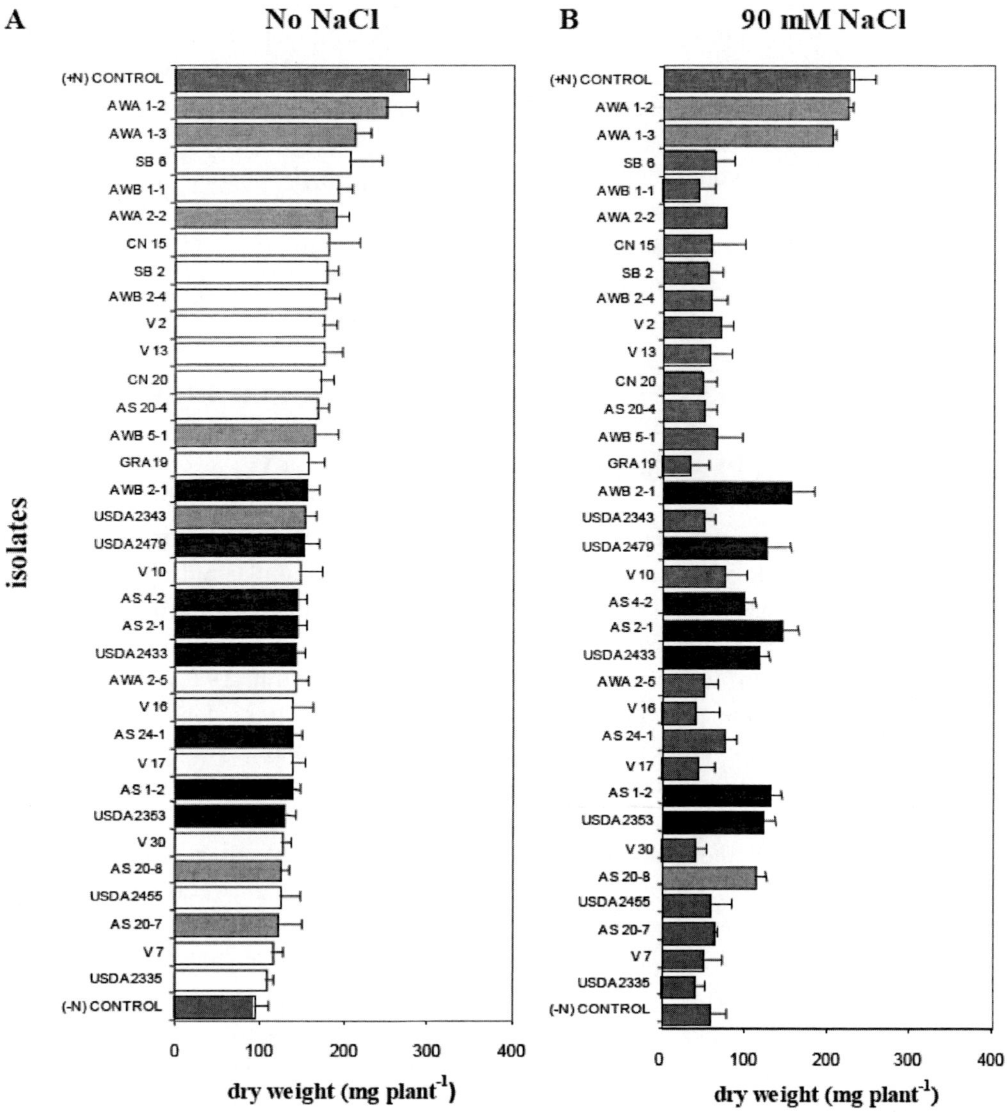

Figure 2.Shoot dry weight of *Pisum sativum* plants grown 30 days under 0 (A) and 90 (B) mM NaCl and inoculated with different *Rhizobium leguminosarum* strains and isolates. Striped bars represent unnodulated plants. Single colour bars represent nodulated plants. Colour is indicative of isolates and strains salt tolerances: white- sensitive; grey- tolerant and black- extremely tolerant.

When no sodium chloride was added, all isolates established efficient symbiosis, as judged by the formation of pink nodules, although growth responses varied greatly (Figure 2A). Shoot biomasses were significantly lower than controls supplied with nitrate. AWA 1-2 yielded a significant higher growth than the other isolates. In the presence of 90 mM NaCl, few isolates were able to establish an efficient symbiosis (Figure 2B). Unsuccessful symbiosis under salt stress may be due to a failure in the establishment of rhizobia in the rhizosphere, or a failure of the infection process due to the effect of salinity (Singleton and Bohlool, 1984).

Under salt stress, none of the sensitive strains was capable to nodulate pea plants, but all extremely tolerant strains were able to establish efficient symbiosis, except for strain AS 24-

1. Surprisingly, shoot dry weights of plants inoculated with extremely tolerant strains AWA 1-2 and AWA 1-3 were not significantly different from plants supplied with nitrate.

The results indicated some success for the establishment of an effective symbiosis under moderate salinity conditions. The significant increase in plant growth provided by the inoculation of the isolates AWA 1-2 and AWA 1-3 may reflect their ability for effective N fixation even under salt conditions. Most tolerant and extremely tolerant strains were able to establish a moderately effective symbiosis under 90 mM NaCl, providing survival in saline soils and increasing the probability of an effective symbiosis under the highest salt concentration tolerated by the legume.

3. Molecular Basis of Salt Tolerance

3.1. Involvement of Plasmids in NaCl Tolerance

Although chromosomal genes are important to bacterial adaptation and survival to environmental constraints, resistance genes are mainly located in plasmids. Their analysis can be a genetic tool to understand the way that bacteria tolerate stress. Furthermore, plasmid-mediated tolerance may be ecologically important because tolerance can be rapidly transferred from tolerant to sensitive bacteria (Gadd and Griffith, 1978). Weaver and Holt (1990) showed that some plasmids influence survival under stress conditions, which open new perspectives for the improvement of the *Rhizobium*-legume symbiosis under salinity. Pereira et al. (2006) also reported the involvement of some plasmids with Cd tolerance in *Rhizobium leguminosarum*.

Table 2 represents the frequency of different plasmids in salt sensitive, tolerant and extremely tolerant isolates. Our results demonstrate striking differences on plasmid profiles of *Rhizobium* isolates. Fifteen plasmids with sizes ranging from 82 to 828 MDa were separated and each strain had between two to six plasmids, except for isolates AWB 1-1, AWA 2-2, AWB 5-1, AWA 1-2 and AWA 1-3, where no plasmids were detected.

When the plasmid profiles were submitted to classification analysis (Bray-Curtis coefficient) four distinct cluster groups were identified (Figure 3). Cluster I included five isolates from Alentejo and one from USDA, all showing moderate (100-500 mM NaCl) or high (600-1800 mM NaCl) tolerance to salinity. Plasmids 828 and 82 were only detected in this group. Plasmid 734 was present in three tolerant strains and additionally a 147 MDa plasmid was present in all Alentejo isolates. Cluster II contained only two strains, one with moderate tolerance (USDA2343) and one tolerant (USDA2433), which had in common the 734 MDa plasmid. Cluster III was the most heterogeneous group, formed by seven strains: five sensitive, one tolerant and one extremely tolerant. Most genotypes of this group shared the 242 MDa, 201 MDa and 170 MDa plasmids. Cluster IV was formed only by salt sensitive strains. Clustering of these strains was due to the presence of the 343, 275 and 220 MDa plasmids. Clusters III and IV exhibited a higher number of plasmids than clusters I and II.

Table 2. Frequency of plasmids in sensitive, tolerant and extremely tolerant *Rhizobium leguminosarum* isolates

	Molecular Weight (MDa)														
	828	**734**	**512**	**427**	**387**	**343**	**315**	**275**	**242**	**220**	**201**	**170**	**147**	**113**	**82**
Sensitive			2	4	4	6	3	8	7	7	7	8	1	1	
Tolerant		1			1			2	1	2	1		3		2
Extremely Tolerant	2	3							2	4			4		

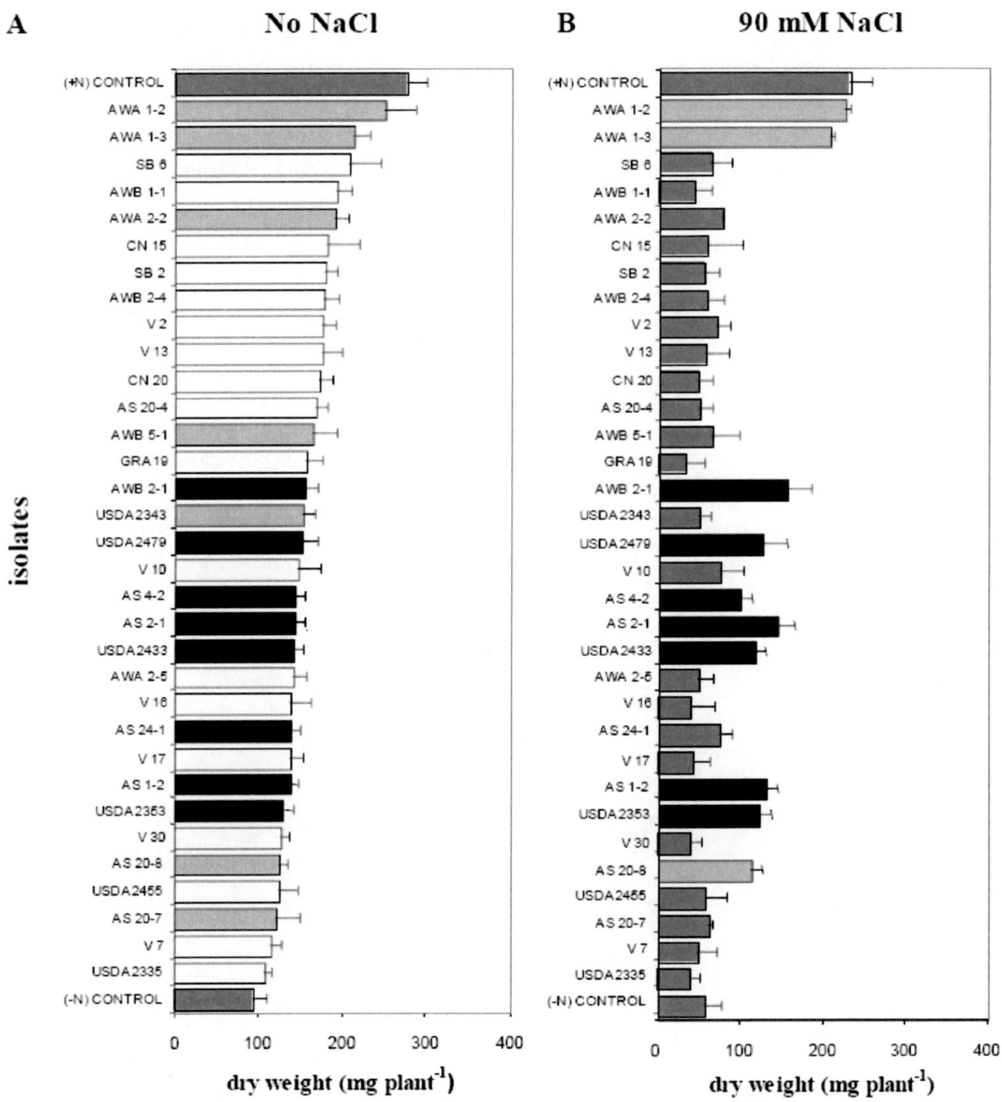

Figure 3. Dendogram representing plasmid profile similarities among *Rhizobium leguminosarum* isolates and strains with different salt tolerances.

Three plasmids were only detected in tolerant and extremely tolerant strains and a plasmid of 147 MDa appeared only in one sensitive strain and in 70% of strains from clusters I and II. According to our results (Table 2 and Figure 3) salt tolerance may be related to the presence of certain plasmids (828, 734, 147 and 82 MDa), however these plasmids *per se* are not enough to bestow salt tolerance, since strains with no plasmids detected were able to tolerate high levels of NaCl. These findings are in agreement with Baldani and Weaver (1992) who observed that plasmid absence in *Rhizobium leguminosarum* bv. *trifolii* did not impair their survival and development under osmotic stress.

The involvement of rhizobia plasmids in the nitrogen-fixing symbiosis is well documented (Martínez et al., 1990; Moënne-Loccoz and Weaver, 1996). Rhizobia plasmids are also related to catabolism of certain carbon compounds, metabolites found in root exudates, chemotaxis and cell motility (Baldani et al., 1992; Boivin et al., 1991; Chen et al., 1993; Hynes and McGregor, 1990). However, the influence that plasmid content of rhizobia has on their ecological competence in the soil is still poorly understood. Moënne-Loccoz and Weaver (1996) suggested that plasmid genes are important to the autecology of rhizobia *ex planta*, and identified two plasmids that contributed to a higher salt tolerance in *Rhizobium*, thus revealing the importance of plasmid genes for the activation of biochemical responses to osmotic stresses. These responses may involve the regulation of intracellular accumulation of certain metabolites (Graham, 1992).

3.2. Salinity Effects on Proteins

Protein synthesis is very sensitive to alterations in cell homeostasis (Yeo, 1998). Therefore, the analysis of protein profile alterations in *Rhizobium* isolates is crucial to better understand the degree of cell stress. The electrophoretic profiles of whole-cell soluble proteins under control conditions and in the presence of salt are shown in Figure 4.

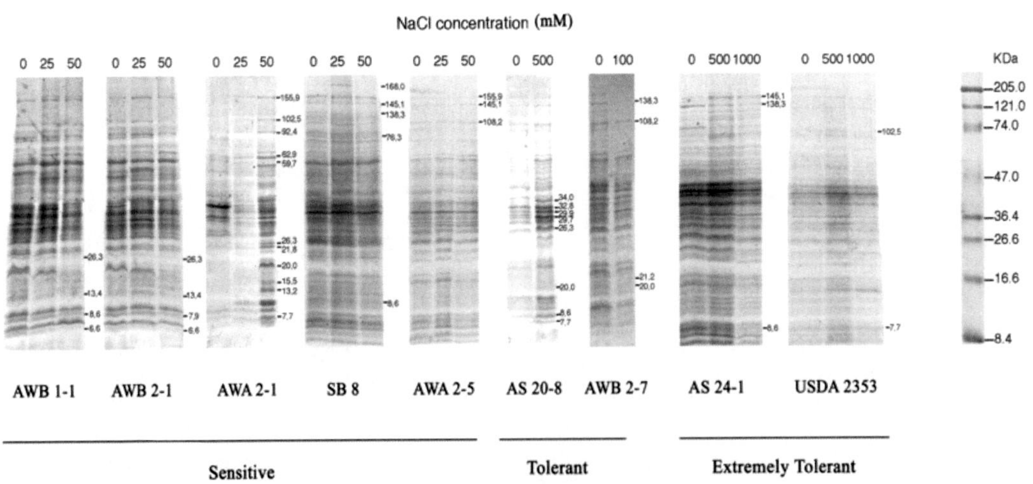

Figure 4. SDS-PAGE analysis of total soluble protein content of *Rhizobium leguminosarum* isolates and strains growing in YEM supplemented with different NaCl concentrations. The molecular mass (KDa) of polypeptides showing salt-induced alterations in band intensity are marked in the gel.

Protein responses to NaCl stress were different even between isolates from the same location. Salt induced increases and decreases of some polypeptides indicating an attempt of cells to adjust to the new adverse conditions. According to Saxena et al. (1996) and Unni and Rao (2001) the imposition of any stress to bacteria results in adaptive responses, which lead to changes in the regular metabolic processes that are then reflected in protein profile changes. In most isolates alterations corresponded to decreases in protein expression, suggesting a deleterious effect on cell basic metabolism, which may have imposed an inhibition of important metabolic pathways hence inhibiting growth and survival of *Rhizobium*. However, in tolerant isolates increments of protein expression were detected, mainly in polypeptides of high molecular weight (> 73.5 KDa), suggesting that these isolates possess mechanisms to minimise stress. Synthesis of stress proteins in order to neutralize the detrimental effects of environmental stresses has been reported (Dubey and Rani, 1987; Joshi, 1987; Saxena et al., 1996; Völker et al., 1992; Zahran et al., 1994).

4. CONCLUSION

In conclusion, our results reveal that in environments with high water availability throughout the year, *Rhizobium* populations displayed a high sensitivity to NaCl and are expected to be highly vulnerable to increases in soil salinity. However, when naturally exposed to large amplitudes of soil water availability and osmotic stress, populations reflect these alterations by seasonal fluctuations in the population's global tolerance. Consequently, the results seem to indicate that the salt tolerance of *Rhizobium* populations was strongly influenced by their origin, which may represent evidence of the population's adaptation to the environmental conditions experienced in the habitat they colonize. This emphasizes the importance of more thorough evaluation of soil bacteria survival when facing to habitat alterations, particularly the predicted climate alterations. *Rhizobium* assumes a double role: 1) it can be used as a sensitive bioindicator of soil microflora when in free-living; 2) when is in symbiosis with legumes it can reflect the influence that these alterations exert on the symbiotic process of N_2 fixation, on which many natural and agricultural ecosystems depend. Bringing into the mind that *Rhizobium* is a ubiquitous bacteria, the analysis of rhizobia populations both as free-living and in symbiotic association can constitute an excellent way to determine the influence of environmental changes, such as climate alterations on different biota.

The establishment of any procedure that may help plant crops to cope with salt stress is of extreme importance in world regions were salinity is a key issue in agriculture. The inhibitory effect of salinity on N_2 fixation and the fact that legumes are generally considered sensitive or moderately tolerant to salinity lead us to the assumption that legume cultivation under saline conditions will only produce economic yields if species and cultivars with enhanced tolerance to salinity are used. Inorganic N use can only be avoided if the host is able to establish an effective symbiosis with a microssymbiont under salt conditions. The results presented in this chapter bring to light new perspectives of agricultural sustainability in salt-constrained soils, since there was a insignificant decrease of plant growth provided by inoculation with two salt extremely tolerant strains when compared to NO_3 fed plants. This is indicative that efficient symbiosis can be established and productivity similar to inorganic fertilization can be obtained under saline conditions.

Plasmid-mediated salt resistance may be ecologically important since resistance can be rapidly transferred from resistant to sensitive bacteria (Lawlor et al., 1999). Weaver and Holt (1990) showed that some rhizobial plasmids influenced survival under stress conditions. Our results also opened new perspectives for the improvement of the *Rhizobium*-legume symbiosis since plasmid profile analysis suggests that the extrachromossomal genome contributes to survival in saline habitats. Protein profile alterations induced by salt did not reflect a common trend. Protein pools of tolerant strains were only slightly affected, suggesting the putative presence of tolerance mechanisms.

ACKNOWLEDGMENTS

The authors wish to thank to Dr. Robert Grace for his revision of the English language. This work was supported by a grant from the Centre for Cell Biology.

5. REFERENCES

Abd-Alla, MH; Vuong, TD; Harper, JE. Genotypic differences in dinitrogen fixation response to NaCl stress in intact and grafted soybean. *Crop Science*, 1998 38, 72-77.

Abedelgadir, AH; Alexander, M. Procedure to enhance heat resistance of *Rhizobium*. *Plant and Soil*, 1997 188, 93-100.

Abdelmoumen, H; Filali-Malouf, A; Neyra, M; Belabed, A; El-Idrissi, MM. Effect of high salt concentrations on the growth of rhizobia and responses to added osmotica. *Journal of Applied Microbiology*, 1999 86, 889-898.

Athar, M; Johnson, DA. Effects of drought on the growth and survival of *Rhizobium meliloti* strains from Pakistan and Nepal. *Journal of Arid Environment*, 1997 35: 335-340.

Bååth, E. Measurement of heavy metal tolerance of soil bacteria using thymidine incorporation into bacteria extracted after homogenization-centrifugation. *Soil Biology and Biochemistry*, 1992 24, 1167-1172.

Bååth, E; Díaz-Ravina, M; Frostegård, Å; Campbell, CD. Effect of metal-rich sludge amendments on the soil microbial community. *Applied and Environmental Microbiology*, 1998 64, 238-245.

Baldani, JI; Weaver, RW. Survival of clover rhizobia and their plasmid-cured derivates in soil under heat and drought stress. *Soil Biology and Biochemistry*, 1992 24, 737-742.

Baldani, JI; Weaver, RWM; Hynes, MF; Early, BD. 1992. Utilization of carbon substrates, electrophoretic enzyme patterns and symbiotic performance of plasmid-cured clover rhizobia. *Applied and Environmental Microbiology*, 1992 58, 2308-2314.

Boivin, C; Barran, LR; Malpica, CA; Rosenberg, C. 1991. Genetic analysis of a region of the *Rhizobium meliloti* psym specifying catabolism of trigonelline, a secondary metabolite present in legumes. *Journal of Bacteriology*, 1991 173, 2809-2817.

Bordeleau, LM; Prévost, D. Nodulation and nitrogen fixation in extreme environments. *Plant and Soil*, 1994 161, 115-125.

Bouhmouch, I; Souad-Mouhsine, B; Brhada, F; Aurag, J. Influence of host cultivars and *Rhizobium* species on the growth and symbiotic performance of *Phaseolus vulgaris* under salt stress. *Journal of Plant Physiology*, 2005 162, 1103-1113.

Brewing, NJ; Ambrose, MJ; Downie, JA. 1993. Root nodules, *Rhizobium* and nitrogen fixation. In: Casey R, Davies DR editors. *Peas: Genetics, Molecular Biology and Biotechnology*. Wallingford: Cab International; 1993; 237-290.

Chen, H; Gartner, E; Rolfe, BG. Involvement of genes on a megaplasmid in the acid-tolerant phenotype of *Rhizobium leguminosarum* biovar *trifolii*. *Applied and Environmental Microbiology*, 1993 59, 820-823.

Cordovilla, MD; Ligero, F; Lluch, C. Effects of NaCl on nitrogen fixation and assimilation of inoculated and KNO_3^- fertilized of *Vicia faba* L. and *Pisum sativum* L. *Plant Science*, 1999 140, 127-136.

Delgado, MJ; Garrido, JM; Ligero, F; Lunch, C. Nitrogen fixation and carbon metabolism by nodules and bacteroids of pea plants under sodium chloride stress. *Physiologia Plantarum*, 1993 89, 824-829.

Dubey, RS; Rani, M. Proteases and proteins in germinating rice seeds in relation to salt tolerance. *Plant Physiology and Biochemistry*, 1987 12, 9-15.

Elsheikh, EAE; Wood, M. Nodulation and N2 fixation by soybean inoculated with salt-tolerant rhizobia or salt-tolerant bradyrhizobia in saline soil. *Soil Biology and Biochemistry*, 1995 27, 657-661.

Fawzy, MH; Kuykendall, D. Plasmid DNA content of several agronomically important *Rhizobium* species that nodulate alfalfa, berseen clover, or *Leucaena*. In: Graham PH, Sadowsky MJ, Vance CP editors. *Symbiotic nitrogen fixation*. The Netherlands: Kluwer Academic Publishers; 1994; 181-188.

Figueira, EMAP. 2000. Aspectos da tolerância salina em *Pisum sativum* L.: influência da nutrição azotada. PhD Thesis. Aveiro University. Aveiro. Portugal.

Fink, CR; Waggoner, PE; Ausubel, F. Nitrogen fertilizer: retrospect and prospect. *PNAS*, 1999 96, 1175-1180.

Gadd, GM; Griffiths, AJ. Microorganisms and heavy metal toxicity. *Microbial Ecology*, 1978 4, 303-317.

Graham, PH. Stress tolerance in *Rhizobium* and *Bradyrhizobium*_and nodulation under adverse soil conditions. *Canadian Journal of Microbiology*, 1992 38, 475-484.

Graham, PH; Vance, CP. Nitrogen fixation in perspective: an overview of research and extension needs. *Field Crops Research*, 2000 65, 93-106.

Hirsch, PR; Jones, MJ; McGrath, SP; Giller, KE. Heavy metals from past applications of sewage sludge decrease the genetic diversity of *Rhizobium leguminosarum* biovar. *trifolii* populations. *Soil Biology and Biochemistry*, 1993 25, 1485-1490.

Hynes, MF; McGregor, NF. Two plasmids other than the nodulation plasmid are necessary for formation of nitrogen fixing nodules by *Rhizobium leguminosarum* strain. *Molecular Microbiology*, 1990 4, 567-574.

Hynes, MF; O' Connell, MP. Host plant effect on competition among strains of *Rhizobium leguminosarum*. *Canadian Journal of Microbiology*, 1990 36, 864-869.

Ibekwe, AM; Angle, JS; Chaney, RL; van Berkum, P. Sewage sludge and heavy metal effects on nodulation and nitrogen fixation of legumes. *Journal of Environmental Quality*, 1995 24, 1199-1204.

Ibekwe, AM; Angle, JS; Chaney, RL; van Berkum, P. Differentiation of clover *Rhizobium* isolated from biosolids-amended soils with varying pH. *Soil Science Society American Journal*, 1997 61, 1679-1685.

IPCC. Summary for policymakers. Climate Change 2007: The physical Science Basis. Contribution of Working Group I to the Fourth Assessment Report of the Intergovernmental Panel on Climate Change. Intergovernmental Panel on Climate Change. 2007.

Joshi, S. Effect of soil salinity on nitrogen metabolism in *Cajanus cajan* L. *Indian Journal of Plant Physiology*, 1987 30, 223-231.

Lakzian, A; Murphy, P; Turner, A; Beynon, JL; Giller, KE. *Rhizobium leguminosarum* bv. *viciae* populations in soils with increasing heavy metal contamination: abundance, plasmid profiles, diversity and metal tolerance. *Soil Biology and Biochemistry*, 2002 34, 519-529.

Lawlor, K; Chaudri, AM; McGrath, SP; Hirsch, PR. Gene transfer in bacteria from soils contaminated with heavy metals. *Letters in Applied Microbiology*, 1999 28, 317-320

Mannion, AM; Future trends in agriculture: the role of biotechnology. *Outlook on Agriculture*, 1998 27, 219-224.

Martínez, E; Romero, D; Palacios R. The *Rhizobium* genome. *Critical Reviews of Plant Science*, 1990 9, 59-93.

Möenne-Loccoz, Y; Weaver, R. Involvement of plasmids in saprophytic performance and sodium chloride of clover rhizobia W14-2 in vitro. *Applied Soil Ecology*, 1996 3, 137-148.

Mohammad, RM; Campbell, WF; Rumbaugh, MD. Acetylene Reduction Activity in Salt-Tolerant Alfalfa/Rhizobium under Drought Stress. *Arid Soil Research Rehab*, 1989 3, 431-438.

Pereira, SIA; Lima, AIG; Figueira EMAP. Screening possible mechanisms mediating cadmium resistance in *Rhizobium leguminosarum* bv. *viciae* isolated from contaminated portuguese soils. *Microbial Ecology*, 2006 52, 176-186.

Rehman, A; Nautiyal, CS. Effect of drought on the growth and survival of the stress-tolerant bacterium *Rhizobium* sp. NBRI2505 *sesbania* and its drought-sensitive transposon Tn5 mutant. *Current Microbiology*, 2002 45, 368-377.

Rozelle, S; Veeck, G; Huang, JK. The impact of environmental degradation on grain productin in China 1975-1990. *Economical Geography*, 1997, 73, 44-46.

Saxena, D; Amin, M; Khanna, S. 1996. Modulation of protein profiles in *Rhizobium* sp. under salt stress. *Canadian Journal of Microbiology*, 1996 42, 617-620.

Shamseldin, A; Nyalwidhe, J; Werner, D. A Proteomic approach towards the analysis of salt tolerance in *Rhizobium etli* and *Sinorhizobium meliloti* strains. *Current Microbiology*, 2006 52, 333-339.

Shannon, M. Breeding, selection and the genetics of salt tolerance. In: Staples RC, Toenniessen GH editors. *Salinity tolerance in plants*. New York: Wiley; 1984; 300-308.

Silver, S; Misra, TK. Plasmid-mediated heavy metal resistances. *Annual Reviews of Microbiology*, 1988 42, 717-743.

Singleton, PW; El-Swaify, SA; Bohlool, BB. 1982. Effect of salinity on Rhizobium growth and survival. *Applied and Environmental Microbiology*, 1982 44, 884-890.

Singleton, PW; Bohlool, BB. Effect of salinity on nodule formation by soybean. *Plant Physiology*, 1984 74, 72-76.

Somasegaran, P; Hoben, HJ. Handbook for Rhizobia. Berlin: Springer-Verlag; 1994.

Soussi, M; Santamaría, M; Ocaña, A; Lluch, C. Effects of salinity on protein and lipopolysaccharide pattern in a salt-tolerant strain of *Mesorhizobium ciceri*. *Journal of Applied Microbiology*, 2001 90, 476-481.

Sprent, JI; Zahran, HH. Infection, development and functioning of nodules under drought and salinity. In: Beck DP, Materon LA editors. *Nitrogen Fixation by Legumes in Mediterranean Agriculture*. Dordrecht: Martinus Nijhoff ; 1988 ; 145-151.

Subbarao, GV; Johansen, C. 1994. Potential for genetic improvement in salinity tolerance in legumes: Pigeon pea. In: Pessarakli M editor. *Handbook of Plant and Crop Stress*. New York: Marcel Dekker; 1994; 581-595.

Unni, S; Rao, KK. Protein and lipopolysaccharide profiles of salt-sensitive *Rhizobium* sp. and its exopolysaccharide-deficient mutant. *Soil Biology and Biochemistry*, 2001 33, 111-115.

van Hoorn, JW; Katerji, N; Hamdy, A; Mastrorilli, M. Effect of salinity on yield and nitrogen uptake of four grain legumes and on biological nitrogen contribution from the soil. *Agricultural Water management*, 2001 51, 87-98.

Völker, U; Mach, H; Schimd, R; Hecker, M. Stress proteins and cross-protection by heat-shock and salt stress in *Bacillus subtilis*. *Journal of General Microbiology*, 1992 138, 2125-2135.

Weaver, RW; Holt, EC. Short-term survival of rhizobia on arrow leaf clover seed at different depths. *Plant and Soil*, 1990 122, 147-150.

Yeo, AR. Physiological criteria in screening and breeding. In: Yeo AR, Flowers TJ editors. *Soil Mineral Stresses – Approaches to Crop Improvement*. Berlin: Springer-Verlag; 1994; 37-59.

Yeo, A. Molecular Biology of salt tolerance in the context of whole-plant physiology. *Journal of Experimental Botany*, 1998 49, 915-929.

Zahran, HH; Räsänen, LA; Karsisto, M; Linddström, K. Alteration of lipopolysaccharide and protein profiles in SDS-PAGE of rhizobia by osmotic and heat stress. *World Journal of Microbiology Biotechnology*, 1994 10, 100-105.

Zerhari, K; Aurag, J; Khbaya, D; Filali-Maltouf, A. Phenotypic characteristics of rhizobia isolates nodulating *Acacia* species in the arid and Saharan regions of Morocco. *Letters of Applied Microbiology,* 2000 30, 351-357.

Zhang, X; Kosier, B; Priefer, UB. Symbiotic plasmid rearrangement in *Rhizobium leguminosarum* bv. *viciae* VF39SM. *Journal of Bacteriology*, 2001 183, 2141-2144.

INDEX

A

B

C

D

E

F

G

J

K

L

M

N

O

P

T

U

V

W

X

Y

Z